Pro Tools for Music Production

Pro Tools for Music Production

Recording, Editing and Mixing

Second edition

Mike Collins

AMSTERDAM • BOSTON • HEIDELBERG • LONDON • NEW YORK • OXFORD
PARIS • SAN DIEGO • SAN FRANCISCO • SINGAPORE • SYDNEY • TOKYO

Focal Press is an imprint of Elsevier

Focal Press
An imprint of Elsevier
Linacre House, Jordan Hill, Oxford OX2 8DP
200 Wheeler Road, Burlington MA 01803

First published 2002
Second edition 2004

British Library Cataloguing in Publication Data
A catalogue record for this book is available from the British Library

Library of Congress Cataloguing in Publication Data
A catalogue record for this book is available from the Library of Congress

ISBN 0 240 51943 4

For information on all Focal Press publications visit our website at:
www.focalpress.com

Typeset by Keyword Typesetting Services
Printed and bound in Italy

Contents

About the author   xv

Acknowledgements   xvii

1 Pro Tools – the World's Leading Digital Audio Workstation   1
Introduction   1
Who this book will appeal to   3
What the book covers   4

2 The Evolution of Pro Tools – a Historical Perspective   6
Introduction   6
How Pro Tools came about – the move to digital   6
The road to Pro Tools   8
The beginnings of Pro Tools   9
Pro Tools III   10
The plug-ins   11
Cubase VST   12
The Yamaha 02R compact digital mixer   12
The ADAT Bridge   13
PCI expansion chassis   13
Twenty-four-bit digital audio   13
Pro Tools|24   14
Pro Tools MIX   14
Hardware controllers   15
Developments for 2000   16
Developments for 2001   16
Developments for 2002   16
Developments for 2003   17
Summary   17

3 Pro Tools TDM Systems   19
Introduction   19
The system   19
Native audio systems versus Pro Tools TDM systems   19
Pro Tools TDM hardware   20

Track counts   20
Pro Tools|HD core systems   21
Pro Tools|HD interfaces   21
    192 I/O   21
    192 Digital I/O   23
    96 I/O   24
    96i I/O   24
Peripherals   25
    PRE   25
    SYNC I/O   26
    MIDI I/O   26
Legacy interfaces   26
Third-party interfaces   27
Expansion chassis   28
Pro Tools custom keyboard   28
DigiTranslator 2.0 option   28
MachineControl option   28
Avid picture options   29
Avid Mojo   29
Do you need to use a mixer with Pro Tools?   29
Pro Tools TDM software   30
Summary   32

4   **The Computer System**   **33**
Introduction   33
What computer should you choose?   33
    The Power Mac G5   33
    The PCs   34
    Computer monitors   35
    The Sony VAIO   35
    The Apple PowerBook   36
Recommended additional software   36
Software updates   37
Leaving your computer 'on' permanently versus switching 'off' when you are
    not using it   38
Hard drives for Pro Tools   39
    Disk requirements and bandwidth   40
        Analogue to digital conversion   40
        Audio data transfer rates   41
        Audio data storage sizes   41
        Multichannel audio data transfer rates   41
    Audio file size limit   42
    Hard disk formatting   42
    Disk failure   43
    Disk maintenance and repair   43
    SCSI interface cards   44
    SCSI hard drives   44
    IDE/ATA hard drives   45
    Firewire hard drives   45
    Hard drives summary   46

Backup systems   46
  Storage drives   47
    DVD-RAM   47
    DVD-RW/+RW   48
    CD-ROM/CD-R   48
    CD-RW   48
    Sony AIT   49
    DLT   49
    Super DLT   49
    Exabyte   49
    DAT   50
  My backup strategy   50
  Another backup strategy   50
Mac OSX installations   51
  New for OSX   51
  Upgrading from OS9 to OSX   51
  Disk formatting/partitioning   52
  Installing Mac OSX operating system   53
  User accounts   53
  The Internet connection   55
  Software updates   56
  Logging in   56
  Forgot your password?   56
  Pro Tools issues   57
  Mac tips   58
  Get info   60
Troubleshooting   60
  Hardware problems   60
  Mac software problems – freezes, foibles and crashes   60
  Start-up problems   61
  A hard drive problem   62
  A true story   62
  Repairing disk permissions and disk volumes   63
  Battery-backed RAM   64
  Zapping the PRAM   64
  Unreadable CD   64
  More help   65
Summary   65

5 **Recording and Editing MIDI**   **66**
MIDI features overview   66
Recording MIDI data   66
  Configuring the Apple MIDI setup for Mac OSX   66
  Setting up to record MIDI into Pro Tools   68
  Recording onto a MIDI track   72
  Loop recording MIDI   73
  Playing back a recorded MIDI track   74
  Assigning multiple destinations to a MIDI track   76
  Recording Sysex into Pro Tools   76
  How to record audio from an external synthesizer into Pro Tools   77

How to record virtual instruments into Pro Tools	77
How to use Groove Control with Stylus	78
How to record virtual instruments as Audio	81
Editing MIDI data	83
Editing issues	83
Graphic editing	85
Event List editing	87
The MIDI menu	87
Applying grooves	91
Drum patterns	91
Editing notes	93
Working with patterns	96
Summary	102

6	**Recording**	**103**
	Introduction	103
	Digitizing audio	103
	Digital audio caveats	105
	Practical considerations	106
	Things you should know about before recording with Pro Tools	106
	Monitoring	107
	Drum punch-ins	109
	The rulers	109
	Edit window view options	110
	Meter and tempo	111
	Setting the default tempo	112
	Manual Tempo mode	113
	Recording to a click	113
	Setting up a click	113
	Destructive Record	115
	Playlists	116
	Punch-in and -out using pre- and post-roll	118
	Voices, channels and tracks	118
	Pro Tools|24 MIX systems	119
	Pro Tools|HD and HD Accel systems	119
	Voiceable tracks	122
	QuickPunch	123
	How QuickPunch affects the available voice count	124
	Loop Recording audio	124
	Half-speed recording and playback	125
	Recording shortcuts	126
	Markers	126
	Time stamping	127
	Importing tracks from CD into Pro Tools	128
	Working on a real recording session	131
	Opening a new session	131
	AutoSave	134
	Recording audio	135
	Topping and tailing	136
	Naming audio files	138

Importing audio   138
A quick edit   141
Setting up markers   142
Using Identify Beat to create a tempo map   144
Inserting Bar|Beat Markers   147
Dragging Bar|Beat Markers   150
Editing Bar|Beat Markers   150
Why not Beat Detective?   151
Pitch shifting to suit a particular vocalist   151
A basic synthesizer tracking session   152
Overdubbing piano, guitar, bass and vocals   154
Summary   157

7   Editing   **158**
Introduction   158
Edit window features   158
Edit window set-up   159
Markers window set-up   160
Scrolling options   162
Playback cursor locator   163
Zooming and navigation   164
Keyboard commands and shortcuts   165
The Edit tools   166
Zoom tool   166
Trim tool   166
Selector tool   167
Grabber tool   167
Smart tool   167
Scrub tool   167
Pencil tool   168
The Transport window   168
Crossfades   169
Auto fade-in and fade-out   169
Editing accuracy   170
Tab to transients   171
Editing regions   173
Grid modes   174
Time Compression/Expansion edits   175
Linked selections   177
Beat Detective   178
Beat Detective basic operation   180
Practical editing techniques   180
Fixing a note or chord played late   180
Editing vocals   183
Zero crossing edits   185
Editing before the downbeat   186
Arranging sections   189
Using playlists   190
Spot mode and time stamping   192
Summary   194

8   Mixing   **195**
Introduction   195
Tracks   195
Sends   196
Output windows for tracks and sends   197
Automation   199
   Automation playlists   203
   Automation write modes   204
   Trim mode   205
   Snapshot automation   206
Mixdown   207
A note on terminology   207
Mixing to analogue tape   208
Mixing to DAT   208
Mixing to digital media   208
Making your own CDs   209
Audio compression   209
Bounce To Disk   209
Recording to tracks   211
When to use the Dither plug-in   211
Mixing precision   212
The Dithered Mixer plug-ins   214
Mixer automation   214
Setting up a mix session   214
   Choosing views   214
   Setting up groups   215
   Using Automation 'snapshots'   217
   Editing and inserting breakpoints manually to create automation   219
   Writing mutes   220
   Alternatives   221
   Setting up Auxiliary routings   222
   Mixing vocals   226
   Panning the instruments   228
   The final mix   229
Summary   231

9   Audio Plug-ins   **232**
Introduction   232
Plug-in types   232
Using real-time plug-ins   234
Plug-in delays   236
Making plug-ins inactive   236
AudioSuite plug-ins   237
Real-time TDM and RTAS plug-ins   238
Multi-channel RTAS plug-ins   239
   D-Verb   239
   Dither   239
   POW-r Dither   241
   Dynamics   242
   EQ   242

Delays   242
  Short Mod Delay II and Slap Delay II   243
  Medium Mod Delay II   243
  Long and Extra Long Mod Delay II   243
  Tempo, meter, duration and groove   243
Mono In, Stereo Out Mod Delays   244
Multi-channel TDM plug-ins   245
  The Real-time Pitch Processor   245
  TimeAdjuster   246
Multi-mono TDM and RTAS plug-ins   248
  Click   248
  DeEsser   249
  Signal Generator   249
  Trim   249
    Using the Trim plug-in   251
Inserting Plug-ins during playback   252
Optional Digidesign plug-ins   253
Digidesign distributed third-party plug-ins   253
Plug-in packs   253
More goodies   253
Summary   254

10 Virtual Instruments   **255**
Introduction   255
Digidesign   255
  Access Virus Indigo TDM Synthesizer plug-in   255
  Waldorf Q TDM   256
  Orange Vocoder   257
Spectrasonics   258
  Stylus   259
  Atmosphere   259
  Trilogy   260
Propellerheads   261
  Reason   261
Bitheadz   263
  Unity Session   263
Native Instruments   264
Native Instruments Studio Collection   264
  B-4   264
  Pro-53   265
  Battery   266
Emagic software for use with TDM hardware   266
  HD Extension   267
  Host TDM Enabler   267
  ESB TDM   268
  Epic TDM   270
  Logic Pro update   270
Summary   270

11   ReWire   **271**
Introduction   271
ReWire in action with Reason   271
Setting up the Reason synthesizers   275
Routing MIDI from Pro Tools to Reason   277
Playing ReWire application sequences in sync with Pro Tools   278
Using Ableton Live with Pro Tools   279
Recording from Ableton Live into Pro Tools   280
Looping playback   281
Caveats   281
Summary   281

12   Midi + Audio Sequencers   **282**
Introduction   282
MOTU Digital Performer   282
Emagic Logic Audio   285
Steinberg Nuendo   288
  Using Nuendo with Pro Tools hardware   290
Steinberg Cubase SX   292
  Using Cubase SX with Pro Tools TDM cards   297
BIAS Deck   297
  Using Deck with Digidesign cards   299
Cakewalk Sonar   299
Ableton Live   300
Sony Acid Pro   302
Summary   302

Appendices   **305**

Appendix 1: Hardware Control   **307**
Introduction   307
Digidesign ProControl   307
ProControl modules   309
  ProControl rear panel   309
  ProControl summary   310
Digidesign/Focusrite Control|24   310
  Control|24 control surface   311
  Control|24 rear panel   313
  Control|24 summary   313
Mackie HUI   314
  Getting into the details   316
  Rear panel   318
  Monitoring   318
  Summary   319
Hardware control summary   319

Appendix 2: File Management   **320**
Introduction   320
The Workspace Browser   321
Volume Browsers   322

Catalogs 322
The Project Browser 323
Unique file IDs 323
Missing files 324
Relinking missing files in an open session 324

Appendix 3: Transferring Projects **327**
Pro Tools transfers 327
Getting ready to make the transfers 327
Transferring MIDI 329
Importing MIDI files 330
Exporting MIDI files 330
Importing tracks and track attributes 330
Time code mapping options 333
Track playlist options 333
Selecting session data to import 334
Transferring audio 335
Preparing for OMFI transfers 336
Exporting OMFI files from Pro Tools 337
Importing OMFI files into Pro Tools 338
Moving sessions between platforms 340
Moving between PT LE and PT TDM 340
Summary 342

Appendix 4: Pro Tools 6.4 **343**

Glossary **345**

Bibliography **361**

Index **369**

About the author

Mike Collins started out in music back in 1965 as a guitarist in the R & B group 'The Atlantics' playing mostly Stax and Motown material – with a sprinkling of blues and rock'n'roll. During the last year of that decade, a move from his home town (Burnley in Lancashire) to Manchester to manage a record shop saw the start of a new career as a club DJ – working at the Blue Note playing soul music and at the Magic Village playing a mix of blues and progressive rock. For 6 years between 1970 and 1976, Mike ran a successful hire company back in Burnley, renting out PA systems and discothèque equipment while working in the evenings as a mobile DJ specializing in dance music throughout the North-West of England. Attracted by the idea of making records rather than simply playing them, Mike spent 3 years studying recording technology and music at Salford University. Soon after graduating with a BSc in Electroacoustics-with-Music in 1979, Mike moved to London to find opportunities to get involved in professional music recording. After spending 14 months designing micro-phones for STC and following a brief 6-month stint in the Planning and Installation Department at Neve Electronics, Mike went into record production in 1981, co-producing a Top 40 single for London's leading jazz-funk band of the time – 'Light of the World' – and working on various recording sessions, remix productions and musical arrangements for EMI and other record companies. After 3 years working as a songwriter for Chappell, Dejamus and other music publishers, Mike briefly moved back into the professional audio field to work as a Film Sound Consultant for Dolby Labs for 6 months in 1984. An attractive offer from Phonogram Records to produce TV recordings for programs including *Top of the Pops*, *Soul Train* and other popular music TV programmes then led to a busy couple of years back in the music studios working with top US Soul and R & B acts including Cameo, Shannon, Joyce Sims, Sly Fox and others. Excited by the emerging new MIDI and computer technologies, Mike wrote a speculative letter to Yamaha. This resulted in a 12-month contract to help get Yamaha's London R & D Studio up and running as Senior Recording Engineer and Music Technology Specialist during 1986 and 1987. During this period, Mike was also involved in choosing the specifications for the world's first affordable digital submixer – the Yamaha DMP7D. The realization that music and audio were moving inevitably toward adopting the new digital technologies provided the motivation for a period of further education at London's City University. This resulted in the award of an MSc in Music Information Technology (with Distinction) in 1989. Around that time, Mike's career as a writer for audio, music technology and computer magazines started to develop, and this has since led to the publication of well over 500 articles and reviews tracking the latest developments in music and multimedia technologies [a listing of these is available on Mike's website at www.mikecollinsmusic.com]. In parallel with this, during the first half of the 1990s, Mike was very much in demand as a MIDI programmer working on records, films, music tours and TV with popular bands such as the Shamen and well-known film composers such as Ryuichi Sakamoto. Increasingly involved in multimedia, Mike worked on a variety of CD-ROM projects for Apple Computers, Canon Cameras and others, and, later, spent two years programming and editing a website for Re-Pro – the Guild of Record Producers and Engineers. In 1997, responding to a growing demand for editing and recording projects using Pro Tools, Mike set up a project studio comprising a Pro Tools system, Yamaha 02R mixer and high-quality ATC monitors. The studio has since been involved in everything from dance remixes to TV ads, background and featured music for TV and video, album editing and compilation, and, most recently, a move back to an early passion – recording jazz. Mike also offers consultancy, troubleshooting and personal tuition to other professionals with Pro Tools-based project studios and has recently been offering seminars on Music Technology, Pro Tools and other music software – and on Audio and Video for the Internet. Mike was Technical Consultant to the Music Producers Guild (MPG), contributing to the Education Group and organizing and presenting Technical Seminars, between 1999 and 2002. Currently, Mike is a Contributing Editor to *Macworld* magazine, writing about software and hardware for music and audio, and a regular contributor of news items and features to *Pro Sound News Europe* magazine.

The author may be contacted via email at 100271.2175@compuserve.com or at mike@mike collins.plus.com, or by phone at +44 (0)20 8888 5318, or by letter at Flat 1c, 28 Pellatt Grove, London N22 5PL.

Acknowledgements

I would first of all like to thank Jenny Welham at Focal Press for believing in this project and backing it so effectively. Thanks to Jenny, you can see the screenshots in full colour, and I could not have wished for a more supportive and flexible publishing editor.

A particular mention must be made of all the editors at the magazines I have written for over the last decade, especially including Ian Gilby, Paul White and all at *Sound On Sound*, Dave Lockwood at *Audiomedia* and *Sound On Sound*, Julian Mitchell at *Audiomedia*, Keith Spencer-Allen (formerly at *Studio Sound*), Phil Ward and Dave Robinson at *Pro Sound News*, Andrea Robinson at *The Mix*, Steve Oppenheimer at *Electronic Musician*, Ted Greenwald at *InterActivity*, Simon Jary and David Fanning at *Macworld UK*, Adrian Pennington at *Video Age* and *Production Solutions*, and the many other personnel in magazines around the world with whom I have been privileged to work.

I would like to thank Su Littlefield and Chas Smith who have regularly supplied me with Digidesign hardware and software to review over the last decade. Thanks also to Paul Foeckler and Nora Hayes at Digidesign who helped by providing information and software during the final stages of preparation of the first edition of this book.

Other individuals whose help is greatly appreciated are the many people from the hardware and software manufacturers, distributors and dealers, including Phil Dudderidge, Rob Jenkins, Giles Orford and all at Focusrite, Ralf Schluenzen at TC|Works, Thomas Lund at TC Electronic, Thomas Wendt and Katherine Kuehler at Steinberg, Gerhard Lengeling and all at the E-Magic team, Ken Giles at Drawmer, Graham Boswell and Ian Dennis at Prism Sound, Andy Hildebrand at Antares, Ken Bogdanowicz at Wave Mechanics, Georges Jaroslaw at HyperPrism, Gilad Keren and Orly Nesher at Waves, Paul de Benedictis at Opcode, Simon Stock and all at MOTU UK distributors Music Track, Martin Warr at Mackie UK, Simon Stoll at Raper & Wayman, and Nick Thomas at Media Tools/Turnkey.

All the clients and friends who have engaged my services to engineer recordings, program, play and record music, troubleshoot their systems, offer consultancy for studio installations, teach and present technical seminars have played a role also – giving me the relevant experience with Pro Tools, MIDI recording systems and Macintosh computers which I can now pass on to others in this book. So thank you to Ryuichi Sakamoto, Gil Goldstein, David Arnold, John Cameron, Francis Shaw, Brian Gascoigne, Mike Batt, Richard Horowitz, Sussan Deyhim, Daniel Biry, Joe Cang, Jerry Dammers, Feargal Sharkey, Ray Davies, Phil Harding, Mick Glossop, Pip Williams, Robin Millar, Gus Dudgeon, Alan Parsons, Dave McKay, Andy Hill, Richard Burgess, Pete Brown, Mitch Easter, John Leckie, Tim Simenon, Jeremy Healy, Lenny Franchi, Nitin

Sawhney, Keith O'Connell, Jim Mullen, Lyn Dobson, Damon Butcher, Marc Parnell, Chris 'Snake' Davis, John Mackenzie, Noel McCalla, Paul 'Wix' Wickens, Blair Cunningham, Hugh Burns, Tony O'Malley, The Shamen, The Style Council, The Chimes, Louie Louie, Dave Dorrell, Dave Angel, The Eurythmics, Zion Train, The London Community Gospel Choir, Juliet Roberts, Courtney Pine, Jimmy Somerville, Barry Manilow, Patsy Kensit/8th Wonder, Ian Prince, Zeke Manyeka, Cameo, Shannon, Joyce Sims, Jermaine Stewart, Aurra, Millie Scott, Nu Shooz, Dr York, Kurtis Blow, Sly Fox, Farley Keith, Darryl Pandy, Mark Shreeve, Daniel Biry, Ravi, Alquimia, Dave Lee, Light of the World, Beggar & Co, DC Lee, Mel Gaynor, Tony Beard, Jon Anderson, Keith Emerson, Mort Shuman, Jeremy Lubbock, Richard Niles, Craig Pruess, Bruce Forest, John Michaelson, John Edward, Keith Van Strevett, Dario Marianelli, Roddy Mathews, Roger Askew, Silja Suntola at the Sibelius Academy in Helsinki, Francis Rumsey at Surrey University, Jim Barrett at Glamorgan University and Dave Howard at The International Youth House in Leicester – to mention but a few.

I would also like to thank my brother, Anthony Collins, my parents, Luke and Patricia Collins, and my girlfriend, Athanassia Duma for all the support they have given me – without which this book would have never been completed.

I must specially thank my friend Barry Stoller for reminding me to back up my work just a short while before a hard drive malfunctioned, losing all my data, halfway through writing the first edition of this book.

The Second Edition

Thanks to Beth Howard and all at Focal Press for all the teamwork that produced the Second Edition of this book.

Thanks to Digidesign for including me in their 3rd-Party Developers category to supply me with the Pro Tools|HD1 Core system and 96 I/O interface at a more affordable price.

Thanks also to Liz Cox at Digidesign UK for supplying NFR software and plug-ins for review.

And thanks to Tom Dambly, Certified Pro Tools Instructor, for supplying additional information.

Thanks to Risto Sampola at Arbiter, Ian Cullen at Sound Technology, Mark Gordon at Music Track, and Mandy Rayment at Time + Space for supplying NFR software and plug-ins for review.

Special thanks to Nick Peart and all at Adobe UK for supplying the Adobe Creative Suite (Acrobat Professional, Photoshop, Illustrator, InDesign, and GoLive), which was of invaluable help during the preparation of this book.

1 Pro Tools – the World's Leading Digital Audio Workstation

Introduction

I believe that Digidesign's Pro Tools system for recording, editing and mixing audio is one of the most significant systems to have been developed within the last century for professional audio work. Computer-based audio recording equipment has been available since the latter part of the 1980s and the Digidesign system was not the first to be developed. It is also true that there are many competing systems available. However, Pro Tools is now undoubtedly the leading such system used for professional music recording around the globe.

So how has it become possible for a desktop computer-based system to rival the high-end professional audio mixers and recorders? This revolution has partly been brought about by the inexorable rises in processing power and storage capacities available in desktop systems allied to the falling costs of these computers, components such as RAM, and peripherals such as hard drives. Developments in DSP technologies have also paved the way. Nevertheless, full credit must be given to the people at Digidesign who had the vision to see the potential in targeting the professional market, allying with industry leaders in the non-linear video editing market – AVID (Digidesign's parent company since the mid-1990s) – and allying with hordes of third-party developers. These third-party developers have developed the incredible range of signal processing software plug-ins along with hardware peripherals such as the Focusrite Control|24 that have helped to take Pro Tools to the top.

This breakthrough has not come without a 'price' though. Producers, engineers, composers and musicians used to working with conventional recording equipment do face a very real learning curve when it comes to making the transition to Pro Tools (or any similar system). Getting used to new ways of working is always a problem for anyone working at full pace on a busy schedule of projects as there is rarely much opportunity to take time out from these. And the change in this case is to working both with digital audio and with desktop computers – both of which can be very alien 'beasts' to those 'reared' on analogue recording technologies. The recording engineer has to get used to the reality that the sound produced by overloads is virtually never going to be subjectively pleasing – unlike the case with analogue recording technologies, where overloads can produce subjectively pleasing distortions which are often actually used to enhance the recorded sound of instruments such as rock guitars and drums. The amount of re-adjustment required to cope with this change is nowhere near that required when learning to work with desktop computers as opposed to conventional mixing consoles and recording machines, however. That is why I have included a lengthy chapter in this book about the computer systems and peripherals typically used with Pro Tools systems. Today's recording engineers definitely need to be computer-literate to run a top-notch Pro Tools system and stay on top of it. Hard disks, like analogue and digital tape recorders, still use magnetic media – but don't care if it is population statistics, digital video downloads from the web, or digital audio that is being recorded. It is all ones and zeros to a hard drive. Software applications and operating systems and their features are a constantly-moving target and hardware upgrades come fast and furious. The 667 MHz G4 computers from Apple were discontinued after just 4 months, for example. You need to know as much as possible about all this stuff to make proper buying decisions, upgrade choices, and even just to 'tread water' with these systems. Hard drives need constant maintenance to keep them in tip-top shape for recording audio, people are increasingly networking systems, and a good understanding of backup systems is a top priority.

Another issue is whether you actually have enough processing power in your system to do everything in real-time. Computers are great for anyone who has the time to wait while the machine processes a file in non-real-time. But in the music-recording world, there are many circumstances where this is just not good enough. When the musicians are waiting (and often being highly-paid) to be creative, the last thing you want to do is to break their creative flow – or to extend the session beyond your budget. 'Can you just thin out the snare drum a little?' says the producer. 'Not right this second' replies the engineer, 'I don't have a plug-in for EQ inserted on that channel yet, you can't insert one on-the-fly, and I don't have enough DSP in my system anyway as we are using lots of plug-ins on lots of tracks already.' The good news is that if you have the budget, you can put together a powerful-enough Pro Tools system to run all but the very largest music recording sessions with good access to real-time EQ and other signal processing. Nevertheless, many studios will still find that combining Pro Tools with a conventional mixing desk that always has EQ and other processing available on each channel – with no argument – is actually still the best way to go.

Pro Tools is most widely used in music recording although it is making serious inroads when it comes to video post-production and film work as well. Surround sound for DVD production is another growing area of application. This book will focus on music recording – which will undoubtedly leave plenty of scope for future books focusing on other applications. This book also includes information about the basic Digidesign plug-ins. However, the number of third-party plug-ins is already so great (and growing rapidly) that to deal with all these effectively could easily fill yet another book [so I wrote that book – *A Professional Guide to Audio Plug-ins and Virtual Instruments* – and it is packed full of info about plug-ins!].

I believe that when considering a complete Pro Tools system you should include not only the software and hardware that Digidesign manufactures and markets but also the many other component parts that go to make up a working system. These may encompass any amount of third-party software and hardware, MIDI devices, computer systems and a wide range of studio equipment. These issues are discussed as appropriate throughout the book. Problems that may be encountered and any limitations that exist are also addressed. This book is about Pro Tools 'the system' and about Pro Tools in the real world – not the world of marketing hype.

The manuals supplied with Pro Tools are first rate, so this book will not attempt to replace these. Instead, the focus will be on helping the reader to understand the whole system; to understand how to set up and run the computer, hard drives and other peripherals; and to draw the reader's attention to a selection of useful tips and hints about using the system along the way.

Who this book will appeal to

OK – you want to know about Pro Tools? Then this book is for you. Maybe you are a recording engineer and you have noticed how many studios are bringing in Pro Tools systems to work with these alongside the tape-based recorders or stand-alone hard disk recorders. Some studios are even replacing their other recording systems with Pro Tools. Pro Tools is ideal for music producers who can now take their work away from the studio and run it on a compact system at home. Using Pro Tools LE with hardware such as the Mbox, for example, they can try out trial mixes and edits in their own time and then go back to the main studio to complete their production.

Project studio owners are increasingly choosing Pro Tools systems as these will easily interface with all the popular mixers and outboard, while providing onboard mixing and effects. Many project studios started out as MIDI programming rooms and have grown into full-blown recording facilities. If you have been using Logic, Cubase or Digital Performer you can add Pro Tools hardware to give you just about all the professional features you will need. These packages provide an alternative 'user-interface' which anyone who has grown up with these can use instead of the Pro Tools software – making it easy to get 'up and running' with the system. Re-mixers and DJs often work with MIDI + Audio sequencers like Logic Audio, so for these people Pro Tools is a natural choice when they want to upgrade to a more professional system.

But you should not overlook the power of the Pro Tools software itself. This book can help you to make the transition to using the full system – software and hardware. Like me, you will probably find that the Pro Tools software is better for some projects, while others, involving lots of MIDI programming or when working to picture, may be better handled using Logic Audio or Digital Performer.

For working to picture, the Pro Tools AVoption systems are a great choice for editors working on video that is being produced on AVID systems. These projects can be brought directly into the Pro Tools environment to allow audio post-production engineers direct access to the video source material and accompanying soundtracks to which additional or replacement dialogue, sound effects and music can be added. You can also record or transfer video from other sources into these systems and you can output your work in a variety of useful ways at the end of the project. Pro Tools lets you import video in QuickTime format so you can actually digitize your video using most popular desktop video systems, convert this to QuickTime format and use it in your Pro Tools session. Low-cost Firewire interfaces are widely available for personal computers to let you bring in video digitally from popular DV video sources that can be converted to QuickTime format. And you can always synchronize playback of conventional video recorders using the Pro Tools nine-pin Machine Control option.

Increasingly, musicians, songwriters and composers need music recording systems to demo their work or to use as part of the compositional process. Using a music scoring package such as Sibelius or Finale, both of which are available for Mac and PC, people who use conventional music notation can write music scores or arrangements, then transfer these via MIDI files into Pro Tools or one of the MIDI + Audio sequencers. Once in Pro Tools or one of the third-party applications, they can get to work right away recording as many of the parts as possible as audio. A trip to a Pro Tools-equipped studio can be made to add 'live' instruments that it is not practical to record at home. The mixes can be completed at home – often to produce the final product that can then be supplied to the client at the music publishing, record or film company.

If you are a student or teacher of Music and Audio technology, a Pro Tools system can provide relatively affordable software simulations of just about every component that you will find in a music-recording studio – although you still need to use real microphones, mic pre-amplifiers and loudspeakers. The hardware supports most other music and audio software and is available for both Mac and PC with entry-level systems such as the Mbox available at very affordable prices. This book will be invaluable to students and teachers alike. Serious music hobbyists will also find what they are looking for here, although the book is not aimed at complete beginners – other books cater specifically for these.

Pro Tools was developed on the Mac and I use the Mac for all my music and audio work, so examples in this book will be Mac-specific. Having said this, there is very little difference in the user-interface between the Mac and PC versions, and the features are virtually identical, so almost everything written here about the Mac version applies (at least in principle) to the PC version as well. Because the vast majority of Pro Tools users choose Mac systems, I have included plenty of information about setting up the Mac for use with Pro Tools and troubleshooting any problems that may occur. All of these same problems can and will occur on PC-based systems. The principles behind finding and fixing (or preventing) such faults are very similar, although the detail will differ according to which operating system and hardware is in use.

What the book covers

First I will tell you how Pro Tools came about from a historical perspective and then I will explain what it consists of – the hardware, the software, and the various peripherals that you can use. Pro Tools is like a kind of 'construction' kit in some ways. You can build your system to suit your needs, whether you are running a small project studio right up to a world-class professional facility for audio and music production.

Once you have chosen the appropriate hardware and software, you need to decide on which computer platform. If you decide to go for the PC, Digidesign recommends various models on its website. I have tested the top-of-the-range Sony VAIO laptop with Pro Tools LE and an Mbox and found this to be a very fast, responsive system – ideal for running plug-ins and virtual instruments and for editing or recording sessions away from the studio. I personally recommend that you use a PowerMac G5 desktop as your main machine, as Pro Tools was originally developed on the Macintosh platform and the PC versions of the Pro Tools software have tended to lag behind in certain respects technically. Also, I believe that the range of compatible software and the levels of integration of this software with Pro Tools are greater on the Mac than on the PC. Nevertheless, if you do decide to use a PC, I would caution you to choose a top-of-the-range model from a well-known manufacturer, selected from the list of qualified machines on the Digidesign website. This will probably cost you more money than the top-of-the-range Apple G5, so it will not be a cheaper solution. However, one advantage would be that you could undoubtedly run more plug-ins and virtual instruments on the host computer with a 3 GHz+ PC than with a 2 GHz G5.

The computer is so integral to the whole system that I have included a detailed chapter all about computer systems and peripherals, covering recommended choices of hardware and software, hard drives for Pro Tools, backup systems, Mac OSX installations, and troubleshooting.

You can use Pro Tools to create music using any MIDI instruments, and an increasing number of virtual instruments are available for the Pro Tools platform if you prefer software to hardware.

Many projects start out as 'sketches' made using MIDI instruments that are replaced later on by real instruments, and Pro Tools is so much better than it used to be for MIDI. So I have included a chapter all about recording and editing MIDI.

Recording, editing and mixing are very big subjects – which is why they get a chapter each. There are so many ways to go about doing these things and you can take many different approaches – depending on the type of projects you are working on. Each of these chapters will offer an overview of the kind of techniques available to you in Pro Tools, followed by practical examples from actual recordings I have worked on.

Audio signal processing is another big topic. A chapter is included to give you an overview of the audio plug-ins that are supplied with the standard system. Many other plug-ins are available from Digidesign and from third-party companies – see my book *A Professional Guide to Audio Plug-ins and Virtual Instruments*, also published by Focal Press, for more details.

Software synthesizers, drum-machines and samplers form a relatively recent, exciting and fast-growing area of development. The chapter on Virtual Instruments offers an overview of what is available here. Again, there is more information in *A Professional Guide to Audio Plug-ins and Virtual Instruments*.

Rewire is an important technology that allows popular software such as Propellerheads Reason and Ableton Live to be used with Pro Tools. A chapter is included to explain how this important technology works.

Many people prefer to use a MIDI + Audio sequencer as the 'front-end' for their Pro Tools hardware so I have included a chapter which presents overviews of the different choices available on the Mac. For more detailed information and tutorials on these, see my book *Choosing and Using Music and Audio Software*, also published by Focal Press.

An Appendix about hardware controllers explains how these offer a viable alternative to using hardware mixing consoles with Pro Tools systems.

File management using the DigiBase browsers and Catalogs is covered in Appendix 2, and lots of useful information about transferring projects between platforms and systems is provided in Appendix 3.

A comprehensive Bibliography lists many books and magazines that you may find useful and a Glossary is provided containing definitions of terms you may encounter when using Pro Tools systems.

2 The Evolution of Pro Tools – a Historical Perspective

Introduction

If you are involved in recording audio today – whether for music or post-production – you cannot fail to have heard of Digidesign's Pro Tools. There are plenty of digital audio recording, editing and mixing systems 'out there', but Pro Tools is just everywhere these days! And not without good reason. It is, in my humble opinion, simply the most well-integrated system available today. Now, chiming with the new Millenium, Pro Tools has truly come of age – bringing just about every conceivable type of recording technology into its ambit, whether by software simulation or by hardware add-ons.

Possibly the biggest breakthrough has been the explosion of software plug-ins which can be used to extend the system to incorporate just about every kind of signal-processing imaginable. And, having 'sucked' the whole studio into the computer, Digidesign and its partners have gone on to develop a range of control hardware which can be connected to the computer to provide the tactile control required in professional studio situations which the QWERTY keyboard and mouse simply cannot provide – faders and knobs to mix with, panels with buttons to allow direct access to editing functions, joysticks for surround panning, and so forth.

The latest development is to provide synthesizers, samplers and drum-machines as software modules from which the audio can be routed directly into the Pro Tools mixer. So now we truly have the whole studio in a computer-shaped box!

Perhaps the next step is to have simulated producers, engineers and musicians in there as well – and let them all get on with it. All that would be left for we humans to do would be to decide whether the results constituted a masterpiece – or not. A bit like those apocryphal monkeys with their typewriters eventually coming up with the works of Shakespeare I suppose!

How Pro Tools came about – the move to digital

When I started out in the recording industry in 1981, the first signs of digital technologies were beginning to appear. I worked briefly at Neve Electronics, planning and installing studios with analogue equipment, and I occasionally popped my head around the door of the room where Neve's team of world-class engineers were putting the final touches to their first digital desk – which was eventually installed in CTS Studios, London. I moved into music production later that same year, and found myself working on Neve analogue desks at Utopia Studios, also in London

– which is where I first worked with fader automation systems. The original Neve 'Flying Faders' system was often switched on and left running with the faders forming moving patterns in one of the rooms and some unsuspecting visitor would be prompted to walk into the room – to marvel (or get freaked out!) at the sight. This system was rather cumbersome to use compared to today's systems and arguments raged among engineers as to whether a mix sounded more 'organic' or natural when created manually as opposed to using the automation.

My first professional recordings were made using analogue tape machines and analogue mixing consoles with analogue synthesizers such as the Prophet 5 and the Oberheim OB8. The exciting news at that time was all about the first digital music production systems from Fairlight and Synclavier – both of which included synthesizer-style keyboards. But they were much more than synthesizers: the Synclavier included a sampler and an FM synthesizer along with sequencer software, and the Fairlight models I, II and III also offered sampling, synthesis and sequencing. However, these instruments were confined to the largest studios and a handful of the most successful musicians and composers. The Synclavier and the Fairlight paved the way for what was to come. But they cost more than the price of a decent home – and they quickly became 'dinosaurs' when the economics of the far larger markets for the more affordable equipment that followed allowed the pace of developments here to quickly outstrip those from Synclavier and Fairlight. These two companies have since regrouped and focused on producing high-end digital recording and mixing equipment, which is mostly used in post-production studios these days.

Another significant development that took place around this time was the 'birth' of the Compact Disc. Michael Jackson's *Thriller* album was released on CD in 1982 and went on to become the biggest-selling album of all time. This album undoubtedly helped to pave the way for the success of CD as a new consumer format for music distribution – and, of course, CDs are digital. Even though most of the early recordings released on CD had been recorded using analogue technology, these all had to be digitized at mastering studios for release on CD.

Over the next few years, a raft of influential new products appeared in the studios – many of these falling into the categories of musical instrument technology or 'prosumer' recording equipment. Most significantly, these new products were affordable by many more people – at least when hired for sessions as needed (as I did), if not to buy outright. Yamaha's DX7, the first commercially available all-digital synthesizer was everywhere, closely followed by rival models from Roland, Korg and others. The SPX90 digital effects unit and the Rev 7 digital reverb also became ubiquitous. MIDI sequencers first appeared around this time – with stand-alone models from Roland and Yamaha being particularly popular. The Linn Drum was the first commercially successful drum-machine featuring digital audio samples of real drum sounds and the Emulator sampler was the first sampling keyboard to appear – quickly followed by the Emulator II, which became the industry standard for some time.

Around 1984, personal computers started to become a serious option for musicians to use – at first for MIDI sequencing and music scoring. A sequencer called the UMI was developed for the BBC Computer that won something of a following in the UK, and there was some software available for the Commodore 64, as well as for the older Apple II models. The Amiga was a much more powerful machine which built on the Commodore 64's early popularity and found favour with some, and the IBM PC and the first Macintosh models came out about the same time – although there was little or no music software available for these extremely-limited machines at first. The Atari ST was also developed around this time – and Notator MIDI sequencer became available very early on, followed some time later by Cubase.

The first Akai samplers also started to appear – from which the ground-breaking 16-bit S1000 series was quickly developed to provide a truly affordable sampler with acceptable professional performance. On the Mac, Performer MIDI sequencing software appeared around 1985, with Opcode's Vision appearing a little later on, and I found myself using these almost exclusively by the end of 1986. There were a couple of dedicated music computers produced by Yamaha in the mid-80s – most notably the C1 which was a type of PC which could also run software such as Cakewalk, which was one of the first sequencers developed for IBM-compatible machines.

Yamaha introduced the first affordable compact digital mixer, the DMP7D, in 1987 – and I was a (very small) part of the R & D team that developed this. Several 'cutting edge' music production studios, such as the BBC Radiophonic Workshop in London, mapped out the direction for today's professional project studios – equipping their studios with DMP7 mixers, banks of synthesizers, samplers and drum-machines, with Macintosh computers running Performer software along with customized Hypercard software to manage all the MIDI devices – saving configurations, patches and so forth to allow near-total recall. As for individual musicians – a DX7, an S1000 and an Atari running Logic or Cubase became their 'industry standard' kit.

By the end of the 1980s, CD players, first developed around the beginning of the decade, had become widely available and Sonic Solutions introduced their first Mac-based digital audio workstation – Sonic Studio. This system 'cornered' the professional CD-mastering market for several years – and is still popular, despite stiff competition from Sadie and others.

DAT had also been launched as a 'wannabe' consumer format in the mid-80s, but had pretty much failed in the marketplace. Then, a strange thing happened! Professional studios, engineers and musicians around the world all bought DAT machines! Even though the most expensive models from Sony still essentially used the same 'prosumer' technology, top professionals quickly discovered that the advantages of digital – flawless copies with no generation loss being one of the most important – far outweighed the fact that the analogue to digital (A/D) converters were rather less than perfect in most early DAT machines.

Meanwhile, in the 'big' studios, open-reel digital audio tape recorders started to appear – including models from Mitsubishi, Sony, Studer and others. Newer large-format digital mixing consoles were also introduced by Neve and several others. These formats were slow to be adopted at first, and have still not completely replaced analogue multi-track recorders and mixing consoles at the time of writing (2004). For orchestral or band recording in large studios, these state-of-the-art designs provided the large numbers of channels and tracks required along with the 'industrial strength' build quality demanded by the rigours of continuous all-year-round operation – with a price to match!

The road to Pro Tools

Digidesign started out in 1983 producing replacement chips for early drum-machines such as the E-mu Drumulator offering alternative sounds for these.

By 1987 they had developed Q-Sheet software – which allowed users to place MIDI events at time code locations so that MIDI synthesizers and samplers could be used to place sound effects to picture – along with Softsynth for additive and FM synthesis. This was followed later by Turbosynth which also offered subtractive and phase-distortion synthesis methods using a graphical programming method – paving the way for today's software synthesizers such as Reaktor.

Digidesign's first digital audio system for the Macintosh was developed in 1988. This system was based around a 'Sound Accelerator' card for the Macintosh which offered 44.1 or 48 kHz audio recording at 16-bit resolution when interfaced with a two-channel A/D converter provided in a separate box (the AD In) connected to the card. The A/D converters were not great, but they were more than good enough when working with demo material. Fortunately, a two-channel digital interface box, the DAT-I/O, with AES/EBU and SPDIF input and output was launched in 1989. The software provided was called Sound Designer II, and the whole system was referred to as 'Sound Tools'. Sound Designer II was an excellent piece of software for editing mono or stereo audio recordings, and the software could also be used to transfer audio to and from popular samplers such as the Akai S1000. This allowed much more detailed sample editing and looping on the much larger computer screen. More importantly for some, this system made it possible to transfer audio from DAT into the computer, edit and compile selections, and put the audio back onto DAT to be sent off to a mastering studio or whatever.

Q-Sheet was upgraded around this time to work with the new Sound Tools hardware. Q-Sheet let you spot audio regions to SMPTE time code, play back MIDI files, and control MIDI equipment such as the Yamaha DMP7 mixers directly from within the software. I recall using this for audio post-production of a *Thomas The Tank Engine* video series in the early 1990s, for example, and it proved perfectly adequate for the task.

Digidesign at that time were perceived as a computer hardware and software company specializing in audio for the emerging 'desktop' audio editing marketplace which was beginning to branch off from the Musical Instrument marketplace – rather than as a professional audio company. However, Sound Tools and Q-Sheet attracted enough professional users to warrant the development of the more professional systems that followed.

The beginnings of Pro Tools

A more professional interface, the Pro I/O, featuring 18-bit converters, was introduced in 1990 along with the SampleCell I 16-voice/16-Mb sampler card, and Mac Proteus – an E-mu Proteus on a card.

Around the end of 1990 Opcode introduced Studio Vision. This was a milestone development which paved the way for today's integrated MIDI + Audio software environments – including Pro Tools itself – by providing multi-track audio recording and editing alongside MIDI sequencing. Multi-track audio recording software appeared on the Mac around the same time in the form of OSC's Deck software – which worked either with the Sound Accelerator card or the Digidesign Four-track Audiomedia card (which was also introduced around this time). Deck featured four tracks of audio and offered both mixing and editing features. One limitation was that these audio cards only had two audio outputs, so the four tracks of audio had to be mixed internally before being output via the two output channels. Nevertheless, the principle was established of having audio recording, mixing and editing using one software application.

Around the beginning of 1991, Digidesign brought out the first Pro Tools system, simply called Pro Tools, although you could refer to this as Pro Tools I. This included a four-channel interface in a 1U 19-inch rack with both analogue and digital inputs and outputs. As with Sound Tools, this used a single NuBus card and Sound Designer II software was supplied for two-channel editing, but these original Pro Tools systems used a new version of OSC's Deck called ProDeck as the mixing software along with Digidesign's new ProEdit software. This first version of Pro Tools was really very basic and most Digidesign customers were still using Sound Tools at this stage.

In 1992, Sound Tools II was launched with a new DSP card – using the new four-channel interface developed for Pro Tools I – and, subsequently, the Pro Master 20 interface was added to the range of hardware to provide 20-bit A/D conversion for this system. Also in 1992, 2 years after Studio Vision appeared, Steinberg added audio features to Cubase and named this Cubase Audio – bringing the concept of MIDI + Audio recording to a much wider group of users.

About a year later, at the beginning of 1993, Pro Tools II appeared, featuring the first release of Digidesign's own Pro Tools software. Four cards could be linked together using Digidesign's newly developed TDM technology to form a 16-track expanded system. This was a major leap forward, allowing Pro Tools to compete with the 16-track analogue systems which were generally regarded at that time as the minimum required for professional multitrack music recording. Grey Matter Response developed a SCSI accelerator card called the System Accelerator to provide the faster data transfer rates needed for the larger number of tracks with the expanded systems. The Video Slave Driver and then the SMPTE Slave Driver synchronizers were developed around this time to allow Pro Tools systems to be synchronized with video or audio tape machines.

During this period, Steinberg shifted the main development of Cubase onto the Mac – while continuing to develop Cubase on the Atari and on the PC. Rival company, E-Magic, also developed their Logic sequencer for the Mac and later for the PC. Early in 1993, Mark of the Unicorn added audio to their award-winning Performer sequencer and called this Digital Performer, while E-Magic followed suit, just a little later in the year as I recall, with the release of their Logic Audio software.

Many people, myself included, chose to continue using their favourite MIDI + Audio sequencer with the Pro Tools hardware at this time – in preference to the Pro Tools software itself which was not as well developed as it is today. Navigation and editing commands were not as highly developed as those in the MIDI + Audio sequencers, and the MIDI in Pro Tools was extremely rudimentary at this time. Digidesign developed their Digidesign Audio Engine software as a separate application that these other software applications could use to communicate with the Pro Tools hardware – making it easier for third-party developers to support the hardware.

Other developments from Digidesign included the Audiomedia II with digital I/O and higher-quality converters and Digidesign's first offering for the PC – the Session 8. This included a hardware controller with faders for mixing – a forerunner of the Pro Control.

Pro Tools III

The next breakthrough came when Digidesign introduced their Pro Tools III system early in 1995. The Pro Tools software was upgraded for the new system and many of the earlier problems with awkward editing and such-like were sorted out. I started to make the transition to using the Pro Tools software more than the third-party MIDI + Audio software around this time – finding the Pro Tools software user-interface to be much clearer and simpler to work with for audio, and realizing that the increased track-count meant that I could get rid of my old TASCAM eight-track multi-track tape recorder.

Pro Tools III featured 16 tracks using a single NuBus card, but allowed you to link up to four cards using TDM for a total of 64 tracks. The DSP Farm card was also introduced for additional processing power. A dedicated fast SCSI card was required for use with multiple-card systems to provide the required speed of data communication with the hard drives, and only the faster 'AV' drives were recommended for use with these systems. The SampleCell II card was also introduced around this time with 32 voices and 32 Mb of RAM and a TDM option to allow the audio output to be routed directly into the Pro Tools mixing environment. To accommodate all these extra cards, Digidesign developed a 13-slot Expansion Chassis – which added significantly to the cost of these systems.

There was plenty of activity from third-party developers around this time as well. Lexicon launched their NuVerb card and Cedar introduced their Noise reduction system for Pro Tools using NuBus cards. However, these were to be extremely short-lived products on account of the demise of NuBus – which took place very soon after the Cedar system was introduced.

In 1996, in response to Apple discarding NuBus in favour of the PCI bus, Digidesign released a version of Pro Tools III using PCI cards. The Disk I/O card incorporated a high-speed SCSI interface along with DSP chips for mixing and processing and up to three Disk I/O cards could be installed for a total of 48 tracks. A second card, the DSP Farm, was provided to handle additional signal processing duties, and multiple DSP Farm cards could be used if required.

The 888 interface was introduced with this system, featuring AES/EBU digital inputs and outputs for professional studio work along with reasonable quality 16-bit analogue converters interfacing via professional XLR connectors. At this time, digital desks were not too common, so most people used the analogue inputs and outputs on the 888 interfaces, and some professional engineers questioned the quality of these, although they were an improvement on the earlier interfaces. The lower-cost 882 interface was also introduced with 1/4-inch jack sockets for unbalanced analogue I/O.

On the software front, Post View was originally developed as a special version of the Pro Tools software with support for Sony nine-pin machine control and QuickTime movie playback. At this time, the video playback using QuickTime with the slower computers and hard drives of that period was restricted in practice to a smaller window – such as a quarter-screen size – so the nine-pin machine control for a conventional U-Matic VCR or similar was essential to keep clients happy! Other Digidesign products developed during this period included MasterList CD to burn CDs or output data to DDP/Exabyte tapes, Post Conform to autoconform standard CMX EDLs, and a version of Session 8 for the Mac.

The plug-ins

The next major development around this time was plug-ins! The first 'plug-ins' for Pro Tools were software modules which provided various signal processing capabilities such as compression, EQ, and so forth using the additional DSP power available on the Digidesign cards. So, the typical studio 'outboard' processors were now being modelled in software running on Digidesign's dedicated DSP hardware and provided for use within the Pro Tools software environment. Actually, the first processors that appeared were not directly modelled on any actual hardware counterparts, although many subsequent plug-ins have been. In fact, a number of features are possible in software which could never be achieved using hardware units, and most plug-ins take advantage of this fact to offer some unique aspects. This development made it possible to emulate just about all the components of a traditional recording studio from within the one integrated software environment – another truly innovative technological breakthrough!

A company called Waves, based in Israel, were the first to develop professional audio plug-ins – originally for Sound Designer II software. Unfortunately for Sound Designer II users, Digidesign were on their way to dropping Sound Designer II development by this time. No new features were being developed – although maintenance releases continued to appear for another couple of years to keep step with updates to the Mac operating system and hardware. By 1997, Waves had shifted the focus of their development to plug-ins for Pro Tools TDM systems and a number of other third-party developers had started to develop plug-ins for TDM – notably Arboretum, Antares and Steinberg.

Cubase VST

Steinberg released their Cubase VST (Virtual Studio Technology) software later in 1996 with the intention of providing a new MIDI + Audio environment complete with signal processing plug-ins which could work either with the built-in audio on the PowerMac computers or with a range of third-party cards – with the notable exception of Pro Tools TDM systems. Cubase VST started out as a hobbyist-level system, but has recently been developed into a much more professional system which has found favour with many professional users. Unfortunately, many of these professional users had already bought Pro Tools TDM hardware to use with Cubase Audio as the 'front-end' and were left 'high and dry' when Steinberg dropped TDM compatibility – especially as Steinberg delayed the final announcement that they would never add TDM compatibility to Cubase VST until December 1999 (while saying right up to this point that they would add TDM compatibility)! A consequence of this was that Cubase VST users could not work with TDM plug-ins. Although by this time a number of plug-ins were being developed for VST-compatible and other popular software such as Digital Performer, the situation between 1996 and 2000 was that the most interesting and high-quality plug-ins were almost always developed for TDM first and were often not available for other systems such as VST.

It is worth noting here that it is possible to route the audio inputs and outputs from Cubase VST via Pro Tools TDM hardware using the ASIO Direct I/O drivers for Pro Tools. You can only access the first 16 direct outputs, so if you have a 24-output system as I do, or if you have 32 outputs or more, you will not be able to use the additional outputs. Also, you cannot access TDM plug-ins from within Cubase VST using these drivers. But you do get 16 channels of high-quality audio input and output – which is obviously very useful if you want to run Cubase VST with your Pro Tools hardware.

The Yamaha 02R compact digital mixer

Another technological breakthrough which impacted on people using Pro Tools (and the many similar systems which were starting to appear by this time) was the launch in 1996 of the first affordable yet professional compact digital mixing console – the Yamaha 02R. Several optional interfaces are available for the 02R, including AES/EBU and ADAT optical versions – both of which can be used to interface to Pro Tools systems digitally – helping to realize the dream of an all-digital system for project studios. The ADAT tape cassette-based digital recorders that had appeared some years earlier used an eight-way optical interface that allowed audio to be trans-ferred digitally between machines. It was quickly realized by other manufacturers that this optical interface provided full digital quality and was cost-effective and convenient to implement, so Yamaha (like several other manufacturers) decided to include this type of interface on their 02R as an option.

The ADAT Bridge

Digidesign themselves had developed an ADAT interface for Pro Tools – originally to allow audio to be transferred into Pro Tools from ADAT recorders. In 1998, Digidesign launched their 20-bit ADAT Bridge interface with two pairs of ADAT optical inputs and outputs at a very affordable price – allowing Pro Tools systems to be interfaced digitally to 02R mixers very cost-effectively compared with using the relatively expensive 888 interfaces via their AES/EBU connections.

PCI expansion chassis

When Apple launched their G3 computers, these could only accommodate three PCI cards rather than the five that had been provided with previous models such as the 9500 and 9600. This meant that users of larger system configurations with several Digidesign cards, a SCSI accelerator card, a second monitor card, a video digitizing card and maybe an ISDN card, were compelled to invest in a PCI expansion chassis which would link to one of the existing PCI slots to provide up to a dozen or more extra slots. Digidesign and various third parties had been offering such devices since around 1995 to accommodate the larger Pro Tools III systems, and, currently, suitable chassis are available from Magma, Bit 3 and other companies. The G4 range originally had just three PCI slots, but the top of the range G4 machines now have four slots – which makes these much more viable for use with Pro Tools systems. With one of these machines you can now install a SCSI card, a couple of MIX cards and a SampleCell card, for example.

Twenty-four-bit digital audio

In response to demand from professional users, Digidesign developed 24-bit systems, which offer significantly better audio quality than the 16- or 20-bit systems. 'Such as what?', I hear someone ask. Well, there are various advantages to working 24-bit. First of all, you get a much better signal-to-noise (S/N) ratio. The rule of thumb is 6 dB per bit, giving you 96 dB for 16-bit, 120 dB for 20-bit and 144 dB for 24-bit. In practice, real-world systems do not achieve the theoretical maximum S/N ratios at these higher bit rates, but they do achieve significantly higher dynamic range than 16-bit systems. Naturally, if you are recording audio working with very wide-range dynamic material such as classical music, you will value the extra dynamic range of 24-bit. But what if you work with pop music, which typically has much less dynamic range – in other words, it is loud most of the time? Well, the higher resolution of the 24-bit analogue to digital conversion will always give you a much more accurate representation of the original analogue audio – so it will sound 'truer' to the original than 16-bit audio. Of course, with more dynamic range available in the system, you can use some of this as 'headroom' above the nominal 0 dB level to cater for unanticipated peaks in the audio that would otherwise clip, which is another major advantage.

This extra resolution is particularly valuable on systems that support extensive mixing, because working with higher bit rates/resolution not only helps to create the most accurate image of the original audio source, but also keeps this image clearer throughout the various stages of processing. Complex calculation algorithms are used not only for mixing, but also for signal processing (EQ and the like) – so the higher the resolution you start out with and maintain, the less rounding will be required, and therefore the less distortion will result from this.

Pro Tools|24

Around the end of 1997, Digidesign brought out their first 24-bit card, the d24, which was basically an upgrade for owners of PT III PCI systems who could use the d24 along with their existing DSP Farm cards. The d24 provided 24-bit input and output to either the older 888 interfaces or to the 888|24 interfaces that were introduced around the same time. It is worth mentioning here that both of these interfaces can handle 24-bit digital audio I/O – the difference being that the analogue I/O on the new 888|24 interface was upgraded to include much higher-quality 24-bit A/D converters, along with 20-bit D/A converters. A new version of the 882 was also released with 20-bit converters – the 882|20.

Digidesign claimed that the converters in the 888|24 surpassed the performance of both the Apogee AD-1000 and the AD-8000 converters – although most people believed that the Apogee's sounded better. The Apogee AD-8000 does have a lot to offer as an alternative to the 888|24 – with eight-channels of 24-bit A/D and two or eight channels of 24-bit D/A and also featuring Apogee's proprietary SoftLimit circuitry along with UV22 encoding for 20-bit or 16-bit audio. The Apogee interface will connect directly to Pro Tools, and various format conversion cards are also available for expansion with ADAT or TASCAM interfaces – making this a very flexible unit.

Pro Tools MIX

At the end of 1998, the Pro Tools|24 MIX system was launched. This featured yet another new card, the MIX card, with six Motorola Onyx chips, along with digital I/O. A variation of this card, the MIX Farm, was available to expand the capabilities of these systems. The MIX I/O card was also available at a low price to provide support for an additional 16 channels of I/O and mixing but with no DSP for plug-in processing. With the MIX I/O installed, the Pro Tools mixer first uses the DSP on the MIX I/O card before claiming any DSP from the MIX Core or Farm cards.

Three Pro Tools|24 MIX configurations were offered: Pro Tools|24 MIX featured the MIX Core card, supporting up to 16 channels of I/O, 64 tracks of simultaneous audio recording and playback. Pro Tools|24 MIXplus included the MIX Core card and a MIX Farm card, with support for 32 channels of I/O, 64 tracks of simultaneous audio recording and playback and twice the DSP power of a MIX system. Pro Tools|24 MIX3 featured the MIX Core card and two MIX Farm cards, supporting up to 48 channels of I/O and 64 tracks of simultaneous audio recording and playback.

These MIX cards can successfully be used alongside the older DSP Farm cards whose DSP is optimized for use with older TDM plug-ins, leaving the DSP on the MIX cards to run the newer plug-ins – which makes a lot of sense for anyone upgrading from older systems.

The Pro Tools MIX systems can run in either 32-track or 64-track mode, providing sufficient audio tracks for most people's needs, and Pro Tools 5 software provides many MIDI tracks to run along with these. Pro Tools software has always included MIDI capabilities. However, earlier versions were only designed to allow simple replay of existing MIDI files that you would prepare previously in a more full-featured MIDI sequencer – although it was possible to do very basic recording and editing in Pro Tools. Pro Tools 5 introduced useful MIDI capabilities for the first time.

Interfaces included the 888|24 I/O, the 882|20 I/O, the 1622 I/O and the 24-bit ADAT Bridge I/O. The 888|24 I/O features eight channels of analogue I/O equipped with 24-bit A/D and D/A converters. The 888|24 I/O also has eight channels of digital I/O for sample-accurate, 24-bit transfers from digital tape or mixing consoles. With a Pro Tools|24 MIXplus or MIX3 you can combine multiple 888|24 I/O units for up to 72 channels of analogue or digital I/O. The 888|24 I/O can also be used as a stand-alone A/D or D/A converter.

The 882|20 I/O featured 20-bit D/A and A/D converters at a lower price to suit project studios. The eight analogue I/O connections on the 882|20 I/O use balanced TRS 1/4-inch connectors and can be globally switched in software for +4 dBu or −10 dBV operation. Two channels of S/PDIF digital I/O allow links to DAT recorders, CD players, and other devices. The S/PDIF ports can also be used to record, write, and playback 24-bit audio files.

The 1622 I/O provides 16 balanced or unbalanced inputs with 20-bit A/D conversion using 1/4-inch TRS connectors, and two analogue outputs with 24-bit D/A conversion. An S/PDIF port capable of 24-bit digital I/O allows you to connect other digital devices. Each input has software controllable gain to accommodate a wide variety of synthesizer, sampler, and effects unit output levels. These levels are stored in memory even when the unit is powered off. The 1622 I/O in conjunction with Pro Tools software also acts as a virtual patch bay, enabling you to change connections with just one click.

The 24-bit ADAT Bridge I/O features 16 channels of ADAT Optical Inputs and Outputs, separate AES/EBU and S/PDIF ports for mastering to DAT, and includes a pair of high-quality 24-bit analogue outputs for monitoring your audio. It supports any device carrying the Alesis 'ADAT Optical' logo – making it an ideal solution for transferring audio back and forth between Pro Tools and ADAT optical-compatible consoles such as the Yamaha 02R.

Hardware controllers

Controlling faders and tweaking signal processing parameters 'on-the-fly' during a mix is possible using a mouse – but far from ideal. Hardware controllers (from J.L. Cooper and others) first appeared in the mid-1990s. These connected to Pro Tools via MIDI and could be used to control faders and mutes, for example. The largest of these controllers still only had 16 faders – although these could be switched to control different combinations of channels in Pro Tools. Both Digidesign and Mackie decided to develop hardware controllers for Pro Tools around the same time toward the end of 1998. The Mackie Human User Interface, known as the HUI (pronounced 'Huey'), appeared first and provided eight faders plus controls for most of the functions in the Pro Tools software – including the plug-ins. Like the earlier controllers, this hooked up via MIDI.

The Digidesign ProControl which arrived in 1999 is a much more expensive modular system which hooks up via Ethernet to provide a better level of control on account of Ethernet's greater-than-MIDI bandwidth. Both of these systems assume that no external mixer will be used, so they will sit in the studio set-up in place of a conventional mixer – simply to provide a hardware control surface for the Pro Tools software. Both the HUI and the ProControl include talkback features with switching controls and suitable outputs to feed monitoring systems and headphones. They also have inputs for microphones or line-level sources.

Developments for 2000

Focusrite designed a 24-channel controller for Digidesign featuring 24 high-quality Focusrite microphone pre-amplifiers. Descriptively named the Control|24, this provided a third option mid-way in price and features between the HUI and the ProControl.

The other significant development for the year 2000 was the low-cost Digi 001, which offered 24 tracks of 24-bit audio using one PCI card along with a 19-inch rack-mountable interface. Featuring two microphone pre-amps plus audio outputs to connect to a monitoring system, this allowed users on a tight budget to dispense with the need for an external mixer. The Digi 001 was popular as a small project studio system used to edit work from a large professional set-up, or to prepare material to take to a large set-up to finish off.

Developments for 2001

The year 2001 saw the introduction of the Pro Tools version 5.1 software with fully integrated surround capabilities – along with a host of new features and enhancements. Of particular interest for music production, the new 'Soft SampleCell' software-based sampler offered even greater integration than the hardware version; and Digidesign also developed a decent reverb plug-in for TDM systems at last – the Reverb One.

Developments for 2002

Digidesign introduced their 96 kHz Pro Tools|HD systems sometime around the second quarter of 2002. Three basic card configurations were made available that used the HD Core card either on its own or together with one or two HD Process cards. A range of matching interfaces and peripherals was also rolled out during the year, including the 192 I/O, the 96 I/O, the Sync I/O, the Pre eight-channel microphone preamplifiers, and the MIDI I/O 10-way MIDI interface.

Digidesign also introduced the Mbox – a low-cost system that includes an audio interface with Pro Tools LE software. The interface connects to the computer via USB and is powered via USB – ideal for use with laptops. Mbox measures just over 6-inches high, three and a half-inches wide and seven-inches deep It has two analogue audio inputs into which you can feed a microphone, a line level signal or a direct instrument signal via the XLR/1/4-inch jack combination connector. It also has two channels of S/PDIF digital I/O. One of the neatest features is the Mix control. This lets you blend the sound of anything plugged into the Mbox with the playback from Pro Tools LE to avoid the latency delays.

In July 2002, Digidesign launched the mid-level Digi 002 featuring a touch-sensitive control surface, 24-bit and up to 96 kHz sample rate operation, with Pro Tools LE 5.3.2 software for both Windows XP Home and Mac OS9.x. Aimed at demo studios and small radio stations, the Digi 002 featured dedicated monitor and headphone outputs and a FireWire interface to the host computer. Borrowing the same technology found in Digidesign's Control|24, Digi 002's array of touch-sensitive faders, rotary encoders and LCD scribble strips provides tactile and visual command over nearly every Pro Tools LE feature and parameter, including plug-ins. And for live applications, Digi 002 can be uncoupled from Pro Tools and used in stand-alone mode as an 8 × 4 × 2 digital mixer with onboard EQ, dynamics, delay and reverb.

Also in 2002, Yamaha introduced their updated 02R mixer – the 02R 96 – along with its siblings, the DM2000 and the DM1000. These all include control software for Pro Tools systems so that the faders can be used remotely to control Pro Tools systems. They also include surround-monitoring facilities that make these mixers ideal partners for many Pro Tools users.

Pro Tools LE 5.3.3 became available with MBox systems for Windows XP, along with ASIO drivers for popular third party applications, in November 2002.

Developments for 2003

Pro Tools 6 for TDM and LE systems was introduced around the end of January 2003. This was the first Digidesign software release for Mac OSX, with the PC software remaining at version 5.3.1 at this time.

In May 2003, Digidesign introduced the Digi 002 Rack. A single FireWire/IEEE 1394 cable connects the Digi 002 Rack to a compatible PC or Mac – making the 002 perfect for laptop or desktop, studio or stage. The Digi 002 Rack features up to 18 simultaneous channels of I/O, 1-in/2-out MIDI I/O, 24-bit/96 kHz A/D/A and four built-in high-quality mic pre-amplifiers. Its Pro Tools LE software interface supports 32 tracks of 24- or 16-bit audio, with integrated MIDI sequencing and real-time mixing and processing.

In September 2003, the HD Accel cards were introduced along with the 192 Digital I/O with its 16 channels of digital I/O and the 96i I/O interface with its 16 analogue line inputs for keyboards and samplers and stereo outputs for monitoring. The HD Accel card has almost twice the DSP power as the HD Process cards – enabling it to support more complex plug-in algorithms and increased plug-in counts.

Summary

Pro Tools 'grew up' in the digital age, starting out from its humble beginnings as Sound Tools and developing into the extremely powerful and all-encompassing music production system that it has become today. As more advanced converters and DSP processors, hard drives and computer systems have become available, Digidesign has always been quick to adopt these. Another key to the company's success is the way that such a wide range of third-party man-ufacturers have been successfully encouraged to develop software that interfaces so effectively with Pro Tools systems. This 'critical mass' of software development is undoubtedly what has helped to position Pro Tools systems right at the cutting edge of today's music technologies. The year 2003 has been probably the most successful ever for Digidesign, with its Pro Tools TDM systems taking over not only in the world's leading music recording studios, but also in the film industry on music scoring sessions – not to mention making serious inroads into the video post-production field.

Despite the fact that Pro Tools is now a cross-platform system with software supplied with every system for both Windows XP and for the Mac, the vast majority of TDM systems are Mac based. This inevitably puts Pro Tools users at the mercy of Apple (and of the Wintel 'axis' if they are PC-based). The saga of OSX, for example, is still ongoing at the time of writing. Apple introduced the first OSX-only computers in January 2003, only to release a further range of dual-bootable OS9/OSX models later in the year to cater for the many users who still could not migrate to OSX. Interestingly, this included the team at Pixar, Steve Job's 'other' company that produces film animations, who found that they could not move to OSX until around October 2003, despite all pressure from Apple to do so, because they were still depending on OS9 applications that were not ready or available for OSX. Other Mac users unable to migrate to OSX for practical reasons included many professional musicians and studio users who were 'wedded' to their OS9 systems on which they had installed their favourite plug-ins. A move to OSX would involve upgrading all the software and plug-ins – and many of the leading software applications such as Digital Performer and Nuendo (and many of the plug-ins) were simply not available for OSX until the third or fourth quarters of the year. So many users 'stayed put' and continued to work with OS9 systems throughout 2003. Other factors were the cost of upgrading not only all the software, but, in many cases, needing to upgrade the computer systems to cope with the ever-increasing demands of the processor-hungry operating systems, application software, plug-ins and virtual instruments. And for some people, this was also the time to consider upgrading existing Pro Tools MIX systems to HD systems. That is a lot of upgrading and expenditure – and consequent studio downtime. I know – I made the transition to OSX more or less completely by about October 2003, at which point I upgraded to a Pro Tools|HD system. But it took another 3 months before I could change my 02R for a 96 kHz DM1000 and yet another 3 months before I could raise the funds for a 2 GHz G5 computer and an HD Accel card. Interestingly, for 6 months of 2003 I worked with a Sony VAIO laptop running Steinberg Cubase and Nuendo, and Pro Tools LE software, with an Mbox hardware interface. It turned out that this computer would let me run twice as many virtual instrument and effects plug-ins than even the fastest G4 Macs!

All this goes to show just how incredibly dependent we have all become on computers and software and the audio cards and peripherals that are used with these. And it shows how significant computer processor and operating system upgrades have become for anyone working in audio or music recording. Oh what a brave new *binary* world we now inhabit!

3 Pro Tools TDM Systems

Introduction

In this chapter I will present overviews of the Pro Tools TDM hardware and software for people thinking about buying or upgrading.

The system

Digidesign's Pro Tools system has always included a software application and various hardware components. The hardware consists of one or more cards for the computer along with one or more audio interfaces, while the software offers a multi-track waveform editor along with a mixing console emulation and controls to let you record and play back audio from hard disk. One of the reasons the system has achieved such widespread popularity is because Digidesign has always encouraged other software developers to include options to use Pro Tools hardware for recording and playback of audio. So plenty of musicians and composers who bought Pro Tools systems chose to control the hardware using the Midi + Audio sequencer of their choice – such as Logic Audio or Digital Performer. There were two main reasons for this. Firstly, if you needed to do any MIDI sequencing, previous versions of the Pro Tools software were so poorly equipped for this that you had to be crazy to even attempt anything much more than a simple bassline or keyboard 'pad'. Secondly, many people who bought Pro Tools systems had previously used one of the popular MIDI sequencers and bought Pro Tools hardware, as this was the first audio hardware supported by Opcode, Mark of the Unicorn, Steinberg and E-Magic – in that order, as it happens. For these users it was much simpler to continue using the software interface they already knew and loved. Recording engineers and producers were always more inclined to use the Pro Tools software – especially if they were working on audio-only projects. With the release of Version 5, Pro Tools 'came of age' as an integrated Audio + MIDI recording environment which can competently handle most straight-ahead MIDI programming 'gigs' – although the accent is still on the audio capabilities. Version 6 has consolidated and enhanced the audio and MIDI features without significantly upgrading these.

Native audio systems versus Pro Tools TDM systems

A question which I am sometimes asked is why you would want a relatively expensive Pro Tools TDM system when so much software is available which will work with the 'native' audio on the Mac or with the increasing numbers of very affordable, or downright cheap, cards available for Mac and PC. The first thing I explain is that professional audio interfaces which provide the very highest audio quality are readily available for Pro Tools systems, but not always for the cheaper cards. Then there is the question of the DSP on the Pro Tools TDM cards. The range of high-quality plug-ins which run on this DSP helps to provide what is probably the most well-integrated system for music production available today. Also, the fact that DSP is provided on the Pro Tools TDM cards means that the computer's CPU is more able to take care of the important recording, mixing and editing functions than if it had to handle all the signal processing as well. It is worth bearing in mind that anyone working with a sophisticated software-controlled system such as Pro Tools will have to invest a great deal of time (and money) in learning to use the system thoroughly. If you start on an entry-level Pro Tools system, you can build up your expertise while you move forward in your recording career so that when you finally arrive at Abbey Road Studios or wherever, you are ready to go into action with Pro Tools 'at the drop of a hat'. These things simply cannot be said of the cheaper entry-level systems that some people choose to use.

Pro Tools TDM hardware

Digidesign introduced the Pro Tools|HD systems sometime around the second quarter of 2002. Three basic configurations were made available that used the HD Core card either on its own or together with one or two HD Process cards. In the final quarter of 2003, the HD Process card was superseded by the HD Accel card, which features almost twice the DSP power as the HD Process cards (and nearly four times that of older MIX Farm cards) with support for more complex plug-in algorithms and increased plug-in counts. The HD Accel card supports Digidesign's next-generation TDM plug-ins, including the Impact mix bus compressor. Third-party companies are also developing more advanced plug-ins that will only run on the HD Accel card.

Pro Tools|HD Accel cards can be added to previous HD systems while continuing to use the original HD cards with no problems. Also, because Pro Tools|HD Accel systems use the HD Core card, existing HD-compatible plug-ins are fully supported. And HD Accel-enhanced systems continue to support TDM, HTDM, RTAS and AudioSuite plug-in formats, allowing users to use HD Accel's dedicated DSP alongside the native resources of the host computer.

Unlike host-only systems, Pro Tools|HD's hardware architecture is designed to handle the majority of the system's audio processing tasks, off-loading that work from the computer so it is left with plenty of processing power to run additional plug-ins or applications.

> Note: At the time of writing, Pro Tools TDM 6.2 software, which ships with all Pro Tools|HD systems and cards, is the first release to support the new HD Accel cards. Pro Tools 6.2 combines the 6.1 feature set with support for the new, 16-input 96i I/O audio interface for Pro Tools|HD, Windows Media Audio 9 and Windows Media Audio 9 Pro import and export (Windows XP systems only), specialized plug-ins, and other features. All Digidesign and Digidesign-distributed Development Partner plug-ins have been updated to support the HD Accel cards. And installers for these are included on the Pro Tools TDM 6.2 software CD-ROM.

Track counts

The number of simultaneous voices that Pro Tools|HD systems can play back depends on the sample rate being used.

At 44.1 or 48 kHz, Pro Tools|HD systems can play 96 tracks while Pro Tools|HD Accel systems can play 192 tracks.

At 88.2 or 96 kHz, Pro Tools|HD systems can play 48 tracks while Pro Tools|HD Accel systems can play 96 tracks.

At 176.4 or 192 kHz, Pro Tools|HD systems can play 12 tracks while Pro Tools|HD Accel systems can play 36 tracks.

Pro Tools|HD core systems

Pro Tools|HD 1 features the HD Core card, supporting up to 32 channels of I/O and guaranteed support for up to 96 simultaneous audio tracks with no stress on the computer. Pro Tools|HD 2 Accel includes the HD Core card and an HD Accel card, offering support for 64 channels of I/O. Pro Tools|HD 3 Accel features the HD Core card and two HD Accel cards, supporting up to 96 channels of I/O. Pro Tools|HD 2 Accel and HD 3 Accel systems both offer guaranteed support for up to 192 simultaneous audio tracks with no stress on the computer. All Core systems feature sample rate support up to 192 kHz and include TDM II, Digidesign's new bus architecture design, which provides support for extremely large mixing configurations. HD Core systems also support Digidesign's new DigiLink interface standard that allows connectivity of HD Core and HD Process cards at lengths of up to 100 feet – although the 100-foot cables only support up to 96 kHz sample rates.

> Note: All Core systems require that you have at least one Digidesign HD audio interface connected before you can use them. These are sold separately so that you can choose whichever interface suits you best.

Pro Tools|HD interfaces

The Pro Tools|HD Interfaces all feature low-jitter clocks but you may still want to consider adding a Sync I/O to your system, or a higher-quality third-party clock source.

192 I/O

The top-of-the-range interface is the 192 I/O which features 24-bit and up to 192 kHz A/D and D/A conversion. The A/D converters offer a 120 dB dynamic range while the D/A converters offer 118 dB. Sample rates of 44.1, 48, 88.2, 96, 176.4, 192 kHz ±10% are supported. The basic 192 I/O supports up to 16 simultaneous channels of analogue and digital I/O. The standard interface has two 25-way D-sub connectors on the rear panel providing 16 channels of balanced analogue inputs either at +4 dBu or at −10 dBV, depending on which of the two connectors you use. A single 25-way D-sub connector provides 16 channels of balanced analogue output. The digital I/O bay is occupied by a 25-way D-sub connector that provides eight channels of AES/EBU I/O, a second 25-way D-sub connector for eight-channels of TDIF I/O, and a pair of eight-way ADAT optical input and output connectors.

A pair of XLR connectors provides two channels of AES/EBU I/O and a pair of RCA/Phono connectors provides two channels of S/PDIF I/O. There is also a pair of optical two-channel S/PDIF I/O connectors. Two pairs of BNC connectors are available to provide loop sync and external clock (Word 1x and Slave Clock 256x) input/output. Additional ports include an Expansion Port that allows for direct connection of another 192 I/O or 96 I/O and a Legacy Peripheral port.

The 192 I/O has one empty I/O bay that can be used to add more inputs or outputs. To expand the analogue I/O capacity, you can add either a 192 AD card, providing eight more analogue input channels, or the 192 DA card, which gives you eight additional analogue output channels. The 192 I/O can also be fitted with the 192 Digital card, which adds a further eight channels of AES/EBU, TDIF and ADAT I/O.

If you fully expand the 192 I/O, it supports a total of 50 channels of I/O, including the 16 basic analogue channels, the 24 digital I/O channels, the eight additional I/O channels, and the two-channel digital I/O.

> Note: One of the most important features of the 192 I/O is that it offers switchable, real-time sample rate conversion on digital inputs on the 192 Digital card, so you can input digital signals at any sample rate and it will convert these to the session sample rate.

> Tip: If the session sample rate is 88.2 kHz or higher, ADAT and TDIF input sources can still be used with Digidesign's 192 I/O, but the sample-rate conversion option must be enabled using the Hardware dialog from the Setups menu.

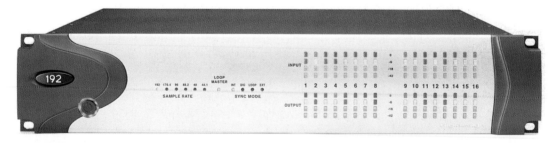

Figure 3.1 192 I/O.

A neat feature provided for the analogue inputs, the Soft-Clip Limiter, allows you to print higher levels to disk for punchier, hotter recordings.

192 Digital I/O

If you do not need analogue I/O, the 192 Digital I/O features a wide range of digital I/O options, including up to 16 channels of AES/EBU, TDIF, and ADAT I/O, along with S/PDIF I/O. This unit supports up to 16 simultaneous channels of AES/EBU I/O at 96 kHz or up to eight simultaneous channels of AES/EBU I/O at 192 kHz. The rear panel has two bays filled with digital I/O connectors. Each bay has a 25-way D-sub connector for AES/EBU I/O, a 25-way D-sub connector for TDIF I/O, and a pair of optical connectors for ADAT I/O. You cannot use all of these I/O connections at the same time – they are provided to allow flexibility when connecting to the different types of digital equipment that you may wish to use.

Just like the 192 I/O, a pair of XLR connectors provides two channels of AES/EBU I/O, and a pair of RCA/Phono connectors provides two channels of S/PDIF I/O. There is also a pair of optical two-channel S/PDIF I/O connectors. Two pairs of BNC connectors are available to provide loop sync and external clock (Word 1× and Slave Clock 256×) input/output. Additional ports include an Expansion Port that allows for direct connection of another 192 I/O or 96 I/O and a Legacy Peripheral port.

Note: The 192 Digital I/O also has switchable, real-time sample rate conversion on digital inputs to allow input of digital signals at any sample rate.

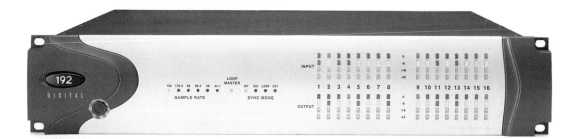

Figure 3.2 192 Digital I/O.

96 I/O

The 96 I/O is an affordable 16-channel audio interface for Pro Tools|HD systems that features a versatile selection of I/O options, including eight channels of high-definition analogue I/O via 1/4-inch TRS jacks and eight channels of ADAT optical I/O, plus two channels of AES/EBU and S/PDIF I/O, and Word Clock I/O. The A/D converters offer a 115 dB dynamic range while the D/A converters offer 114 dB.

> Note: The 96 I/O audio interface does not have any sample rate conversion capability, so the ADAT port will always go off-line with sample rates above 48 kHz.

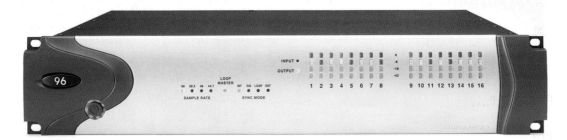

Figure 3.3 96 I/O.

96i I/O

The 96i I/O provides 16 analogue inputs to enable you to easily connect keyboards, samplers, effects, and other line-level equipment to your Pro Tools system. Its 16 analogue inputs provide balanced or unbalanced connections using tip-ring-sleeve (TRS) 1/4-inch connectors. A 24-bit capable S/PDIF port allows you to connect professional DAT recorders, CD players, and other digital recording devices. The A/D converters offer a 111 dB dynamic range while the D/A converters offer 113 dB. Six-step software-adjustable level controls are provided for inputs 1–4, with two-step software-adjustable level controls provided for inputs 5–16. Outputs are software switchable between +4 dBu and −10 dBV. This unit also has two pairs of BNC connectors for loop sync and Word clock input/output and an Expansion Port to connect additional units.

> Note: The 96i I/O does not have sample rate conversion, it does not have AES/EBU or ADAT I/O, and it does not have a Legacy Port.

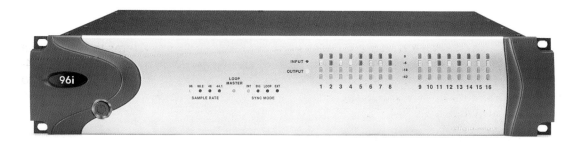

Figure 3.4 96i I/O.

Peripherals

PRE

One of the most interesting new peripherals that Digidesign introduced with their HD systems is the 'PRE' – an eight-channel remotely controllable microphone unit featuring eight discrete, matched-transistor, hybrid microphone pre-amplifier circuits. You can connect almost any type of input signal including microphones via XLR inputs, and line or direct instrument (DI) level inputs via 1/4-inch jacks to any of the eight channels. Each channel includes a high pass filter, a phase reverse switch, 48 V phantom power, and a −18 dB pad. You can control any of its settings remotely using the Pro Tools software or any of the Digidesign control surfaces – so

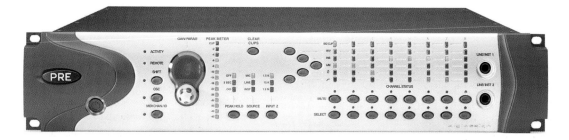

Figure 3.5 Digidesign PRE eight-channel Microphone Pre-amplifier.

you can position the PRE anywhere within reasonable distance of your main Pro Tools system, such as out in a studio area close to your audio sources. The PRE can also be used as a stand-alone device and can be controlled remotely using any standard MIDI controller for non-Pro Tools applications.

SYNC I/O

To allow near sample-accurate synchronization to time code or bi-phase/tach signals, Digidesign offers the SYNC I/O peripheral. This features a 192 kHz-capable low-jitter Word Clock that can be used to provide the stable clock signals that every digital system needs to maintain the highest audio fidelity. The SYNC I/O supports all the industry-standard clock sources and time code formats, including the most widely used pull-up/pull-down rates for film and video, and two nine-pin ports are included for use with the Digidesign MachineControl option. This unit also features AES/EBU clock I/O, video reference in/thru, and video program in/out connectors.

Figure 3.6 SYNC I/O.

MIDI I/O

Digidesign also offers a 10-way multi-port MIDI interface, imaginatively named the 'MIDI I/O'. This has 10 MIDI inputs and 10 MIDI outputs, each of which can carry 16 MIDI channels independently for a total of 160 MIDI channels. A hardware 'thru' mode allows you to patch any number of inputs to any number of outputs without using the computer – enabling the unit to function as a MIDI patchbay. The MIDI I/O connects to your computer via a self-powered USB connection and includes support for Digidesign's MIDI Time-Stamping technology to provide superb timing accuracy and precision. Ports 9 and 10 are mirrored on the front panel for convenience and Input 1 can be patched to all outputs in Hardware Thru Mode for stand-alone operation.

Legacy interfaces

The Legacy peripheral port on the 192 I/O, 192 Digital I/O and 96 I/O interfaces allows you to connect any of the following audio interfaces to your HD system: the 888|24, the 882|20, the 24-bit ADAT Bridge I/O, the original 20-bit ADAT Bridge I/O or the 1622 I/O.

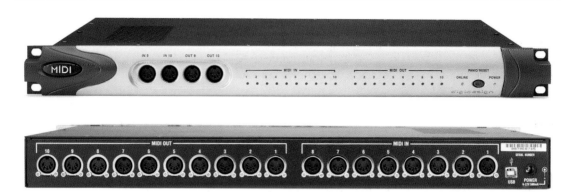

Figure 3.7 MIDI I/O.

Note: You cannot use the 888 I/O or the 882 I/O with HD systems.

There are some restrictions when using the Legacy Peripheral port. The biggest of these is that you can only work on sessions at up to 48 kHz. At higher sample rates, the Legacy port ceases to function. Also, the maximum number of Legacy peripherals that you can hook up to an HD system is eight – and for this you would need to have four HD-series audio interfaces with four 16-channel Peripheral Cable Adapters ('Y' cables).

Third-party interfaces

You may also encounter a couple of high-end third-party interfaces. The PrismSound ADA-8 features converters from the UK's leading converter manufacturer, while the Apogee AD8000 is available from one of the USA's leading converter manufacturers. The Apogee AD8000 offers appreciably higher quality than many of the Digidesign converters, along with a host of other useful features.

The Prism Sound ADA-8 offers the highest-quality conversion commercially available – matching or exceeding that of any converters you will find elsewhere. The Prism Sound ADA-8 multi-channel ADDA converter range adds a Firewire (IEEE 1394) interface module compatible with Apple's OSX 'Panther' operating system. Support is also planned for Windows XP. In this configuration, the ADA-8 can provide eight channels of digital AES/EBU and analogue input and output at sample rates up to 96 kHz with software such as Emagic Logic Pro, Apple's Final Cut Pro and many other applications. The ADA-8 provides maximum flexibility and top performance for a wide range of PC and Mac-based recording, editing and sequencing systems including Digidesign Pro Tools|HD and MIX systems, Logic Audio, Final Cut Pro and many others. In addition to Firewire, ADA-8 interfaces include AES, DSD, Pro Tools|HD and MIX. Connection to Pro Tools is direct to the audio card, eliminating the need for Pro Tools I/O hardware. The ADA-8 also allows conversion between its various interface formats, such as for example interfacing to and from DSD (Direct Stream Digital) and PCM systems such as Pro Tools or Logic Audio, or simply converting between AES and DSD.

Expansion chassis

Most Macs and PCs will have at least three PCI slots available to take up to an HD3 system – and you can use up to six Pro Tools cards in some computers. However, if you want to add extra cards, or to use SCSI cards or video cards, you may need to use an expansion chassis.

> Note: To achieve maximum track counts you may need to use a Digidesign-qualified SCSI Host Bus Accelerator (see the Digidesign website for details).

An expansion chassis connects to your computer via a cable to its own PCI card that sits in your computer. Thirteen-slot expansion chassis support up to ten Digidesign cards although the number of HD or MIX cards is limited to seven – the other cards could be the AVoption video capture cards or the discontinued SampleCell cards, for example. Seven-slot chassis support up to seven HD or MIX cards. Some two- and four-slot CardBus expansion chassis have been approved for use with laptops and Pro Tools TDM (again, see the Digidesign website for details).

Pro Tools custom keyboard

If your Pro Tools system uses a computer that is dedicated to running Pro Tools, it makes sense to add Digidesign's custom USB keyboard. This has all the Pro Tools keyboard commands and shortcuts printed directly onto the keys alongside the normal QWERTY keyboard letters, numbers and symbols. Various keys and groupings of keys are colour-coded to help to further distinguish keys according to function.

DigiTranslator 2.0 option

The Pro Tools 6.1 and later versions, with the DigiTranslator 2.0 option installed, support the new Advanced Authoring Format (AAF) that allows file interchange between compatible systems. At present, this mainly facilitates easy interchange of projects between Avid video systems and Pro Tools audio systems. As more third parties adopt and implement this format, this will improve interoperability in general. DigiTranslator 2.0 also supports the OMF interchange file format that is now widely used by third parties such as Steinberg and Emagic.

MachineControl option

If you need to control the transports of external video (or audio) recording and playback equipment using Sony nine-pin or V-LAN protocols, you can add Digidesign's MachineControl software option for Pro Tools. With Sony nine-pin devices, this also allows you to control the Pro Tools 'transport' from a remote device. This means that you can press Play, Record, Stop and so forth and arm tracks for record either using Pro Tools or using an attached VCR. The 'Remote LTC Mode' allows you to chase LTC while receiving nine-pin record and track-arming commands. An Offset setting for MachineControl is also provided that allows you to compensate for any differences in source tape versions when chasing timecode and MMC.

Avid picture options

If you regularly work on audio post-production in collaboration with Avid picture editors, it makes a lot of sense to add the Avoption|XL to your Pro Tools|HD system. Avoption|XL allows conversion-free capture, import, and playback of up to uncompressed Avid video and audio media directly within Pro Tools. Pro Tools 6.1 offers cross-platform support for AVoption|XL and allows recording and playback of up to 1:1 uncompressed JFIF resolution video. Pro Tools 6.1 TDM software also allows recording, playback and seamless interchange of audio and video files via Avid Unity MediaNetwork shared storage systems with Pro Tools|HD. On Windows XP systems, support for Avid Unity MediaManager even allows users to drag and drop Avid sequences directly from Avid Unity MediaNetwork into Pro Tools sessions – without having to copy a single file.

AVoption|V10 is a professional 10-bit video I/O peripheral that is fully compatible with files created by Avid's Media Composer Adrenaline, providing high quality projection in the highest profile environments. Adding AVoption|V10 to a Pro Tools|HD system enables real-time playback of Avid video directly from the Pro Tools application. AVoption|V10 can play back video content created by a broad range of Avid products, past and present. Supported video resolutions include all Avid uncompressed standard definition video, JFIF, AVR, and DV formats. Because AVoption|V10 works with the same video files as those on an Avid system, bringing a project into a Pro Tools|HD/AVoption|V10 system is as easy as importing an AAF sequence. AVoption|V10 includes the full complement of video inputs and outputs including component, composite, and SDI, and comes with custom Avid software for automated audio/video capture and conform, import, rendering, export, and playback.

Avid Mojo

The Avid Mojo video I/O peripheral delivers cost-effective, integrated Avid video playback – turning Pro Tools into the perfect audio companion for Avid Xpress Pro projects using 1:1 uncompressed 601 or DV25. It can also play several JFIF resolutions from Media Composer systems, including 15:1s, 14:1p (24p), 35:1, and more.

Do you need to use a mixer with Pro Tools?

The Pro Tools Mix window has its own very useable built-in EQs, dynamics, and delays and also lets you access many plug-in software versions of conventional effects processors. Truly total automation and recall of all the mix parameters, including the effects, gives you more detailed control than you can get on most other systems. However, low-level microphone or instrument signals will not drive the line-level inputs on the various Digidesign interfaces successfully, so you will need DI boxes or pre-amplifiers for these. One option is to use the Digidesign PRE, which provides eight high-quality channels of microphone preamplification. This makes a good choice for professional studios that need to use lots of microphones. For smaller project studios, it can be more cost-effective to use a small mixer with suitable microphone and instrument-level inputs – such as the Mackie 1604 or similar. You may also need to monitor other external equipment such as a CD, DAT, cassette, or other instruments, and some kind of small mixer is pretty much essential for this.

Already, many artistes and producers in both the semi-professional and professional fields with Pro Tools systems no longer use large mixing consoles – or use these for monitoring purposes only. This kind of usage is increasing quite dramatically, with the Mackie HUI, Digidesign's Pro Control and Focusrite Control|24 now well established in professional project studios. These controllers make Pro Tools' mixing features much more easily accessible – giving engineers and producers the type of console control surface that they are already used to. It does make increasing sense to use Pro Control instead of a compact digital mixer for the bigger project and post-production studios – and it certainly looks impressive! Similarly, for a project studio recording bands where lots of microphones are required, the Control|24 makes a lot of sense, while for smaller studios working with lots of MIDI tracks and normally overdubbing just one or two tracks at a time, the HUI is a better choice. With an expanded Pro Tools HD3 (or larger) system that has plenty of DSP available for plug-ins, you can have enough signal processing available to produce very ambitious mixes with Pro Tools. And if you only need to record or overdub instruments in mono or stereo, then the pair of mic pre-amps in the HUI would be fine. You can always add extra mic pre-amps if you need more – or go for the Focusrite Control|24 instead.

So would I use Pro Tools with a controller rather than an external mixer? Well, this would depend on what Pro Tools system I could afford and on the kind of work I was doing. If my work was mostly post-production and editing with occasional overdubs and if I had plenty of DSP available, then I would be happy to work with just a hardware controller, but in my situation using a mixer as well makes much better sense. I have a Pro Tools|HD1 system with one 96 I/O interface at present. I plan to expand this by adding an HD Accel card. This system will let me use a fair number of plug-ins – but never enough when I have, typically, 32 tracks of audio in a mix session. So I route several of the more important tracks to my Yamaha DM1000 so that I can EQ, compress or add reverb or delay effects from the DM1000 without needing to use Pro Tools plug-ins. This also has the advantage of allowing me to use physical faders on the DM1000 for any tracks that need the tactile control that only real faders can provide.

Pro Tools TDM software

The biggest of the changes that came with Pro Tools 6.x software versions was the move to Mac OSX. Yes, it is true that the user-interface has a more colourful, three-dimensional 'look' than before, and there are some changes to the way you access some of the controls and settings – but, by and large, the new features added were incremental improvements rather than any revolutionary redesign of the software. Specifically, the way you select which views and which rulers are displayed in the Edit window, the Tab to Transients, Commands Focus and Timeline Selections buttons, and the grid and nudge values have been tidied up into a convenient area – just above the rulers display. Everything else is more or less where it was and as it was before – so you should not have too much trouble finding your way around if you are upgrading. Digidesign has also made great efforts to achieve parity between the Mac OSX versions and the Windows XP versions – which are now within 'spitting' distance of each other features-wise.

Recent Pro Tools versions feature four types of so-called DigiBase browsers that let you organize your files better. When using the Workspace browser to display all your mounted disk drives, for example, you can simply drag and drop any audio file or any CD track that is available in your Workspace onto any audio track in the Edit window of the current Pro Tools session – or into the Audio Regions list. The user-interface has also been improved by the addition of a 'Playback Cursor Locator' that lets you quickly navigate to wherever the Playback Cursor is positioned on-screen, and there are several useful new editing features. There is also a powerful new Relative

Grid Mode that lets you edit audio and MIDI regions that are not aligned with Grid boundaries as though they were. The MIDI features were 'supercharged' for version 6, so, for example, Pro Tools sessions now support up to 256 MIDI tracks. The supplied set of plug-ins has also been improved, by the inclusion of the basic D-Verb reverberation plug-in, for example, and ReWire technology has been implemented to allow Pro Tools users to work with Propellerheads Reason and Ableton Live.

The Edit and Mix windows are the main Pro Tools work areas. The Edit window provides a timeline display of audio, as well as MIDI data and mixer automation for recording, editing and arranging tracks. Each track has controls for record enable, solo, mute and automation mode. And, using the Display menu options, you can also reveal the input and output routing assignments, the inserts, the sends, the comments, or any combination of these – which makes it possible to work with just the Edit window for most of the time. The Edit window lets you display the audio and MIDI data in a variety of ways to suit your purpose, and you can edit the audio right down to sample level in this one window.

In the Mix window, tracks appear as mixer channel strips with controls for signal routing (inserts, sends, input and output assignments), volume, panning, record enable, automation mode, and solo/mute. From the Display menu you can choose to display all of these items, or just some of them. Pro Tools provides three different types of audio tracks – mono, stereo and multi-track Audio tracks; Auxiliary Inputs; and Master Faders – as well as the MIDI tracks. The Mix window lets you balance all your levels, pan the sounds to create a stereo or multi-channel effect, and process the audio using software plug-ins or using external devices connected via the audio interface. Pro Tools lets you use up to five Inserts on each audio track, Auxiliary Input, or Master Fader. Each insert can be either a software plug-in or an external hardware device. You can also use up to five Sends on each track. Sends let you route signals via internal buses or to audio interface outputs so that one plug-in or one external signal processor can be used to process several tracks at once. Track Input Selectors let you choose the input source for audio tracks and Auxiliary Inputs. Track input can be a hardware input, or bus. Track Output Selectors let you route the post-fader signals to the assigned output or bus paths. The Output Selector routes the main track output to the chosen main or sub-path. Tracks can be routed directly to hardware outputs, or to internal bus paths for sub-mixing. The Pro Tools mixer also has comprehensive automation facilities including both 'snapshot' and real-time automation. There are four real-time automation modes – Read, Touch, Latch, and Write – plus a Trim mode.

Surround mixing features come as standard, with support for all currently defined surround formats up to 7.1. So you get a choice of mono, stereo or multi-channel tracks to work with. Not only can you mix in every popular surround format, including LCRS, 5.1, 6.1, 7.1, you can also work in several surround formats at the same time by assigning multiple outputs and send destinations for each audio channel. The plug-ins also work in these multi-channel formats so you can add signal-processing in surround. Once your surround mix is complete, you can deliver several versions simultaneously by assigning tracks to multiple output paths at the same time. For example, if you're working on a 7.1 mix, you can set up Pro Tools to also give you outputs for, say, a 5.1 and a stereo mix.

Digidesign's original audio file format, the Mac-only Sound Designer II format, is fast becoming history. The default file format now is BWF (.WAV). Improved cross-platform support allows both the Mac and PC versions to record and play both AIFF and WAVE files, and the Pro Tools session file format can be opened on either platform without needing any conversion. Also, you can import any track, complete with all its parameters and assignments, from any other Pro

Tools session. Another extremely useful feature allows you to open and work with offline media. Pro Tools can open and modify a session even if all the audio or video files for that session are not currently available. And any edits that you make to tracks containing offline media are reflected in the session when the files are available again. Also, to conserve DSP resources in a session, tracks, I/O assignments and plug-ins can now be set to 'inactive'. Inactive items retain their various settings, routings and assignments, but are taken out of operation – freeing the DSP they were using for other uses. The original settings will remain saved so you can always see what you have deactivated and return to these at any time. Even better – when you move a Pro Tools session to a system that has different plug-ins and I/O configurations, Pro Tools will automatically deactivate tracks, plug-ins, sends, or I/O channels as necessary while letting you preserve your original session settings so you can return to these when you move back to the original system.

All these features, and the others that you will learn about in this book, have helped Pro Tools to reach the top spot as software of choice for the world's top recording engineers.

See Appendix 4 for details of Pro Tools 6.4.

Summary

Pro Tools TDM systems can be built up to suit just about any application using some combination of cards, interfaces and peripherals. With the release of version 6.1, the user-interface has been significantly enhanced – particularly by the inclusion of the new DigiBase features. Upgrades from older systems are now essential in order to use the latest G5 computers with OSX and to get the benefits of the latest features for Windows XP. And if you are still using another system, you should strongly consider changing to Pro Tools.

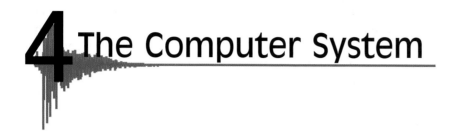

4 The Computer System

Introduction

Every Pro Tools system runs on a computer system, so you will need to know how to choose the best computer system for your application. Pro Tools was developed originally for Macintosh computer systems – which I prefer to use – so much (although not all) of this chapter will be Mac specific. You also need to know about disk maintenance and repair and at least the basics of troubleshooting – when the computer won't even blink at you this can be a real session-stopper! Even if the studio has a maintenance engineer or support engineer to keep the Pro Tools systems running, and assuming he is always on hand, you should still be aware of what the issues are and what can and cannot be fixed if there is trouble. Recording engineers and producers, composers and musicians, especially if they have been brought up on analogue technologies, will probably be unfamiliar with much of this stuff. Nevertheless, if you can handle an SSL console or a Yamaha 02R, or use Logic Audio or Sibelius, you should not have too much trouble with anything here. But don't even dream of skipping this chapter if you are working professionally with a Pro Tools system, as having a grasp of the information presented here will 'save your bacon' time and time again.

What computer should you choose?

Dealers often recommend the basic or intermediate-level models for use with Pro Tools systems, but I always recommend that you get the most powerful computer you can afford. It is possible to use a basic or intermediate level model if you are using Pro Tools software with TDM plug-ins, because the processing for the plug-ins and mixing is provided by the Digidesign HD or MIX cards while the computer's CPU mostly takes care of the user-interface and file handling – tasks which are not too demanding of CPU power. Nevertheless, if you want to use AudioSuite or RTAS signal processing, this uses the computer's CPU. Also, many third-party software applications run on the computer's CPU and virtual instruments, for example, can be very demanding of CPU time. You should definitely choose a dual processor model if you are using Logic Audio, Cubase VST, Digital Performer or the various software synthesizers on Mac OSX or Windows XP systems (these operating systems automatically spread the processing load between dual processors).

The Power Mac G5

Apple PowerMacs typically come in three basic models – basic, intermediate and high-level configurations to suite various requirements. These are mainly distinguished by the speed of the

CPU, the size of the internal hard drive, and the amount of RAM fitted as standard. Other differences may include the size of the power supply, speed of the internal busses and caches used, and additional CD-ROM or DVD drives fitted. Some models can be ordered with dual processors.

Figure 4.1 Apple Power Mac G5.

Technically, the Power Mac G5 is superb – It is the first personal computer with a 64-bit processor and it can use up to 8 gigabytes of main memory. Apple claims that the G5 is the fastest personal computer available, but in reality there are faster PCs. Nevertheless, the G5 is no slouch – especially if you go for the top-of-the-range Dual 2 GHz model. The Power G5's system architecture features a 1 GHz front-side bus on each processor, providing ultra-high bandwidth, and the G5 has three PCI-X slots – the latest advance in PCI technology. The PCI-X protocol is perfect for high-performance PCI devices, increasing speeds from 33 MHz to 133 MHz and throughput from 266 MBps to 2 GBps. The physical design of the Power Mac G5 is also first rate. It has a great-looking chassis constructed of anodized aluminium and the enclosure features an easy-to-open access panel for speedy access to internal components. And with front-mounted FireWire, USB, headphone ports and handles, the Power Mac G5's enclosure certainly delivers on convenience. Apple divided the inside of the Power Mac G5 into four discrete thermal zones, compartmentalizing the primary heat-producing components – processor, PCI, storage and power supply. This allows the system to increase or decrease the temperature of a single zone without affecting the others. Each of the four thermal zones is equipped with its own dedicated, low-speed fans. Apple engineered seven of the nine fans to spin at very low speeds for minimum acoustic output. And Mac OS X constantly monitors component temperatures in each zone, dynamically adjusting individual fan speeds to the appropriate levels for the quietest possible operation. As a result, the Power Mac G5 runs three times quieter than the previous Power Mac G4 enclosure – and much quieter than most PCs – giving it a distinct advantage if it is used in a music studio control room.

The PCs

There are lots of PCs 'out there', but the buyer definitely has to beware! Top models from Hewlett-Packard, IBM, and Dell are a good bet, and there are custom machines available (from companies such as Carillon in the UK) that are guaranteed to work with Pro Tools systems. The bottom line is that you do have to check Digidesign's compatibility recommendations on their website to see which computer models they have qualified for use with Pro Tools. This list is updated as new models are released, but there is always something of a lag between a new model appearing and its passing through Digidesign's qualification program.

Computer monitors

I recommend that you get the largest monitor you can afford and have room for. And, ideally you should have two of these. A third would not be completely out of order, either. I can highly commend the LaCie 22-inch monitors that I am currently using – although Sony, Mitsubishi, Formac and others also make excellent large monitors. Thin, flat monitor panels such as the Apple 23-inch Cinema Display look very 'sexy' and take up much less space than conventional monitors. However, these cost much more than the price of an equivalent conventional monitor – and the picture quality is not always as good. A minimum recommendation would be a pair of 17-inch monitors – one for the Pro Tools Edit window and one for the Mix window.

The Sony VAIO

Now if you need a laptop, why not try the Sony VAIO range – especially the top-of-the-range models with clock speeds greater than 2 GHz, 16.1-inch 1600 × 1200 pixel LCD display screens and DVD-R/CD-R combo drives. The VAIO is fast! And, unlike the Apple PowerBook, the VAIO doesn't get too hot either! It has a very efficient cooling system that keeps the CPU, chipset and graphics chip wonderfully cool. Also, it is very quiet in operation compared with most other laptops. With its 16.1-inch screen, it is possible to regard this laptop as an alternative to a desktop if you are limited for space – or if you value the extremely cool-running and quiet operation of the VAIO.

Figure 4.2 Sony VAIO GRX616SP.

The Apple PowerBook

Of course, diehard Apple Mac users will probably prefer an Apple laptop, such as the top-of-the-range 1 GHz PowerBook G4 with its 17-inch screen. This offers the same viewing area as a 19-inch CRT monitor and supports a resolution of 1440 × 900 pixels. This size of screen really does make this laptop a viable alternative to a desktop model. However, the Sony VAIO GRX616SP with its 2.4 GHz clock speed will run more than twice as many plug-ins and virtual instruments.

Figure 4.3 17-inch PowerBook G4.

Recommended additional software

Every Power Mac comes with easy-to-use software tools for creating music CDs, interactive DVD videos and Desktop Movies. Every model includes iTunes software and a CD-RW drive. Besides letting you listen to CDs and hundreds of Internet streaming radio stations, iTunes enables you to create, play and organize MP3 files into playlists, then burn these to CD. iTunes is certainly a useful item to have access to, but professional users will find that Roxio's Toast software actually has the features they will need most. Power Mac models with the built-in SuperDrive come pre-loaded with iMovie, which lets you digitize and edit video in a basic fashion. They also include Apple's iDVD software, which lets you create basic DVD titles. Professionals are likely to prefer Apple's award-winning Final Cut Pro software (for sophisticated video editing, compositing and special effects) and DVD Studio Pro (a complete set of interactive authoring and production tools for producing professional-quality DVDs from start to finish). The important point here is that you can not only burn a CD to play on your hi-fi or in your car, but you can also burn a DVD on your desktop that you can play on your home-cinema system. Now that's progress! Mac users can choose from a range of video applications. After Effects, again, is the popular choice for motion graphics along with PhotoShop for still images. Bias Peak is a popular audio application used to extract audio from video files and replace it with edited or newly created audio. Avid editors often choose Avid Xpress Pro software for video editing, appreciating the compatibility with the high-end Avid systems. Most other video editors are choosing Final Cut Pro – Apple's award-winning video editing software. And for DVD authoring, Apple's DVD Studio Pro is the software of choice on the Mac.

For the PC platform, the Adobe Video Collection includes Premiere Pro for video editing; Audition (formerly Cool Edit Pro) to let you edit audio for the video; After Effects Professional for motion graphics and visual effects; PhotoShop for working with still images and effects for video; and Encore DVD which provides professional DVD authoring facilities. Adobe Creative Suite Premium, available for both Mac and PC, includes PhotoShop with ImageReady for working with bitmapped images and Illustrator for working with vector graphics – great for creating diagrams with lines, circles and other geometric shapes. The Suite also includes a desktop publishing package called InDesign. This is just what you need for preparing anything from a promotional leaflet to a studio brochure or instruction manual. The big new thing in publishing, of course, is the World Wide Web, so the Suite includes GoLive which lets you put together and manage anything from a single web-page to a full-blown interactive website. Most professional musicians and studios now have websites – and GoLive is a pretty essential piece of software to have if you want to put a website together yourself. Adobe's Creative Suite also includes the full professional version of Acrobat that lets you both read and write PDF files. Acrobat at first sight appears to be a simple utility. It lets you read and create Portable Document Files (.pdf) on your computer. These so-called 'Acrobat' or 'PDF' documents contain all the information necessary to recreate the layout and font images used when these documents are opened on any other computer that has Acrobat installed. This format is often used for the 'Help' files and manuals that come with software packages. Acrobat files are also common on the Internet as a format to enable downloading of articles and other information that you may need.

> Note: Acrobat Reader is supplied free with Pro Tools systems but you may need to download updated versions from Adobe's website from time to time to keep up with upgrades, as older versions will not always work with files saved using newer versions.

Microsoft Office is the standard package that includes a word processor, spreadsheet and so forth. This is ideal for letters, accounts, billing, and the like – and you sometimes come across help files supplied in Microsoft Word format. There are cheaper alternatives, such as Microsoft Works, but I recommend sticking with the industry standard here to avoid compatibility problems. FileMaker Pro is the most popular database software for the Mac, and is also available for the PC. Again, this is useful to have around – not only for its powerful database capabilities which can be used to keep lists of clients and so forth, but also because useful technical information is occasionally supplied in this format.

You also need a suite of hard disk utilities such as Norton Utilities or Tech Tool Pro. Norton checks your drives for faults and fixes these wherever possible. It also lets you de-fragment and optimize your drives and carry out various other maintenance and file recovery tasks. Tech Tool Pro does all this and more – it also lets you check out the status of your hardware and will report if you have hardware faults in RAM or CPU or peripherals or whatever. You also need a virus checker such as Norton Anti-Virus. The Mac is not as bad as the PC when it comes to viruses – but they certainly do exist and can lead to inexplicable crashes and software malfunctions.

Software updates

Don't forget to update your software as appropriate. Often, this can be done via the Internet, which is a good reason why you should have an Internet connection available to your computer system. Utilities such as Norton need to be updated regularly to keep step with new peripherals and CPUs, as do Virus checkers (new viruses appear all the time), application software and operating system software.

It is a question of judgement as to when to upgrade to a new operating system. Often, the first release of a major new version will be buggy and application software may not yet be available which has been tested and 'tweaked' for compatibility with this. It is often wise to wait for a subsequent release – and to check the technical information posted on your application software's website for news about compatibility. Digidesign, for example, publishes comprehensive compatibility charts on their website for this purpose.

> Note: Another problem is that some older software will not always work properly with a newer operating system. This can provide another good reason for staying with an older, but known-to-be stable, operating system. Of course, eventually you will want to upgrade to take advantage of new features that only work with newer operating systems.

With application software, the same comments apply in general, and some users may decide that they do not require the extra features available in a new version. However, new versions often include fixes for previous buggy behaviour, compatibility with the latest operating systems, and other enhancements to the existing feature set – so I recommend that you keep up to date with new releases wherever possible. You can get up-to-date information from other users, from user-groups on the Internet, and from popular magazines.

Leaving your computer 'on' permanently versus switching 'off' when you are not using it

Many people ask whether it is better to leave your computer (and any other electronic equipment) switched on when you are not using it. There are two things to consider here. Firstly, when the equipment is on it gets warm. Heat ages electronic components, so the longer it is left on, the more chance there is of component failure. Keep in mind, however, that cold and damp don't help either. The second factor is that turning on and off can lead to stressing of various components due to power surges. If a component is going to fail, it is more likely to do so when the equipment is turned on or off.

Computer (and most electronic) equipment is designed to be left running day and night year in, year out and will last for several years with no breakdowns if you are lucky. Of course, you cannot realistically expect to never have a power supply breakdown or a monitor screen burn out – or whatever.

I generally leave most of my equipment turned on all the time – always ready for action – unless I specifically know that I will not be using it for some time.

I have been assured by several of the most well-informed technical engineers that the pros and cons for leaving electronic equipment on permanently against turning it off regularly just about balance out – so there is no great advantage or disadvantage either way.

Having the equipment ready for action at all times is the advantage which tips the balance for me toward leaving much of my equipment turned on at all times.

Hard drives for Pro Tools

The hard drives you use for recording are extremely important components of any Pro Tools system. Remember that these drives use disks coated with magnetic material that can suffer physical damage just as magnetic tapes can. Not as easily as open-reel tapes perhaps, but they can get damaged nonetheless. A small area of damage would correspond to a tape drop-out while major damage would correspond to a crumpled or broken tape. Just as tape machines can run at different speeds, so can hard drives. Spin speeds of 7200 rpm are commonly used for audio drives and 10 000 rpm drives are used for more demanding situations and for use with digital video. Another measure of disk drive performance is the Average Seek Time, i.e. how fast the read head can get to the data. Ten milliseconds would be about the slowest Average Seek Time that would provide acceptable performance for Pro Tools systems.

Before choosing hard disks to use with your Pro Tools system, you should always check the Digidesign compatibility guidelines on their website at www.digidesign.com for recommendations regarding these. Drives other than those recommended may work – but they may not. The exact model of the drive, and even the exact version of the firmware in the drive, can be absolutely crucial to the successful operation of the drive with a Pro Tools system.

Storage requirements for Pro Tools systems depend on how many tracks you will be recording and at which sample rate and bit resolution. As with all Pro Tools systems, drive performance depends on a number of factors, including track count, edit density, CPU speed, single or dual processors, and the use of cross-fades or processing like Beat Detective. Digidesign advises that you should use a dedicated internal or external audio drive or drives with Pro Tools systems. Specifically, recording to or playback from system drives is not supported. Digidesign also recommends that you keep the Pro Tools TDM application software on your start-up drive – i.e. the drive that contains your operating system and other system-related files. Your data files – session files, audio files and fade files – can be located on any compatible drive.

A common question is 'Can you use the internal hard drive to record audio?' The short answer is 'Yes – but do so at your peril!' There are several reasons why this is generally not a good idea. When you work with digital audio your hard drive can quickly become fragmented – especially if you make a lot of edits. You will need to regularly defragment the drive using Norton Utilities or similar software, or back-up your files, initialize (or sometimes completely re-format) the disk, and restore the files before each important new session. This is not so easy to do with your boot drive as this may contain copy-protection keys for your music software that need de-installing before you can initialize the drive and re-installing afterwards. So a dedicated audio drive is always the best idea. Nevertheless, you can record to system drives when necessary – for example, if your computer system has just one hard drive or if your other hard drives have run out of space.

For maximum performance, SCSI drives are recommended for Pro Tools TDM systems and these can be connected to the internal SCSI bus or the external SCSI bus of your computer (on older models), or to a SCSI host bus adapter card in your computer. FireWire and ATA/IDE drives are also supported.

Apple computers now use fast, large capacity internal ATA/IDE drives like the drives fitted to Windows PCs. There is plenty of space on these drives to record audio, so you can use the internal drive to get started. However, you should partition the drive first, with, say, one partition

> Note: You cannot spread audio files from one session across different types of drives with Pro Tools|HD Systems. For example, if you have a session with audio files on SCSI drives, you cannot record or play back Pro Tools files from an IDE/ATA or Firewire drive. (This applies to audio files only, video can be on a different type of drive, or the same type of drive as the audio files in a session.)

> Tip: When using QuickTime movies with Pro Tools|HD systems, for reliable performance of 16 or more audio tracks, video should be stored on a separate drive.

for your software, one for your audio, and one for any video. Most IDE drives are fast enough to let you work with 24-track sessions, as long as you don't push these to the limits with too many tracks or edits and cross-fades. You can always add an external SCSI hard drive as soon as your budget allows and as your demands on the system grow. Many of the popular SCSI cards have an internal connecter that can be used to hook up an internal SCSI drive, and there is usually space inside the computer to add one of these. However, it is more flexible to use an external drive as you can always connect this to another computer, or just take the drive to another studio, or to the repair shop or wherever.

The advantage that SCSI hard drives have over even the latest, fastest IDE drives is that the faster SCSI drives still have the edge when it comes to performance – especially when recording to a large number of tracks. If you try to record too many audio tracks with an ATA/IDE drive you will get a short delay before recording begins – which is definitely not what you want when the musicians are out there ready to play in the studio. Also, bear in mind that when you choose to work with 24-bit files you can only record two-thirds as many tracks for two-thirds the recording time as when you are working at 16-bit resolution – onto the same size of hard disk. If you want to be able to use more audio tracks, your storage requirements will obviously increase. And your bandwidth requirements will also increase substantially when working with 24 or more tracks of 24-bit audio. In this case, you will almost certainly need to use a SCSI card with large-capacity *fast* and *ultra-wide* hard disk systems to achieve the required data throughput from the drives.

Disk requirements and bandwidth

Analogue to digital conversion

If you are sampling your audio at 44.1 kHz, this means that the A/D converters are measuring the amplitude of (i.e. 'sampling') the incoming analogue audio signal 44 100 times a second. The bit-depth of the A/D converters dictates the fineness of these measurements. Sixteen bits gives you a range of 65 536 (i.e. 2 to the power 16) numbers that you can use to measure these amplitude values. The measured values are represented using binary numbers and these numbers form the 'digital' output from the converters.

So, the situation is this: each 44 100th of a second, a 16-bit number representing the amplitude value of the analogue signal is generated by the A/D converter and is output as 16 binary digits – i.e. 16 bits.

Audio data transfer rates

To work out how many bits are output from the converter each second, you simply multiply the number of bits output each 44 100th of a second by 44 100 – in this case, $16 \times 44\,100 = 705\,600$ bits per second (bps) or 705.6 kilobits per second (kbps).

[Note that 1000 is 10 to the power 3 and is represented by the abbreviation kilo, or k for short.]

It is often more useful to convert this figure to bytes, where there are eight bits in every byte. In this case the data rate would be $705.6/8 = 88\,200$ Bytes per second or 88.2 kiloBytes per second (kB/s or kBps)

The rate per minute is 60 times the rate per second, i.e. $60 \times 88.2 = 5.292$ MegaBytes (MB).

Audio data storage sizes

If you want to work out how much hard disk space will be needed to hold this audio data, then you need to make yet another conversion – this time to take account of the way hard disk drive sizes are quoted in multiples of kiloBytes, where 1 kiloByte is equal to 1024 Bytes – commonly referred to as 1 K of storage.

[Note that 1024 is 2 to the power 10 and is represented by the abbreviation (uppercase) K.]

In this case, the data rate of 88 200 Bytes per second must be divided by 1024 then multiplied by 1000 to get the number of kiloBytes of disk storage required to hold this data, i.e. $(88\,200/1024) \times 1000 = 86.13$ K for 1 second of audio (KB/s).

Again, the rate per minute is 60 times the rate per second, i.e. $60 \times 86.13 = 5.168$ MB.

> Note: This data rate is often quoted as (approximately) 5 MB/minute for a single channel/track of 16-bit, 44.1 kHz audio. This figure is a reasonable practical approximation to use for either data transfer rates or disk storage sizes.

Multichannel audio data transfer rates

Now let's look at the data transfer rate required for two channels, i.e. stereo. If we take the rate in kiloBytes based on powers of 10 and multiply this by 2 we get $88.2 \times 2 = 176.4$ kiloBytes/sec. So if you are working in stereo at 44.1 kHz sample rate and 16-bit resolution, you have to be able to transfer information around your system at the rate of 176.4 kiloBytes per second for everything to work properly.

The computer's PCI bus must be able to transfer data at this rate, the CPU/RAM must be able to cope, and the hard disk must be able to handle this data throughput. And don't forget that as you increase the number of audio tracks, you increase the amount of bandwidth required of the system. There are limits to the speed of data transfer (i.e. the bandwidth) of any system. Four tracks of audio would require 352.8 kbps, eight tracks would need 705.6 kbps, and so forth. Every link in the signal chain must be capable of handling what you are asking it to handle, or your audio will get messed up along the way to its destination and will sound distorted or stutter.

Audio can also be quite 'hungry' for disk space – depending on the sample rate and the bit depth. As we know, 44.1 kHz, 16-bit files need 5 MB per minute for each mono track. At 48 kHz each mono track requires 5.5 MB/minute, at 96 kHz it requires 11 MB/minute, and at 192 kHz it requires 22 MB/minute. If you are recording 24-bit files, at 44.1 kHz this requires 7.5 MB per minute per mono track. At 48 kHz it requires 8 MB/minute, at 96 kHz it requires 16 MB/minute, and at 192 kHz it requires 32 MB/minute.

> Note: If you choose to use 24-bit files, you will quickly realize that you need additional hard disk space to accommodate these, as they require 50% more hard disk space than 16-bit files. For example, mono, 16-bit 44.1 kHz audio needs about 5 MegaBytes/minute of disk space, while mono, 24-bit, 44.1 kHz audio needs 7.5 MegaBytes/minute.

Audio file size limit

Pro Tools has a single audio file size limit of 2048 MB. This equates to about 4.5 hours at 24-bit, 44.1 kHz.

Hard disk formatting

Mac hard disk drives can be formatted using the older Hierarchical Filing System (HFS) or the newer HFS+ that was designed to better cope with lots of small multimedia files. If you are working with relatively long audio files there is no major advantage to HFS+, but if you are working with lots of small samples it will definitely be advantageous.

Digidesign recommends using the Mac OS Extended (HFS+) formatting option for use with Pro Tools, although Mac OS Standard (HFS) can also be used. All drive types, including IDE, FireWire and SCSI drives, should be formatted with the Apple Disk Utility that comes with Mac OSX. You can find this in the Applications/Utilities folder on your OSX hard drive, or in the Apple Menu when you boot from the OSX installation CD.

> Note: It is not necessary to use the ATTO ExpressStripe disk-formatting software (included on the Pro Tools 6.0 CD) with SCSI drives and Pro Tools 6 – the Apple Disk Utility works just fine and is recommended by Digidesign.

The PCs running Windows XP can use the NTFS file system. This removes the restrictions on file sizes that made older file systems unsuitable for use with video systems that often create files greater than 2 GB in size. However, NTFS formatted drive partitions cannot be 'seen' from older operating systems such as Windows 98. So, for greater compatibility, especially on a dual bootable Windows XP/Windows 98, SE or Millenium computer, the FAT 32 file system is the one to choose.

Disk failure

In case you were wondering, hard drives can and do break down. If you are lucky, it might just be the power supply. This can be easily fixed in a repair shop and your data should be intact. But the drive itself can suffer various kinds of malfunction. The bearings that spin the disk platters can break down or the read/write head can crash into a spinning disk platter and cause severe mechanical damage. In such cases there is much less chance that you can ever recover any data, and even to try you will have to send the drive to a data recovery specialist. Such specialists do exist, but the price you will be asked to pay if they can recover your data may be prohibitively high – hundreds or even thousands of pounds. So if you really value your data, here's what you have to do: first you backup; then you backup again and keep this copy off-site; and, finally, make a safety copy for luck.

Disk maintenance and repair

In the case of a catastrophic hardware failure, you have no choice – you have to take the machine back to the shop to have a replacement drive fitted, unless you feel confident to do this yourself. Luckily, most everyday drive faults are much less severe, mostly involving the disk directories and corrupted files, and can be fixed using software. Perhaps the most essential utility that Pro Tools users will need is a suite of disk utilities such as Norton Utilities. This type of software lets you diagnose and repair damaged disks, optimize their performance, recover accidentally deleted files, recover data from an accidentally initialized or crashed disk, and keep your computer virus-free. And, before you ask, yes, you will need to do all of these things at some time or other – and you can never know when! A disk can go down the day you buy your computer or several years later – or at any, and possibly many, times in between. Viruses are generally not as common or as destructive on the Mac as on the PC, but you may well encounter one or more of these at some point – even on the Mac. Norton can be set up to check every file that exists on or is introduced to your disk drives to help prevent any such unwanted occurrence. But even if none of these things ever happened, Norton is absolutely essential for Pro Tools users – in order to keep your audio disk drives optimized.

Under normal conditions of usage, the files on hard drives become 'fragmented' as you edit these and save new versions. The operating system writes the edits to any place on the hard drive where they will fit – and not necessarily next to where the original file is located – so you get a fragmented file. The problem is that the drive's read head may have to move around very rapidly to try to continuously read out audio data from a fragmented file – and there is a limit to how fast the head can move. At some point, even with relatively low amounts of file fragmenta-tion, the read head cannot keep up, and the audio playback will stutter.

To keep your drives in optimum condition, you should defragment them after every heavy editing session. An alternative is to back up all your Pro Tools sessions and any other audio files, then wipe your audio hard drive clean – i.e. re-initialize it. Then restore your Pro Tools sessions and audio files. These will be written as contiguous files onto your newly initialized, empty disk partition – with no fragmentation, by definition.

Small disk faults will also occur with normal usage during the working life of any drive and these faults can lead to corrupted files – especially if any of the bits get 'flipped'. Norton's Disk Doctor examines the disk directories and files on your storage drives and fixes any small faults it finds automatically (it asks you first). If you run a full check on the disk media, it will check for faulty areas on the disk and will attempt to move any affected files away to god areas then 'map out' the faulty sectors so that you avoid them in future.

As I am sure you will agree, this is all essential stuff that you need to be aware of and able to sort out yourself as you work with your Pro Tools system. Even if you have the luxury of a dedicated maintenance engineer in your studio, you just know that problems are going to arise when he or she is not around!

SCSI interface cards

If you are working with 24 or more tracks of 24-bit audio, a SCSI accelerator card with the fastest SCSI drives is essential for best performance with the HD and MIX cards. The data rate required for 24 tracks of 24-bit audio is around 150 kBytes/sec per track which adds up to about 3.6 MBytes/sec of sustained throughput – which can only be achieved using very high-performance drives. Disk storage for 24-bit audio also runs at about 7.5 MBytes/s per mono minute of recording at 44.1 kHz and is correspondingly greater at higher sample rates – so hard disk space requirements have increased fairly substantially compared with the original 16-bit systems. Fortunately, prices of hard drives have come down and hard drive capacities have gone up. For all Pro Tools systems, you need to check the Digidesign website at www.digidesign.com to see what the recommendations are for your particular system.

Personal computers rarely come with SCSI interface cards as standard these days, so you will probably need to add a SCSI card to your system if you want to use SCSI drives. Digidesign offer their SCSI-128 Kit, based on the ATTO Express PCI UL3D SCSI card, which is qualified for 128 tracks at 48 kHz. The older SCSI-64 Kit (still available at the time of writing) based on the ATTO Express PCI-DC, is qualified for 64 tracks at 48 kHz with the G4, but is not compatible with the G5.

> Note: The optional SCSI HBA cards bundled by Apple with some G4 computers are not compatible with Pro Tools due to the lack of NVRAM to store the desired SCSI transfer rate. This means that these cards cannot be configured for use with Pro Tools. Also, Adaptec 2940UW and 3940UW cards with G4s are unsuitable for full Pro Tools performance. Due to PCI bus related issues the Adaptec 2940UW and 3940UW SCSI HBA cards are unable to provide full Pro Tools performance – and Adaptec has discontinued these cards.

SCSI hard drives

One SCSI drive guarantees you 32 tracks at 44.1/48 kHz, 16 at 96 kHz and 6 at 192 kHz. Two SCSI drives guarantees you 64 tracks at 44.1/48 kHz, 32 tracks at 96 kHz and 12 tracks at 192 kHz. Four SCSI drives guarantees you 128 tracks at 44.1/48 kHz and 64 tracks at 96 kHz. If you are using four SCSI drives, you are likely to be using an expansion chassis. The track count at 192 kHz with a 64-bit chassis is 24 – although this reduces to 16 with a 32-bit chassis.

SCSI drives for use with Pro Tools should have either the wide single-ended or low-voltage differential (LVD or Ultra 160) SCSI interface type. The disk drive rotational speed should be 10 000 rpm or faster and the buffer size should be 512 K or larger.

IDE/ATA hard drives

Apple has been using IDE drives since the G3 series and all G4 and G5 models are fitted with IDE internal drives. PCs also use IDE drives instead of SCSI – although high-end machines such as the IBM Intellistation M-Pro can be supplied with SCSI as standard.

IDE drives cost less than SCSI drives but they are not generally as capable. The data transfer rates you can get range from 3.3 to as much as 66 MB/s with the latest drives. With SCSI, data transfer rates range from 5 up to 160 MB/s. So SCSI drives are typically faster than IDE drives, although this is not always the case – it depends on the actual model of drive. IDE works with up to four internal hard drives and/or CD-ROM drives, but not with external devices such as MO drives, tape drives and so forth. SCSI works with either seven or 15 devices depending on whether you are using 'narrow' 8-bit or 'wide' 16-bit devices – and these can be internal or external. Also, SCSI drives use dedicated controllers, allowing greater efficiency, while IDE drives use the host computer for handling data.

Users of older Macs or PCs should note that the built-in SCSI controllers or older SCSI cards are likely to be too slow for best results with Pro Tools and similar digital audio systems, so these models will still need a dedicated SCSI card.

Compatible IDE/ATA drives are qualified for up to 32 tracks at 44.1/48 kHz with Pro Tools TDM systems.

Firewire hard drives

Firewire is a digital interface, also known as IEEE 1394, which was originally developed by Apple and is now featured on both Macs and PCs. It is intended to allow data to be transferred digitally between computers and peripherals such as hard drives, digital cameras and camcorders, and so forth. The G4 and G5 PowerMacs no longer have a built-in SCSI interface – they use Firewire instead.

The Firewire bus has a data transfer rate of 400 Mb/s – which is equivalent to (400/8) or 50 MB/s – so Firewire drives are not as fast as the faster SCSI drives at present. Another issue with Firewire is that all the connected devices share the available bandwidth. This means that if you are using several Firewire devices at the same time the data transfer rate achievable per device will be reduced as a consequence of this sharing. The latest implementation of Firewire has a theoretical data transfer rate of 800 Mb/s (or 100 MB/s) – which is still not as fast as the faster SCSI implementations.

The DigiDrive FireWire 80 and Avid MediaDrive rS80 have been qualified with Pro Tools|HD and Pro Tools|HD Accel systems for up to 24 tracks per drive at 48 KHz/24-bit at 500 ms edit density, with a maximum of eight drives for a total of 192 tracks at 48 kHz/24-bit at 500 ms edit density.

Using one of these 80 Gb Digidesign FireWire drives with a Pro Tools|HD system guarantees you 24 tracks at 44.1/48 kHz, 12 tracks at 96 kHz and 6 tracks at 192 kHz. Using a fully expanded Pro Tools|HD Accel system with eight 80 Gb Digidesign FireWire drives guarantees you 192 tracks at 44.1/48 kHz, 96 tracks at 96 kHz and 36 tracks at 192 kHz.

Hard drives summary

The IDE internal drives fitted to today's computers will play back 24 or more tracks from Pro Tools successfully. However, for maximum track counts and especially where you are using lots of edits, a dedicated SCSI card will definitely be required. At the time of writing, SCSI is the clear winner for digital audio as long as you are using Ultra2 or Ultra3 devices with suitable SCSI cards.

You can also increase data transfer rates by using SCSI RAID systems. These are often used to achieve the higher data transfer rates required for digital video systems. RAID stands for Redundant Array of Inexpensive Devices. If pairs of SCSI drives are arranged such that each alternate byte of data is written to the alternate drive, the overall data transfer rate to and from the computer is doubled.

> Note: RAID systems can also be used in a different type of configuration that writes the same data to different drives. In this case, if one drive crashes your data is still available from one of the other drives.

Backup systems

Backup is a vital area to address when your precious audio recordings are reduced to strings of 1s and 0s being whizzed around on a hard disk platter at up to 10 000 rpm.

Backing up can simply mean making a copy of your files onto another storage system. Alternatively, you can buy specialized backup software which will let you organize your files effectively onto the backup medium and will let you schedule automatic backups. The two most common packages used on the Mac, for example, are Retrospect and Mezzo. You can use Retrospect to make backups to most types of backup system including tape drives and CD-R and you can schedule automatic backup for whenever is most convenient – such as overnight. Grey Matter Response, Inc.'s Mezzo software (see www.mezzogmr.com) has similar features but also includes specialized capabilities for dealing with Pro Tools projects, Digital Performer projects, Media 100 projects and other specialized formats. Mezzo's interface is easier to work with than Retrospect's, especially when it comes to the specially supported software. For example, Digital Performer users can simply drop their project folders into Mezzo and Mezzo will do the rest – scanning and collecting all of the projects' related files (audio files, analysis files, crossfades, clippings, etc.), even if they are spread across multiple storage devices. This process can run as a background operation once initiated, and Performer users can continue to work on their projects throughout Mezzo's backup operations.

Software, of course, is only part of the story. You need hardware and suitable media to backup onto. Some computers have a backup drive (or drives) installed as standard, but for most computers, backup drives are an optional extra. They can be fitted internally (depending on your computer model) or they can be supplied as stand-alone units. Drives can be categorized as random or linear access types. Linear access devices use tape, usually contained within a cassette of some form or other. To read data some way along the tape, you have to wind through the tape until you get to the part where the data is – and this takes time. A tape can be thought of as a line, if you lay it out flat, which is why tape is said to be 'linear', i.e. 'line-like'. So linear access means you have to traverse the line of tape in a linear (line-like) fashion until you reach what you want. In other words, you have to wind the tape backwards or forwards to find the spot you want – which can take time. Optical discs, CD-Rs and similar devices offer random access. The disc drive's read head can be repositioned and the disc spun so quickly that the read head can access any part of the disc in just a few milliseconds. For practical purposes, this means instant random access to the data on the disc – i.e. without having to wait while a tape is wound forwards or backwards.

Another way of categorizing the various storage types is according to their suitability for short-term backups as opposed to long-term storage, or archiving. Re-writable formats are intended for work in progress, where you are likely to want to change the data during the course of a project. DVD-RAM is designed specifically for storing data so that it can be randomly accessed whenever necessary. The DVD rewritable formats are almost, but not quite, as convenient. Write-once formats, such as CD-R and DVD-R, are useful when you want to prevent accidental over-writing of data. Arguments rage as to the longevity or otherwise of data on CD-R or DVD-R with estimates ranging from as little as a couple of years to 25 years or more. Inks or other materials on the disc can eat into the disc surface until they cause a problem – so it is best to avoid writing on the disc surface. If in doubt, make copies of your copies from time to time – checking to make sure that the copies are successful. Tape-based systems such as AIT or DLT are often used for archiving purposes – where the linear access is not a problem.

> Note: Any hard disk system can be used to make additional copies of your data. This then forms a 'backup' of this data – but not one that you can trust very much! Hard drives can easily develop faults at any time during their lifespan, and should only be regarded as temporary storage.

> Tip: Remember this 'golden rule': 'If it only exists on a hard drive, it doesn't really exist at all – and you will wake up one morning to find that it is not there any more!'

Storage drives

There are many different types of optical discs, removable hard drives, and tape-cassette systems that can be used to store computer data. Let's take a look at popular choices here.

DVD-RAM

The RAM designation means that these discs can be written to as many times as you like – providing random access to stored data that can be changed or overwritten. The current range of PowerMacs can be ordered with a DVD-RAM drive fitted instead of a CD-ROM drive. The DVD-RAM drive looks like a CD-ROM drive when you open it, with a tray available to take a CD. When you look more closely you see a spring-loaded plastic piece which pushes back to make room for a Type I/Rewritable DVD-RAM disc. DVD-RAM is a special type of DVD in which the disc is contained in a fixed 'caddy' to protect it – so it cannot be played in a normal DVD player. The drives can read DVDs and CDs and use 4.7 Gb capacity discs. These can be read by the latest generation DVD-ROMs and can store 2 hours of MPEG-2 video. Discs of 9.4 Gb are also available. Double-sided DVD-RAM discs will hold either 4.27 or 2.3 Gb of data on each side (depending on which type you buy) when formatted for the Mac. The DVD-RAM format is currently the lowest cost per megabyte removable storage solution.

DVD-RW/+RW

Rewritable DVD-RW (and the rival format DVD-+RW) drives are now commonly available. Some drives can work with either type of disc. The discs will hold 4.7 Gb of data on one side, which means they are the right size to store the data for a DVD-Video disc.

> Note: DVD-Rewritable discs certainly make a lot of sense as a backup medium for Pro Tools sessions, especially if you are working with many tracks of 24-bit audio – in which case projects can easily require more than a couple of gigabytes of space. However, I would still recommend the use of an additional archiving or long-term storage medium for important projects – possible using AIT, DLT or Exabyte.

CD-ROM/CD-R

The ROM designation means that these discs can only be read once they have been written. This type of drive is fitted to most personal computers today and can also be bought as a stand-alone device. The original drives have been superseded by drives that can read and write much faster. Speeds have been increased to 48× or more in the latest models. Blank CD-R discs can cost as little as 50 pence each for unbranded types, or up to around five times this amount for well-known brands guaranteed to write at the faster speeds. I recommend using good brands such as TDK and HHB for best results. There are two main types of disc, Silver and Gold, which use different types of dyes in their construction. The Gold types are said to offer greater longevity and stability. CD-ROM makes good sense as a backup medium for Pro Tools users, and the drives can also be used to produce audio 'reference' discs for playback in any CD-player. CD-R discs are also accepted by most CD pressing plants.

> Note: There are differences between models that you need to be aware of, such as some drives will not support ISRC or other PQ codes, and the situation is constantly changing as older models are dropped and newer models introduced. Software such as Roxio Jam and Roxio Toast also has to be updated to add compatibility with new machines that have appeared, and there can sometimes be a long wait between new CD-ROM or CD-RW machines becoming available and the software updates arriving from Roxio. Digidesign publishes a comprehensive list of which drives have been tested to work with which software and hardware combinations. You will find this on their website and I strongly advise you to check this list carefully before buying a drive.

CD-RW

CD-RW drives can use two types of discs – standard CD-R discs or the newer CD-RW discs. These let you erase data and add new data whenever you wish – as with any other random access storage media. CD-RW makes a lot of sense for Pro Tools users, allowing both temporary storage onto rewritable discs and more permanent storage on the extremely affordable write-once CD-R discs. As with CD ROM, these drives can be used to produce audio 'reference' CDs or discs to send to CD pressing plants.

Sony AIT

AIT is yet another backup format from Sony that currently comes in two sizes, AIT 1 with 35 Gb and AIT 2 with 50 Gb. AIT3 is coming soon with 100 Gb. The AIT mechanism uses a rotating head helical scanning technology developed from video recorder designs. The benefit is that you can record high densities on small cartridges. The disadvantage is that the drives need to use precision mechanics and tape guides. So the mechanics are more sensitive to wear over time. These drives are very fast for tape – offering speeds of around 43 Gb per hour uncompressed. Data can be compressed to double the capacities stored and this format offers an extremely fast average file access time of 20 seconds. The drives cost around £1500 or £2600, respectively, and tapes cost between £50 and £75. These drives are gaining in popularity within broadcast, mastering and post-production studios – possibly on account of the Sony brand name and the company's reputation for high-quality manufacturing.

DLT

Digital Linear Tape (DLT) uses a linear tape mechanism, similar to conventional digital audio tape recorder mechanisms – all the tracks run in parallel to the edge of the tape. The benefit of this compared with rotating head designs is that the mechanics around the tape are simple and reliable. There are very few moving parts to wear out. Yet another format using 40 Gb or 80 Gb tape cartridges, data on DLT can be compressed to double the capacity. Priced fairly expensively, but comparably with Sony AIT drives, these drives offer fast transfer times of around 360 Mb/minute – with average file access time of 60 seconds. A good example is the Tandberg DLT VS160. This offers 160 GB of compressed capacity and a 16 MB/second (compressed) transfer rate. Although you can halve these figures for uncompressed data, they are still good.

Super DLT

SDLT was developed co-operatively by Tandberg and Quantum and uses a modified DLT recording scheme to offer capacities of 160 Gb uncompressed and 320 Gb compressed. The speed is 16 MB/s uncompressed or 32 MB/s compressed. This translates to a mightily impressive 58 Gb/hour uncompressed. The Tandberg Super DLT 320 drive, for example, is currently one of the fastest in this market segment. The drive is quite expensive to buy, and the tapes are not cheap. You also need a suitable LVD Ultra 160 SCSI controller card such as the Adaptec 29160. Nevertheless, the cost per megabyte of storage on these tapes is relatively low, and the speed they work at means that it will take around 2 hours to back up a 120 Gb drive onto the 160 Gb tape – with room to spare. Although these SDLT drives may appear to be a costly solution, when you consider how much faster they are than most other drives you will realize that you may save much more money in the time you will save while making backups and restoring data.

Exabyte

Exabyte is an older type of computer backup drive that uses tape-cassettes similar to DAT – but the tapes are physically larger. These tapes normally hold more data than the equivalent DAT backup system. A popular model during the 1990s was the 8505 that was later superseded by the Model 820 which used the same tapes. These tapes typically store between 10 and 20 Gb of data. Exabyte also offer the Mammoth and Mammoth II, both of which use bigger tapes and hold more data – up to 60 Gb with the Mammoth II. Unfortunately, the original tapes will not work in these newer drives. The original Exabyte tapes were widely used in CD-Mastering and were one of the standard formats to use for supplying albums to pressing plants. They were also widely used for storing digital video from non-linear editing systems such as the Media 100.

DAT

Although less common today, you may still come across DAT drives for backing up computer data. These use readily available and relatively affordable tape cassettes similar to those used for audio but optimized for computer data backup. The amount of data you can store on a tape ranges from 4 Gb up to about 20 Gb. File-compression can be used to approximately double the capacity, so an 8 Gb drive might be advertised as a 16 Gb drive – which assumes that you would be using this file compression feature. DAT drives were offered with various different capacities, for example, DDS2 with 4 Gb, DDS3 with 12 Gb and DDS4 with 20 Gb. Transfer rates were 47, 72 and 144 Mb/minute, respectively, while average access times were 35, 45 and 55 seconds respectively. While DAT drives offered a relatively affordable backup solution, they were slow to use. It really could take all night to back up several gigabytes of data.

My backup strategy

I make various backups at different times while I am working on projects. While recording instruments and vocals, I often make a backup copy more or less right away onto another hard drive for safety. At the end of each day or of each prolific session, I might create a CD RW or DVD-RW for further safety – or I might put these copies onto DVD-RAM. This way, I can easily update these copies throughout the life of the project by overwriting the original files, or by saving the updated files separately. When I have completed a project, I create at least one, and sometimes two, copies onto CD-R or DVD-R for long-term storage. In the case of the most important files, I keep a copy off-site in case some disaster should strike my studio.

I am also looking into the possibility of backing up all my hard drives automatically each day to DLT or other tape format, or possibly across the Internet – which is appealing as it would ensure that an off-site backup always exists. One problem with the tape-based solution is that I have more data on hard drives than would fit onto one DLT tape, so I would have to feed extra tapes into the drive as necessary – so I couldn't just go to sleep and leave it. And with the Internet backup solution, I would have to feel absolutely confident that the service provider would be in business for many years and would totally guarantee the integrity and privacy of my data before I would entrust it to any third-party.

Another backup strategy

If I am recording in stereo or in two-channels of mono, which is the case on some jazz projects with solo performers or duets, and is also the case with various dialogue recordings, I often feed the audio simultaneously into Pro Tools and onto a DAT recorder. I can leave the DAT recorder running more or less all the time if I need to, the tapes are not too costly – and most studios like mine have a DAT machine available for this purpose. There are times when a computer will crash or malfunction in some way or other and having an alternative recorder running can get you out of a lot of trouble! Of course, all equipment can fail at times – even DAT recorders.

Mac OSX installations

New for OSX

OSX is a very different 'kettle of fish' compared to OS9 and previous Mac operating systems. In 'geek-speak', OSX is a UNIX-based operating system that offers fully pre-emptive multi-tasking (so if one application crashes it does not bring down the rest) and advanced multi-threading capabilities (which take advantage of multiple processors and allow many different processes to run simultaneously). This means that OSX will provide a more stable platform for your software and will allow the computer to do more things at the same time than OS9. OSX also brings many specific advantages for those making music on the Mac platform. The operating system includes the Core Audio drivers that connect audio applications to audio hardware; Core MIDI, which does the same thing for MIDI; and Audio Units, which provide a potential new standard for plug-ins. One of the main benefits of Core Audio is that it makes extremely small latency times possible. Audio Units plug-ins are available system-wide for all applications – so managing these is much easier. The MIDI functionality makes extensions like OMS or FreeMIDI unnecessary and facilitates quick and uncomplicated installation and configuration of MIDI interfaces and settings – since all MIDI applications and devices report to one central location. Ah!, the benefits of standardization! Apple hired top programmers such as Doug Wyatt (developer of OMS for Opcode) to produce the Core Audio software as part of OSX to ensure that the new operating system would provide a standardized interface for both audio and MIDI applications that could be used reliably for professional applications. All the sample rates up to 192 kHz are supported at 24-bit resolution and internal processing is 32-bit floating-point. And you can use as many channels as you need for surround applications – there is no restriction on the number of tracks. A special software application is provided to let you configure the audio and MIDI for OSX – the aptly named Audio MIDI Setup utility. The MIDI section of this utility lets you configure your MIDI setup in a similar way to OMS – you add 'devices' corresponding to each hardware device in your MIDI setup and configure each to reflect the number of MIDI channels used and so forth. The Audio setup section lets you make settings for the internal audio system and for any additional audio devices that you are using.

Upgrading from OS9 to OSX

If you are upgrading from OS9 you may be wondering what to do about setting the memory allocations for your different software applications. One thing you won't have to configure with OSX applications is the memory allocated to individual software applications – unlike OS9, OSX takes care of this for you. And you won't need to worry about whether or not your software applications work with dual processors – OSX automatically shares the workload between processors on dual-processor machines. As a direct consequence of this, I was able to run almost double the number of plug-ins in Logic Audio under OSX than under OS9. OSX is also much more stable than OS9.

You will still need to make sure that the version of OSX that you are using is compatible with the application software that you want to use. There is often a time lag between Apple releasing a new OS version and Digidesign and others updating their software to work properly with this. Neither Digidesign nor any other third party can guarantee the compatibility of automatic updates of Mac OSX or any updates to system software components. So you should disable 'Automatically check for updates when you have a network connection', in the Software Update System Preferences.

Also, be aware that installing OSX on your Macintosh may erase any currently installed OS9 key disk authorizations (such as SoundReplacer for Pro Tools 5.1.1), so you should uninstall all key disk authorizations before installing OSX. Most Digidesign and third-party key disk authorized software can be upgraded to iLok authorization.

Disk formatting/partitioning

Ideally, you will use external drives (or additional internal drives) for your audio and for any video. These should be 10 000 rpm models – and you need to use a SCSI system for best results, especially when working at higher sample rates and bit-depths or with large numbers of tracks.

Nevertheless, the internal hard drives fitted to most recent computers are fast enough to use with Pro Tools as long as you are not trying to push your system to its limits.

Digidesign recommends that you normally use the standard Apple Disk Utility when formatting drives for use with Pro Tools. Apply the Disk Utility software on their System (start-up) disks. This lets you partition drives and choose the formatting type for each disk partition. The original Mac disk format was called the Hierarchical File System, or HFS. HFS+ was developed to handle the large numbers of very small files that became commonplace with the growth of the Internet. Mac OSX refers to these as Mac OS Standard and Mac OS Extended.

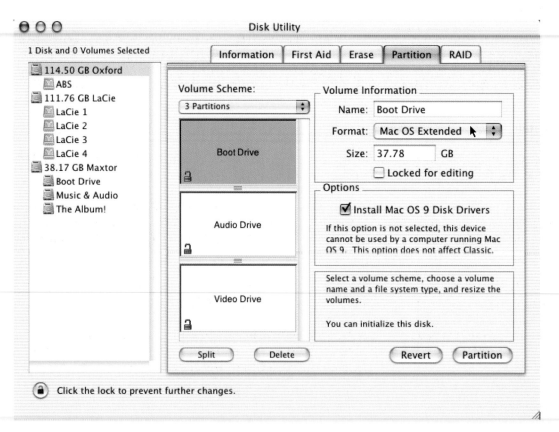

Figure 4.4 Apple Disk Utility – partitioning a hard drive.

Mac OSX 10.3 incorporates a 'journaled' file system that will automatically log any file modifications. If your computer crashes badly enough to require a restart (or a restart from a power failure), the OS will then use the journal to aid in fixing any disk problems caused by the crash. Because any write to the disk will also incur a journal update, this will slightly slow down disk performance.

Mac OSX 10.3 ships with journaling on by default. Digidesign recommends that customers format their media drives with journaling turned off. If you are using Apple's Disk Utility program to format drives, choose 'Mac OS Extended' format, instead of 'Mac OS Extended (Journaled)'.

I recommend that you immediately partition the internal drive on your computer when you are installing this. The current 2 Gb G5 comes with a 150 Gb internal drive so you might create three partitions of 50 Gb each – one as your 'boot' drive partition to contain the operating system, applications and general files; one to hold your audio projects, and one to hold video files.

You can select each partition in turn and choose whether to format with journaling off or on. Keep journaling off for your media drive partitions and turn it on for the boot drive partition.

If you wish to turn off journaling on your boot drive as well, run the Apple Disk Utility, select your boot volume in the main window, then choose 'Disable Journaling' from the File menu. Be aware, though, that this will slow recovery if your system has to be rebooted to recover from a kernel panic or power failure.

> Note: Drive partitions are also referred to as disk 'volumes', by analogy with the different volumes of a book.

Installing Mac OSX operating system

Installation is very straightforward. Insert the Mac OSX Install Disc 1 CD and double-click Mac OSX. The Installer software then guides you through the process.

You can choose which of the connected disk volumes to install to. When you have made your choice, the Installer 'decides' how to install the software. If Mac OSX is already installed on the volume, the Installer will upgrade the software. If Mac OSX isn't installed on the volume, the Installer will install the software for the first time.

With both of these options you are able to choose to 'Erase and Install'. This erases the destination volume first, then installs a brand new copy of OSX.

If there is an existing System installed on the disk volume, you can choose to Archive and Install. This renames the existing System folder as Previous System, thus preserving any extensions, preferences, plug-ins and so forth, then installs a brand new copy of OSX.

User accounts

When you first install Mac OSX you are asked for a name and a password for your user account.

Tip: If you are re-installing a System that has got messed up, you may prefer to keep the previous System folder around until you have tested all your installed applications. Several of these may have needed items that are kept in the System folder, such as plug-ins, preferences, and so forth. If you completely replace the System folder, your applications will not be able to access these items, so these applications will have to be completely re-installed as well. If you keep this previous System folder when you install a new System, then you can search through this, albeit carefully and painstakingly, to identify any additional files required by your applications then drag these across into the new System folder. This can take a while to do, and is only really for the advanced user to attempt, but it can be the best way to do this at times. The only other alternative is to get out all your original program disks and re-install any applications that no longer work properly.

Each user is allocated a Home folder, named after the user and kept in the Users folder on your hard drive. You can store your personal documents, music files, pictures, movies and so forth in folders within this folder. The Library folder contains the preference settings for all the programs you use, any fonts or programs that you have installed for your personal use, and so on.

The Mac assumes that the person who installs Mac OSX for the first time is the 'administrator', i.e. the only user allowed to install new software into the Applications folder, change the System settings and set up other user accounts on this computer.

If other people will be using your computer you can set up new accounts for each user, each with their own password and Home folder.

Figure 4.5 Home folder.

Figure 4.6 Accounts Control Panel accessible from the System Preferences window.

The Internet connection

Once you have created a User Account, the Installer helps you to set up your Internet connection. It helps if you have the details to hand when installing – such as the telephone number of your ISP, your user account name and password, and any other information such as IP addresses or incoming and outgoing mail server details.

If you already have an ADSL or DSL router installed and configured to work with other Macs or PCs in your studio, then all you need to do is hook up the computer you are working with to the router via Ethernet (or via an Airport or wireless network) and it should work straight away.

But why would you want an Internet connection on a computer dedicated to recording audio? The main reason is because so many of the software companies now expect you to register software online – often to email a unique identifying number for your computer so they can supply you with a response number that fully activates the software. Also, you can buy additional software and updates online – any time you like – and you can download software drivers and other useful items whenever you like. Varying levels of technical help are available online, depending on the company's policy, and this can be particularly invaluable when you have a problem in the studio at midnight on a Saturday. You can also swap information and keep in contact with your creative collaborators, your clients, your friends – and the tax man, if you like! You can even find love! I rest my case!

Software updates

Immediately after installing any operating system or application software, you should check to see if there are updates already available that you can download directly to your hard drive.

Apple's Software Update utility, available from the System Preferences window, can be set up to automatically check for updates at daily, weekly, or monthly intervals. This is a great convenience – but you do need to be sure, whenever possible, that updates are compatible with all the software you are using.

One thing to watch out for here is that Digidesign do not normally keep their Pro Tools software updates in step with Apple's operating systems updates – Pro Tools typically lags behind by one, or even two, versions. So make sure you check the Digidesign website for the latest news on compatibility of versions and, if in doubt, don't update.

Nevertheless, the sheer convenience of being able to keep your software regularly updated directly from your computer is absolutely tremendous! No more trekking to the nearest city centre dealer to pick up a copy on CD – only to find they are out of stock!

One potential problem is the possibility of version conflicts – but this possibility exists with all software updates anyway. The other issue is the speed/bandwidth of your Internet connection. Downloads of 50 Mb are fine if you have a DSL or ADSL broadband connection – but a complete pain if all you have is a 56 kbps modem.

Often, you will find information about updates in popular magazines such as *Sound On Sound*. You can also join various user-groups on the Internet – although these tend to be full of messages from people 'sounding off' about their favourite topics. It can be tedious to look through hundreds of messages to find just a couple that are extremely relevant and accurate. Sometimes the best way (or the only way) is to go ahead and install the updates, while being fully prepared to re-install the previous software versions in case of trouble. Just don't try it 5 minutes before an important recording session!

Logging in

'Logging in' involves having to enter the username and password of a bona fide User Account before being allowed access to the computer.

Mac OSX is set up to log in automatically using the User Account that you create when you set up Mac OSX. Obviously, this is no good if you want to stop anyone else gaining access. In this case you need to turn off automatic login by deselecting the checkbox for this in the Login Options in the Accounts window that you can access from the System Preferences.

A number of users working on the same computer at different times can each log in to their own account and save their files into folders reserved for use with this account.

Forgot your password?

If any user forgets his or her password, the Administrator can open up the System Preferences, go to the Accounts window, select the name of the person who forgot their password and click Edit to access a dialog where the password can be verified and entered.

Of course, if you are the Administrator and you forget your password, this can be a bit more of a problem! In this case you will need to restart your Mac using the Mac OSX Install Disc 1. Hold the 'c' key on your computer keyboard with this CD in the tray until the Mac OSX Installer appears on-screen. Go to the Installer menu and choose Reset Password. Select the hard drive partition that contains the Mac OSX System that you want to fix and choose the name of your account from the first popup menu. Choose a new password, type this into both boxes and click 'Save'. Then close the window, quit the installer and restart. Easy enough to do really – when you know what to do!

Pro Tools issues

To ensure proper operation and file management, Pro Tools 6 software must be installed in an administrator-level account in Mac OSX. In addition, to ensure access to all plug-ins from within Pro Tools, it is recommended that the Pro Tools application and all Pro Tools plug-ins be installed and used from the same administrator-level account. Do not install or operate Pro Tools while logged in as a root-level user. File permissions of a root-level user make it possible to perform actions that may conflict with Pro Tools file management tasks.

If the Energy Saver Control Panel is set to switch your computer into Sleep mode this can disrupt your session if it happens when you don't want it to – so you should disable this. Open the Macintosh System Preferences, launch Energy Saver, click the Sleep tab, and disable all the Sleep features. For example, set 'Put the computer to sleep when it is inactive' to 'Never'.

If you are using a Power Mac G5, you must also click the Options tab and set 'Processor Performance' to 'Highest' for optimal Pro Tools performance. Apple's explanation is as follows: 'The PowerPC G5 processor in your Power Mac G5 can run at various speeds. The default

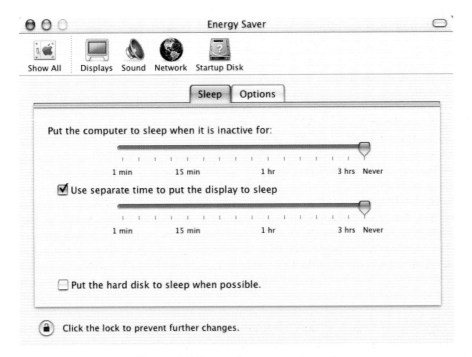

Figure 4.7 Energy Saver control panel.

setting for this feature is "Automatic", which allows the system to run the processor at full speed when required, and at a lower speed during periods of light processing activity. Certain real-time applications, such as audio creation or gaming, may perform optimally when using the highest processor performance setting. You can change the processor performance setting in the Energy Saver pane of System Preferences.'

Similarly, screensavers have been known to disrupt MIDI playback at critical moments. This happened to a friend of mine on a very important session in a major London studio a couple of years ago. The screensaver kicked in just as they were recording with the management and record company representatives all in the control room watching. My friend has almost got over the shock – but still won't use a screensaver!

> Note: There is a 31-character limit on for all Pro Tools file names. Trying to open Pro Tools files with names longer than 31 characters will result in a DAE error –9035.

Mac tips

I recommend that you neatly arrange the folders on your disk drives to let you more speedily identify which folder is which. If you establish a standard set of positions for your folders and group similar types together, this will help you to distinguish between the different folders more quickly. The Panther operating system (like OS9 and earlier) also allows you to distinguish items by allocating any of seven different colour labels – providing yet a further visual 'cue' that helps you find your folders and files more quickly. You should try to keep to a maximum of, say, seven (or at the most 10) folders – putting all your software applications within one folder, all your utility software within another, all your letters and notes in another, and so forth. This neatness can really give you a speed advantage when it comes to working with your system – and it is well worth the trouble of setting up and sticking to.

You should keep all your music project files on a separate hard drive or hard drive partition and arrange these neatly, inside informatively named folders. So how can you keep things neat? For a start, you can make a habit of resizing all your folders to just encompass all of the items inside these by clicking in the 'size' control (the right-most green button on each folder window's title bar). And you can quickly tidy up the positions of the items within a window using the Clean Up command from the Finder's View Menu. If you have a folder open which has several files scattered around inside and some of these are positioned somewhere out of sight, you can bring all the files within the visible area of the folder by choosing any of the 'Arrange By' options from the View menu in the Finder. I usually find that I also need to move icons around manually to be able to get them arranged exactly as I want them – especially if they have long names. Still, it's definitely worth taking the trouble to arrange your files clearly and logically within their folders – or on the desktop – as you will save time over and again when you come back to look for your files in future.

To help you to keep your desktop tidy when you are opening and closing lots of folders, there are a couple more shortcuts. Holding the Option key when double-clicking a file or application causes the folder which the file or application is in to close once the file is open. And if you hold the Option key when closing any open folder in the Finder, all the open folders on your desktop will close along with it.

Learn the standard editing commands – Command X, Command C and Command V for Cut, Copy and Paste work in the Finder and in just about every application. Command-A for Select All works in the Finder and in most applications – although not all. Command-Z lets you Undo your last action, and Command-F brings up a Find dialog in the Finder (or – if there is one – in the application).

If you want any programs or utilities to launch immediately when you log onto your computer, simply drag these to the Login Items window – available from the System Preferences window. To switch off Log In items that you don't need, open System Prefs and go to the Log In items pane to see the list of applications that automatically open when you log in. Select recently added items, click 'Remove', then restart the computer.

To make an alias of a file in one folder and put this into another folder, hold the Command and Option keys simultaneously, while you drag the file's icon into the new folder. Use this method instead of Command-M (Make Alias) to avoid having the word 'alias' in the name.

Finally, if you like working fast, hit Command and the backspace key to instantly put any selected item into the trash.

Figure 4.8 File Info Dialog showing Ownership and Permissions.

Get info

Remember that selecting any file and choosing Command-I brings up an info window for the file. Here you can lock a file to prevent it from being changed, or check its date of creation or modification, or check the application version number. This can be very useful if you are trying to track down problems that may have occurred because you have two versions of what look like the same file on your disk, for example.

Troubleshooting

Hardware problems

As far as hardware failure is concerned, both Apple and Digidesign equipment have proven to be extremely reliable over the 14 years or more that I have worked with their products. I have had three hard drives crash irretrievably, including a 3-month old internal drive in a G4, a main logic board in a PowerMac 8100 develop a major processor fault, on-board RAM on the 8100 logic board become faulty (after several years' use), a 20-inch Apple monitor that once blew up, and a couple of power supplies that have burnt out during that time. And a Lexicon audio card once burned out the first time I switched the computer on with the card newly installed. These types of faults can happen with any electronic audio or computer equipment – but are, fortunately, generally quite rare occurrences. Normally if something is going to go wrong with your hardware it will happen very soon after you buy it, or sometime (hopefully a very long time) after the warranty has run out.

The most common hardware faults I have come across are broken or bent pins on connectors, cards not being seated properly in the computer, cables not being connected properly and so forth. For example, I recall that a video card that had come partly out of its socket once prevented my computer from starting up – even though the computer monitor screen appeared to be working. And sometimes the Pro Tools cards can look as though they are fully inserted even when they are not. These problems are easy enough to fix by checking carefully for these possibilities. Of course, if a connector pin is broken, you may be stumped until you can order a new one from Digidesign. These use special connectors which you are unlikely to have spares for – unless you are really thinking ahead and determined to keep your system up and running at all times. Now if you really want to be sure, maybe you should buy two complete Pro Tools systems and keep one available as a spare at all times. I am sure that Digidesign would heartily approve of this tactic. On the other hand, thinking more practically, a couple of spare cables would probably not go amiss.

Mac software problems – freezes, foibles and crashes

The first thing you can try if an application freezes or misbehaves in some way is to quit the program then re-launch it. You can use a handy keyboard command, Command-Option-Escape, to force the application to quit if you cannot get to the normal Quit command. Select the application you want to quit from the list that appears; and click 'Force Quit'. If it doesn't work at first, try again and it may well succeed. Because the memory space used by each software application running on OSX is quite separate from that used by any other, no other application will be affected by one application crashing. And once you have removed the problem application from the computer's memory, you can safely restart the application. And if you are aware that a particular sequence of events caused the crash, hopefully you can avoid this situation next time around.

The second thing you can try is to trash the preferences file for the problem application. These preference files sometimes, mysteriously, get corrupted. There are at least two places where folders containing these preferences files could be located on your boot drive. The first place is in the Library folder in the main hard drive window. The second place is in the Library folder within your Home folder, which is within the Users folder in the main hard drive window. This is the most likely place. Put the preference file that you think is causing the trouble into the Trash, but don't empty it – yet. Run the program again and it will build a brand-new, uncorrupted, preferences file. If this solves your problem, then you can get rid of the old preferences. If not, you still have the option of reinstating the original preferences that you put in the trash. (There is also a PreferencePanes folder within the Library folder within the System folder, but you are not allowed to trash these files.)

> Tip: The preferences files to remove if you are having trouble with Pro Tools are called 'DigiSetup.OSX' and 'Pro Tools v6.1 Preferences' – or version whatever you are using. When you next launch Pro Tools, new versions of these files will automatically be created with their default settings. Don't forget that you will have to re-set any preferences that you previously set up within the Pro Tools software and reset the hardware settings that are stored in the DigiSetup file.

A great 'quick fix' that you can always try in OSX is to simply log out then log in again. This ensures that any software applications that were running are removed – even if you weren't aware what was running. It is a bit like a 'soft reset' for your system – enabling you to start over with a 'clean sheet'.

If all else fails, you will just have to 'bite the bullet' and restart your computer. To restart, press Command, Control and the Power button (if your computer has one), or press Command, Control, and the Eject button (if your computer has an eject button).

> Note: Sometimes the computer completely 'locks up' and you cannot get the Log Out, Restart or Shut Down commands to work. In this case, you can try pressing the small Reset Button that you will find on the front panel of many Macs. Alternatively, you can press and hold the main power button for at least 5 seconds. Occasionally, none of these methods will work, and you will simply have to pull the power cable out. When you restart, the hard disk drive is automatically checked for problems and these are fixed as necessary.

Start-up problems

If the computer will not start up properly, you will need to insert the Apple System Installation disk which is supplied with all new computers, restart and hold the 'c' key until the Mac OS splash screen appears.

> Note: Don't forget to keep your Apple System Installation disk close to your computer at all times.

Once the computer has started you can re-install the System software – either into the original System folder, in which case all your installed program extensions will continue to operate, or into a new 'clean' System folder. In this case you will either have to manually copy the required extensions and other necessary files into the new System folder from the old one – which is automatically renamed 'Previous System' – or you will have to re-install any of your application programs that require additional files to be installed into the System Folder. Be aware that most Audio and MIDI software applications do require various extra files to be installed.

> Tip: It is worth bearing in mind that no software tools will fix hardware faults – and these can and do arise on drives at any time in their 'lives' – although most typically when the drives are relatively new or relatively old.

Your hard drive is not a perfect piece of machinery. It is tempting to think that it is – especially if you have not encountered any hard drive problems yet. Nevertheless, problems can occur at any time that will prevent your computer from starting up and the potential to lose your precious data is always there. So backup, backup, then backup some more!

A hard drive problem

Recently, I came across a computer that would not start up properly and discovered that the hard drive had zero free space. This prevented the machine from booting properly and the fix was simple – start up from CD-ROM and move some files to another disk or backup to free up some space. The rule of thumb for best results is to always leave at least 10% of your hard drive free. Many programs need to create temporary files on the internal hard drive to operate correctly, so even if the computer appears to work when the drive is almost full, you can still encounter problems with particular application programs.

A true story

While I was writing the first edition of this book, the hard drive on my G4 suffered a mechanical failure! I lost 2000 words that I had written earlier that day about the importance of backup and such like for this very book. Talk about ironic! I also lost 3 months' worth of emails which I had overlooked when doing my backups, and I lost two-days' worth of synthesizer recordings which I had not backed up as I had run out of blank CD-ROM discs at the time. I should also confess that had it not been for a friend who had reminded me a couple of weeks prior to this to back up the words I was writing for the book, then I might have lost the entire previous 3 months' work! For the first 3 months while working with this new machine with its 40 Gb of hard disk space, I did not need to make space for new files as I never filled up the 40 Gb, and I was thinking (not very clearly and definitely very wrongly) that the new hard drive should be safe to work with – so there was not much reason to back up just yet. So I got stung yet again – despite the fact that two or three years prior to this event I had lost 3 months' work through hard drive failure, so I should have known much better! On that occasion, two drives failed on me within a couple of days of each other. They were a pair of identical 4 Gb drives acquired at the same time. When the first one went, I breathed a sigh of relief as all the data on this was 'backed up' – on the other drive! I made a mental note to order some CD-Rs and back up the other drive on a more permanent medium. But I felt no sense of urgency about this as I said to myself – 'lightning does not strike twice in the same place' and 'I am a lucky kind of guy'. Well I was wrong on both counts that time. My luck definitely ran out because the second hard drive died two days later – along with 3 months' worth of recordings and samples. Yet 3 years later, there I was in danger of falling right into the same old trap. Be warned!

Repairing disk permissions and disk volumes

OSX is a Unix-based operating system and, like Unix, it expects various permissions to be set for all files and folders on attached disks. Sometimes, these permissions can be wrongly set, especially after installing new software. If you are encountering strange behaviour of any sort, it is worth using the Apple Disk Utility to fix these disk permissions. You can always use the copy of Disk Utility that can be found in the Utilities folder within the Applications folder on your boot drive. Alternatively, you can boot up from your Mac OSX Install disc and use the Disk Utility. When the Installer screen appears, don't click Continue. Instead, choose Open Disk Utility from the Installer menu. When the Disk Utility opens, select the First Aid tab, choose the disk you want to repair, and click 'Repair Disk Permissions'.

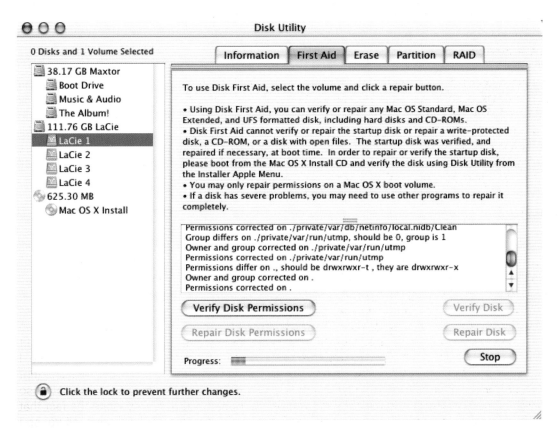

Figure 4.9 Repairing disk permissions.

Battery-backed RAM

One thing worth keeping in mind with Macs is that there is a small battery on the computer's main logic board which provides power for the Parameter RAM (PRAM) when the computer is disconnected from the main electrical supply – in other words when it is turned off. The PRAM holds the clock settings and various other parameters such as the baud rate (i.e. the communication speed) of the comms ports, and other hardware settings that need to be remembered until the next time you switch the computer on. If the battery runs low, these settings will be lost each time you switch on and off.

An easy way to tell if this is happening is to check the date displayed on the computer. If this says that the year is 1956, or some other arbitrary date chosen for the Mac's system clock as the date the computer's time is counted from, then you know that this battery has failed and the system clock has reset to its initial value. These batteries can be bought from photographic suppliers who often keep these in stock to use with popular photographic equipment or from electronics suppliers such as Maplin in the UK or Radio Shack in the US. The battery is a 3.6 or 3.7 volt lithium type, size 1/2 AA.

Zapping the PRAM

Another problem that sometimes occurs with the PRAM is that the data can become corrupted. In this case you need to clear the PRAM and reset it. To clear the PRAM and NVRAM, restart your computer while holding the Command, Option, P, and R keys until you hear the System chord at least twice on start-up. This is well worth trying if you are having mysterious problems that you cannot sort out using other methods.

Unreadable CD

If you insert a blank CD or DVD into your Mac, by definition this will not have any disk storage volumes present on the disk that your Mac can read – because, of course, it is blank and has nothing on it. The Mac knows this, and assumes that you may be unaware of this – so it helpfully posts an error message to inform you of the situation.

Of course, it does make some assumptions here. The first, and biggest, assumption is that it expects that you will understand what is meant by the term 'volume'. Now you probably associate this with the loudness level of those great honking notes you blow on your saxophone or those thrashing chords you play on your guitar. But, of course, as with many words in the

Figure 4.10 Mac OSX disk error message.

English language, there can be other meanings associated with the word 'volume'. The meaning that is intended here is similar to the usage of the word 'volume' as in a book 'volume' – Volume I of the book, Volume II of the book, Volume III of the book, and so forth. But instead of being a volume of a book, we are talking about a volume of a data storage disk – such as a hard disk drive or, in this case, an optical disc – namely a DVD or CD disc. Data storage drives can be 'partitioned' to form two or more 'volumes' – which is what I have done with my internal drive to keep my audio data separate from my applications. My drive is 60 Gb total, partitioned into three 20 Gb volumes. (By the way, note that a floppy or hard disk is spelt with a 'k' while a disc such as a gramophone record or a CD should be spelt with a 'c' – purely by conventional agreement, and as a way of further distinguishing these types of disc(k). The Mac error messages, however, do not go as far as to observe this convention, and refer to any type of storage disk as a 'disk'.) Now, if you are still awake after this lengthy spiel, you will fully appreciate why the Mac has told you 'You have inserted a disk containing no volumes which MAC OSX can read. To continue with the disk inserted, click Ignore.'

More help

If you need more in-depth help, I can recommend *Mac OSX Disaster Relief* by Ted Landau, Peachpit or *Mac OSX: The Missing Manual* by David Pogue, O'Reilly.

Summary

There are three golden rules that you need to observe with computer systems:

1. The hard drives have to be working correctly or problems may arise due to corrupt or missing data on the drives. This can affect the operating system software, the application software or the session and data files. You cannot 'see' what is happening on your hard drives, so the first you may know about these problems is when you lose a recording. Use a good utility software package such as Norton Utilities to help prevent drive problems. And, if in doubt, initialize or re-format your hard drives – or replace them.
2. The operating system software has to be working correctly or the application software can be affected and, again, your recordings are at risk. Don't forget, data on any hard drive can become corrupted and you cannot 'see' this or know about it until problems start happening. If in doubt, re-install 'clean' system software.
3. The application software obviously has to be installed and working correctly, and, as with the system software, if you suspect that it may be corrupted in any way, you have no choice other than to re-install 'clean' software.

Think of it this way, if you are 'standing on shaky ground' in any way, you are likely to start sinking at some point or other. Just as you have to keep analogue mixers and recorders maintained and in good working condition, you have to do this with your hard drives and software as well. Nothing changes!

So – learn how to take care of your computer system, keep the operating system and applications software installation discs to hand at all times, keep your files and folders tidy, and get 'religious' about backups. Follow these basic rules and you won't go too far wrong. Good luck!

5 Recording and Editing MIDI

MIDI features overview

Pro Tools has always been able to record and playback MIDI data – but the MIDI features used to be extremely basic. The latest versions of Pro Tools have all the features you need to work effectively with MIDI data without the more sophisticated features you would find in Digital Performer, Cubase SX or other advanced MIDI software. For example, Pro Tools still does not have the music notation and advanced MIDI processing features that you will find in most MIDI + Audio software applications. However, these features are much less important for straight-forward MIDI recording, editing and playback. Pro Tools sessions now support up to 256 MIDI tracks and have four virtual MIDI inputs, called Pro Tools Inputs, which let you receive MIDI data from other supported applications (such as Ableton Live). Also, if you are using the Digidesign MIDI I/O or any other supported MIDI Time Stamping-capable interface, you can achieve up to sub-millisecond-accurate MIDI timing – allowing Pro Tools to compete effectively with any other sequencer available.

This chapter contains lots of useful tips and hints about how to set up Pro Tools sessions to record MIDI and how to record these MIDI tracks as audio. Using Virtual Instruments is covered in some detail. There are also several sections on editing – Editing Issues, Graphic Editing, Event List Editing – and so forth. The final section, about working with patterns, was originally inspired by a call I got from producer Phil Harding who asked if I could show him how to set up Pro Tools to work with MIDI more like he was used to working in Cubase – with lots of short sections that he could move around on-screen to map out his arrangement.

Recording MIDI data

This section will help you to set up your Pro Tool session to work with MIDI instruments and will show you how to record these as audio when you are satisfied that they sound the way you want them to sound.

Configuring the Apple MIDI setup for Mac OSX

If you are going to use MIDI with Pro Tools, you need to set up your MIDI connections first. With Mac OSX, you can use Apple's Audio MIDI Setup (AMS) utility to identify which external MIDI devices are connected to your MIDI interface and then to configure your MIDI studio for use with Pro Tools. You can open the Audio MIDI Setup utility from the computer's Utilities folder,

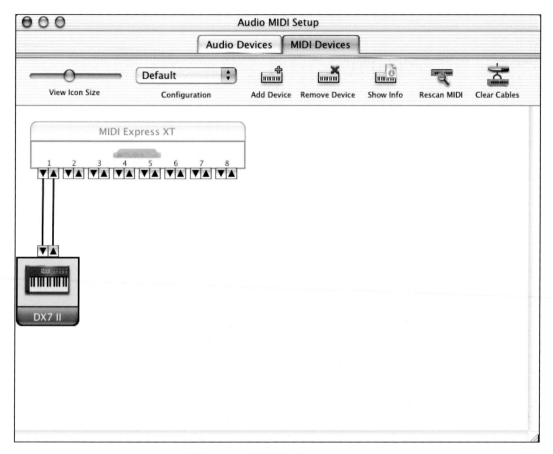

Figure 5.1 MIDI Devices setup.

which resides inside the Applications folder, or by choosing 'Edit MIDI Studio Setup' from the Pro Tools Setups menu. There are two pages, one for Audio Devices and one for MIDI devices – each with its own selection tab at the top of the window. When you click the MIDI Devices tab, AMS scans your system for connected MIDI interfaces. When it finds your interface, this appears in the window with each of its ports numbered. Then you need to add devices corresponding to the actual devices that are connected to your MIDI interface. When you click Add Device, a new external device icon with the default MIDI keyboard image appears. You can drag this icon around the screen to place it wherever you find most convenient. You can connect the MIDI device to the MIDI interface by clicking the arrow for the appropriate output port of the device and dragging a connection or 'cable' to the input arrow of the corresponding port of the MIDI interface.

When you double-click on any device icon that you have inserted, a dialog window opens to let you set various parameters for the device. If the device is a popular model that has been available for some time, it will be listed in the Model pop-up menu in this dialog. If not, just type your own names and set the 'Transmits' and 'Receives' and other parameters manually. When you're done, click 'OK' to close this dialog and save your changes. And that's it – not too difficult really!

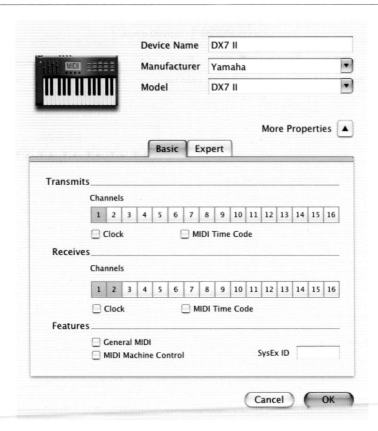

Figure 5.2 Device parameters.

Setting up to record MIDI into Pro Tools

Before you can record any MIDI into Pro Tools you need to make sure that the device you are using, such as your MIDI keyboard or MIDI control surface is enabled as an Input Device. Choose Input Devices from the MIDI menu and make sure that your input device is selected in the MIDI Input Enable window that opens.

> Tip: If you wish to sync to Midi Machine Control (MMC), you must make sure that the MMC source (the device that the MMC messages are coming from) is enabled in the Input Devices dialog.

To record MIDI, you need a MIDI track to record onto. To set one up, choose 'New Track' from the File menu, specify 1 MIDI Track then click 'Create'.

You also need to make sure that the MIDI track's Input and Output are correctly configured in the Mix window. The MIDI Input Device/Channel Selector defaults to 'All', so it 'listens' for MIDI data coming in from any of the inputs — so you don't actually need to change this. But if you prefer, you can use this to select a specific MIDI device for input. To set the Output, click the

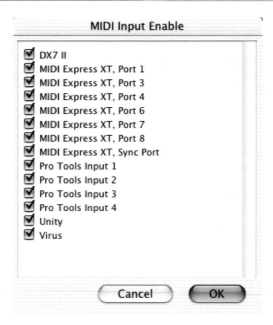

Figure 5.3 MIDI Input Enable dialog.

track's MIDI Output Device/Channel Selector and assign a device and channel from the pop-up menu. Typically, you will see your MIDI master keyboard (I use a DX7II), any MIDI modules (I sometimes use an Access Virus), and any virtual instruments (such as the Digidesign Access Indigo) available here.

Figure 5.4 New Track dialog.

Note: Be aware that you need to insert a virtual instrument onto an Aux track (or onto an Audio track in the case of some virtual instruments, such as Spectrasonics Stylus) before any virtual instruments will appear as available MIDI Output Devices.

You can assign a default program change to the track by clicking the Program button in the Mix window to open the Patch Names window.

In the Patch Names window, make your selections for program and bank select then click 'Done'. Files containing the actual Patch Names used in your synthesizers may be available

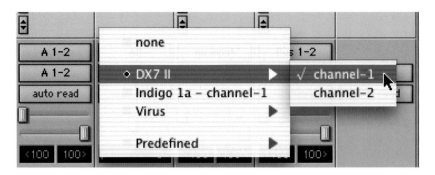

Figure 5.5 MIDI Device/Channel Selector.

on your system, in which case you can select these using the 'Change...' dialog so that they are displayed here.

Now that you have everything set up the way you like it, you can go ahead and record-enable the MIDI track in the Mix window. This is the small button marked 'rec' just above and to the left of each fader.

Figure 5.6 Clicking the Program button in the Mix window.

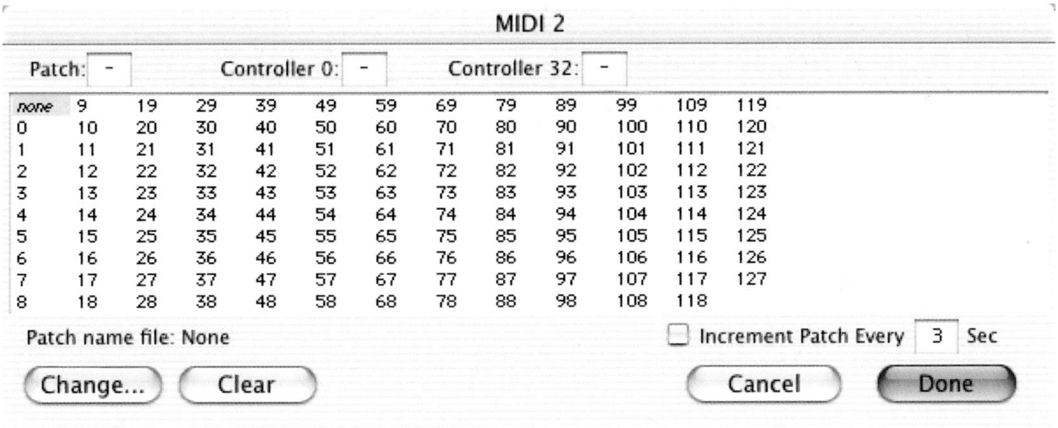

MIDI 2													
Patch:	-		Controller 0:	-		Controller 32:	-						
none	9	19	29	39	49	59	69	79	89	99	109	119	
0	10	20	30	40	50	60	70	80	90	100	110	120	
1	11	21	31	41	51	61	71	81	91	101	111	121	
2	12	22	32	42	52	62	72	82	92	102	112	122	
3	13	23	33	43	53	63	73	83	93	103	113	123	
4	14	24	34	44	54	64	74	84	94	104	114	124	
5	15	25	35	45	55	65	75	85	95	105	115	125	
6	16	26	36	46	56	66	76	86	96	106	116	126	
7	17	27	37	47	57	67	77	87	97	107	117	127	
8	18	28	38	48	58	68	78	88	98	108	118		

Patch name file: None   ☐ Increment Patch Every 3 Sec

(Change...)   (Clear)   (Cancel)   (Done)

Figure 5.7 Patch Names window.

> Tip: Don't forget to make sure that MIDI Thru is selected in the MIDI menu.

Now play some notes on your MIDI controller. The MIDI instrument assigned to the track should sound and the track's meter should register MIDI activity – you should see the meter light up green at the right of the fader.

> Note: Why is it necessary to enable MIDI Thru? Well, if you are using a multi-timbral synthesizer as your MIDI 'master' keyboard, you will normally turn Local Control off on this synthesizer and route the MIDI data from the keyboard into your MIDI sequencer (in this case, Pro Tools) then back out from the track you are working on into the synthesizer section of your MIDI synthesizer. This way, the MIDI keyboard plays the multi-timbral 'part' in the synthesizer that corresponds to the MIDI channel selected as the output from the track you are working with in your MIDI sequencer (in this case, Pro Tools). If you leave Local Control on, the MIDI keyboard will always play one of the sounds in its synthesizer, and this will often be a different sound than the one you are working with in your sequencer – so it will all sound a mess. So you should turn Local Control off on your synthesizer and enable MIDI Thru in your sequencer to avoid this.

> Tip: It can be awkward to have to record-enable a MIDI track that is routed through to a particular MIDI device and channel just so that you can play an attached synthesizer. If there is one main device that you like to use, perhaps on a channel that plays a basic piano or pad sound, you can set up a 'Default Thru Instrument' routing in Pro Tools so that this will always play without you having to record-enable an appropriately routed track. You can set this using the pop-up provided in the MIDI preferences window.

Recording onto a MIDI track

When you have finished setting up to record MIDI, make sure that the MIDI track you want to record onto is record-enabled and receiving MIDI – as indicated by activity on the track meters.

If you are ready to record, all you need to do now is to click 'Record' in the Transport window to enable record mode then click 'Play' in the Transport window or press the space-bar to actually begin recording.

In the Transport window, click Return To Zero to start recording from the beginning of the session – or simply hit the Return key on the computer's keyboard. Alternatively, you can start recording from wherever the cursor is located in the Edit window.

> Tip: If you select a range of time in the Edit window, recording will start at the beginning of this selection and will automatically finish at the end of the selection.

The Transport window can be expanded to display extra controls for MIDI sequencing using the Display menu options. These include controls to set the Meter, to set the Countoff, to enable the Click, and to 'Wait for Note' before recording. A small slider lets you manually adjust sequence tempo or you can enable the Tempo track by clicking on the Conductor button. The expanded controls also include two counters – each of which can show your choice of Bars/Beats, SMPTE Timecode, Feet/Frames, Mins/Secs or Samples.

If you click on the 'Wait for Note' button (the one with the MIDI socket icon), the Stop, Play, Record, and Wait for Note buttons will all flash until the first MIDI event is received. So as soon as you play a note on your MIDI keyboard, for example, Pro Tools will start recording MIDI data from that point onwards.

Figure 5.8 Expanded Transport Window showing the Wait for Note button engaged with the Stop, Play and Record buttons flashed on. The Wait for Note button is the button with the MIDI socket icon, located to the left of the metronome button.

If you click on the Countoff button (the one marked '2 Bars'), then when you click Play, the Record and Play buttons will flash during the Countoff. When the Countoff is finished, recording will begin. The default Countoff is 2 bars. To set any other number of bars, double-click on the Metronome button to open the Click/Countoff Options dialog where you can enter whatever number of bars you prefer.

When you have finished recording, click 'Stop' in the Transport window or press the space-bar. The newly recorded MIDI data will appear as a MIDI region on the track in the Edit window as well as in the MIDI Regions List.

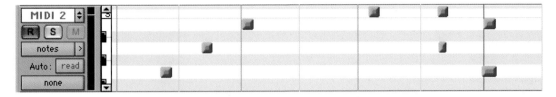

Figure 5.9 MIDI Data in the Edit window.

Loop recording MIDI

There are two ways to loop record with MIDI – either using the normal Non-destructive Record mode with Loop Playback and MIDI Merge enabled for drum-machine style recording or using the special Loop Record mode to record multiple takes on each record pass, as when loop recording audio.

To set up drum-machine style loop recording, where each time around the loop you record extra beats until you have constructed the pattern you want, you need to enable the MIDI Merge function by clicking on the icon at the far right in the Transport. This looks a bit like a letter 'Y' on its side. Deselect 'QuickPunch', 'Loop Record' and 'Destructive Record' in the Operations menu, but select 'Loop Playback' so that the loop symbol appears around the 'Play' button in the Transport window. Select 'Link Edit and Timeline Selection' from the Operations menu, then make a selection in the Edit window to encompass the range that you want to loop around. If you want to hear the audio that plays immediately before the loop range as a cue, you will need to set a Pre-Roll time. So, for example, you would hear your session play back from, say, 2 bars before the loop range then it would play around the loop until you hit 'Stop'. Each time through the loop you can add more notes until it sounds the way you want it to, without erasing any of the notes from previous passes through the loop – just like with a typical drum-machine.

If you record MIDI using the Loop Record mode instead, new regions are created each time you record new notes during successive passes through the loop.

This time, select 'Loop Record' and 'Link Edit and Timeline Selection' from the Operations menu, deselect 'Loop Playback' and check that 'QuickPunch' and 'Destructive Record' are disabled. With Loop Record enabled, a loop symbol appears around the 'Record' button in the Transport window. Make a selection in the Edit window to encompass the range that you want to loop around and set a Pre-Roll time if you need this. Start recording and play your MIDI keyboard or other MIDI controller. Each time around the loop, a new MIDI region is recorded and placed into the Edit window, replacing the previous region. When you stop recording, the most recently recorded of these 'takes' is left in the track, and all the takes appear as consecutively numbered regions in the MIDI Regions list.

The easiest way to audition the various takes is to make sure that the 'take' currently residing in the track is selected, then Command-click (Control-click in Windows) on the selected region

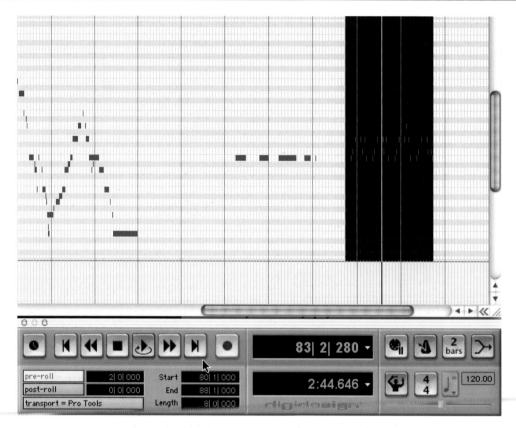

Figure 5.10 Loop recording – drum-machine style.

Note: Loop recording audio is similar, but with the following difference: when you record audio using Loop Record mode, Pro Tools creates a single audio file containing all the takes (which appear as individual regions in the Audio Regions List).

with the Selector tool enabled. A pop-up menu appears called the 'Takes List'. This contains all your recorded 'takes'. Choose whichever you like to replace the 'take' that currently appears in the track.

Tip: Whichever loop recording method you choose, you should disable 'Wait for Note' and 'Countoff' in the Transport window.

Playing back a recorded MIDI track

This is simple enough – just click 'Play' in the Transport window or hit the space-bar on your computer keyboard. The recorded MIDI data will play back through the track's assigned instrument and channel, and sound will be produced by the connected MIDI device.

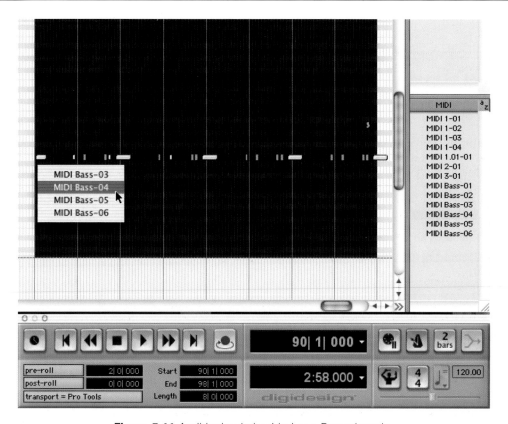

Figure 5.11 Auditioning 'takes' in Loop Record mode.

Note: If you have just finished recording onto the track, you should click the Record Enable button to take the MIDI track out of Record mode for safety.

Of course, there are a few things that you need to watch out for to make sure you will actually hear something. For example, the audio outputs from your MIDI playback device (your synth, sampler, drum-machine or whatever) need to be connected to an external playback system (such as an external mixer and monitors) or they need to be routed into the Pro Tools mixer.

Tip: To monitor your MIDI instrument's audio outputs, you can route these to your Pro Tools audio interface and use Auxiliary Inputs into the Pro Tools mixer. Auxiliary Inputs function as inputs for both internally bussed signals and external audio sources.

Note: Auxiliary Inputs cannot record audio to disk – they just monitor the input and feed this into the mix.

To make sure that you play back from the beginning of the track you can either hit the Return key on the computer's keyboard or click the 'Return to Zero' icon in the Transport window. Of course, you can always play back from wherever the cursor is located in the Edit window. And if you select a range of time in the Edit window, playback will start at the beginning of this selection and will automatically finish at the end of the selection.

Assigning multiple destinations to a MIDI track

MIDI programmers and synthesizer players often like to layer up synthesizer sounds – either before they are recorded or afterwards when working on the mix. There are several ways to do this – starting with the synthesizer itself. Nevertheless, a convenient way can be to simply assign multiple destinations to a single MIDI track. So, for example, you might route a keyboard pad to the Virus Indigo plug-in and the Spectrasonics Atmosphere plug-in at the same time, choosing suitable patches on each to create your layered sound. This is easy to do in Pro Tools – just Control-click (Macintosh) or Start-click (Windows) on the MIDI Output Selector and use this pop-up selector to choose additional destinations from any of the devices and channels available on your system.

Recording Sysex into Pro Tools

Why might you want to record SysEx data? Well, it's often a good idea to have one or two (or more) bars at the start of a MIDI session that you will keep free of music so they can be used for storing set-up and configuration data for your MIDI equipment. Many MIDI devices can use System Exclusive messages (exclusive to that particular manufacturer/device) to transfer useful data such as the contents of any on-board memory to and from devices. So, for example, you can store the actual synthesizer patch data that you are using for your song, or the individual configurations of your MIDI instruments including MIDI channel settings, transposition settings and such like. System Exclusive data takes precedence over MIDI note data and often uses a large amount of MIDI bandwidth, so it is not wise to send too much (if any) SysEx data around your system while your music is playing. It is feasible to send parameter changes or possibly single memory patch information during a sequence, but even in these cases you would have to take care to send this data in between any note data to avoid disruption of playback.

To record SysEx into Pro Tools: enable a MIDI track to record; make sure that the SysEx data is not being filtered out by Pro Tools' MIDI Input Filter; click 'Wait for Note' in the Transport window; make sure that you are at the beginning of your track by hitting the 'Return' key; then initiate the SysEx transfer from your MIDI device. When this has finished, hit the space-bar or click 'Stop' in the Transport window to stop recording. The SysEx data will appear as a MIDI region in the track's playlist and in the MIDI regions list. SysEx messages are also displayed in the MIDI Event List where they can be copied, deleted, or moved as necessary.

Figure 5.12 MIDI Event List window showing a SysEx Event.

Note: Although Pro Tools will record and playback SysEx data it will not let you edit this data or write in data manually – unlike Digital Performer, for example, which does offer these more advanced features.

How to record audio from an external synthesizer into Pro Tools

It is a good idea to record the audio from your synthesizers into Pro Tools as soon as you are satisfied with what you are hearing. This way, you don't have to worry about setting the synthesizers up correctly when you want to mix your music and you don't need to take your synthesizers with you if you take your project to another studio.

First, connect the outputs of your synthesizers to available inputs on your Pro Tools interfaces. If you have an external mixer, you can route signals from this into your Pro Tools interfaces. If you don't have an external mixer, you may need to use outboard microphone pre-amplifiers (which often have instrument-level inputs as well as microphone-level inputs) to raise the output level of the synthesizers to a level that suits the inputs on your Pro Tools interfaces. Of course, if you have the Digidesign Pre, you already have eight high-quality microphone pre-amplifiers with direct instrument level inputs available.

Next, you need to set up a new audio track or tracks, using whatever combination of mono or stereo tracks works for you, and configure the inputs to these to take the audio outputs from your synthesizers. You also need to make sure that your MIDI tracks are playing back correctly.

Once everything is set up, record-enable the audio tracks for the synthesizers, hit Record and Play in the Transport Window, then hit the space-bar to stop recording whenever you like. And that's it – you're done.

Tip: Don't forget to save your session right away in case your computer crashes!

How to record virtual instruments into Pro Tools

Before you can play and record a virtual instrument in Pro Tools, you will need to set up a MIDI track and an Auxiliary Input or Audio track. First, add the Auxiliary Input or Audio track to your Pro Tools session and insert an instance of the virtual instrument plug-in that you want to use. The audio from the plug-in will be monitored via this track or input and you will typically set the output to your main stereo output pair so that you can hear this audio through your main monitors.

Note: It makes best sense to choose an Auxiliary Input for this purpose, rather than an Audio track, to avoid using up any of your available track playback 'voices'. However, you cannot insert RTAS plug-ins onto Aux Inputs, so you have to use Audio tracks for these.

Choose your MIDI keyboard or other MIDI control device as the input to the MIDI track, and choose the virtual instrument that you want to play as the output from the MIDI track. This ensures that when you record onto this MIDI track, MIDI data from the selected keyboard is stored onto this MIDI track, and that on subsequent playback the track routes this MIDI data to the selected virtual instrument.

> Note: To route the 'live' input from your MIDI keyboard through to the virtual instrument while you are playing this keyboard, you have to record-enable the MIDI track and make sure that MIDI Thru is enabled in the MIDI menu. If you don't do this, the virtual instrument will not receive any data from the MIDI keyboard – so you won't hear anything!

Now, when you play your keyboard (with the MIDI track record-enabled and MIDI Thru enabled), you will hear the virtual instrument play back through the Audio track or Auxiliary Input. When you are ready to record your part, press 'Record', then 'Play', in the Transport Window.

> Tip: You may need to adjust the metronome and count-off settings using the controls accessible from the Transport Window prior to recording.

Once you have recorded MIDI data to control your virtual instrument, you can edit this any way and any time you like – change a note here, add a note there, or whatever. Of course, this means that you have to leave the virtual instrument active – which does use up your available DSP resources, leaving less available for playing other virtual instruments, using signal-processing plug-ins, or adding more tracks.

How to use Groove Control with Stylus

Spectrasonics Stylus is available for Pro Tools as an RTAS virtual instrument plug-in. Stylus has a great selection of drum loops and percussion loops that are organized into folders with different BPMs. If you know that you want a drum loop running at exactly 110 BPM, for example, then you can use any of the Stylus loops marked as 110 BPM right away.

But what if there are no loops running at exactly the tempo you want – or if you like a particular loop that is not available at the correct tempo? In this case, Spectrasonics supply special versions of their loops to allow 'Groove Control'. The way this works is that the loop in question is split up into individual beats assigned to play back at their original pitches by different MIDI notes. All the standard drum kits and percussion loops are also available as Groove Control versions along with MIDI files that you can import into any Pro Tools MIDI track.

If you load Stylus with the corresponding 'Groove Control' version of a loop that you like, import the MIDI file version to a Pro Tools MIDI track, then set the output of the MIDI track to play Stylus at the original tempo of that loop, then you will hear the original loop playing back. The neat thing is that if you increase or decrease the tempo, the beats within the loop are played back more quickly or more slowly by the MIDI file – so the loop speeds up or slows down without any need for time stretching.

Choose a MIDI track in Pro Tools, or insert one if necessary, and set up a corresponding Audio track with a Stylus RTAS plug-in inserted so that you can monitor the audio output from Stylus. Set the output of the MIDI track to play Stylus, and record-enable the MIDI track so that you can use a connected MIDI keyboard to play Stylus. Don't forget to make sure that the input selector for the MIDI channel is set correctly to receive data from your connected keyboard.

Try the various loops in Stylus, playing these back using middle C on your keyboard to hear each loop play at its original pitch. When you find a loop that you want to use, such as 72-Slow Jammies c, even if it is not running at exactly the correct tempo, you then select the Groove Control (GC) version within Stylus – for example, 72-Slow Jammies c GC.

Next you need to load the corresponding MIDI file to play the GC version of the loop correctly. Go to the File menu and choose 'Import MIDI to Track'. You'll need to browse to find your Stylus MIDI Files folder.

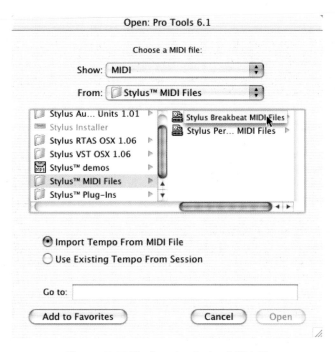

Figure 5.13 Finding the Stylus MIDI Files.

Tip: If the Stylus MIDI files folder is not on your hard drive, you can find this on Stylus Disc 1.

Once you have found the Stylus MIDI Files folder, locate the 72-Slow Jammies c MIDI file, select this and click Open to import it.

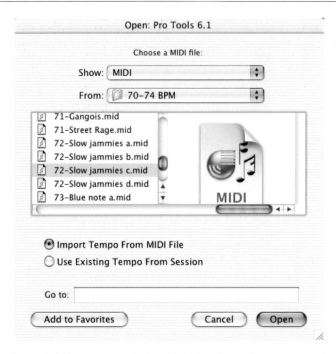

Figure 5.14 Importing a Stylus MIDI File into a Pro Tools track.

Note: One thing to watch out for when importing any MIDI file into Pro Tools is the Tempo option. The default setting is to import the tempo from the MIDI file – which is fine if this is what you want. But I often want to keep the existing tempo – especially if I have spent time creating a tempo map in Pro Tools. It is easy to overlook this and over-write an existing tempo map in Pro Tools. I have managed this a couple of times myself and ended up having to re-create a couple of very time-consuming tempo maps!

Once you have the Groove Control MIDI file in a track in Pro Tools, put the MIDI track into Regions view and then you can slide the MIDI data around in Grid mode to position it at the start of the section where you want to use it. When you have it at the start of the section, it will last for, say 2 bars, but your section may be 16 bars, or it may be the whole song of, say, 80 bars. To fill the time, select the 2-bar MIDI Region and use the Repeat command from the Edit menu, typing 8 to fill 16 bars or typing 20 to fill 40 bars – or whatever. See Figure 5.15.

Tip: With a Groove Control MIDI file controlling a set of Groove Control samples in Stylus, you can simply raise or lower the Tempo in Pro Tools and the tempo of the sampled 'groove' will follow.

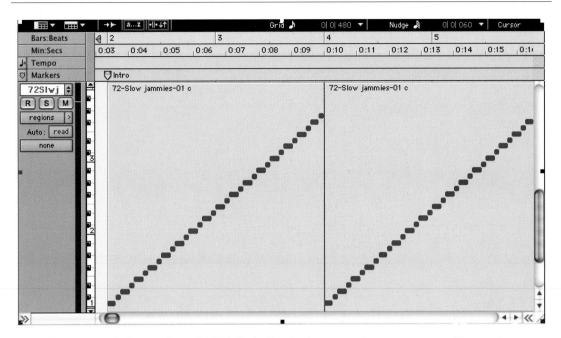

Figure 5.15 A Groove Control MIDI file in Pro Tools repeated as necessary to fill a section.

How to record virtual instruments as Audio

It is a good idea to record your virtual instruments as audio as soon as possible in the production process. There are two reasons for this. Firstly, you can then remove the virtual instrument so that it is no longer using DSP resources. Secondly, and possibly even more importantly, if you take your Pro Tools session to another studio that doesn't have the particular virtual instrument that you used, you would not be able to re-create the original sound that you had – but if you have recorded the audio, then you have it!

You may think that the way to record a virtual instrument is to simply use an Audio track instead of an Auxiliary track to monitor your virtual instrument – so that you can just put this Audio track into record while playing back the MIDI into the virtual instrument. Unfortunately, Pro Tools does not make it as simple as this. If you try this scenario, you will soon discover that the audio from the virtual instrument cannot be recorded onto the same Audio track that is monitoring the virtual instrument.

The solution is to route the output from the Auxiliary (or Audio) track used to monitor the virtual instrument to the input of a separate Audio track using an internal bus – then use this separate Audio track to record the audio playing back through the Auxiliary (or Audio) track.

Take a look at Figure 5.16 to see how this works with a MIDI track sending data to a Virus Indigo that is inserted into an Aux track so we can monitor it (in other words, hear it). When you are ready to record this, route the output from the monitor track via a stereo bus to the input of an Audio track, record-enable the track, then hit Record and Play to 'print' the audio from the virtual instrument onto your hard drive.

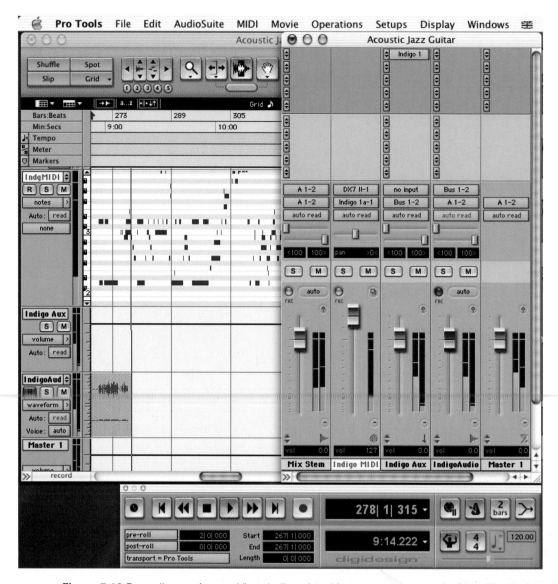

Figure 5.16 Recording an Access Virus Indigo virtual instrument onto an Audio track.

Note: Virtual instruments only available as RTAS plug-ins (such as Spectrasonics Stylus, Trilogy and Atmosphere) cannot be inserted onto Aux tracks for monitoring – only onto Audio tracks that can accommodate RTAS plug-ins.

Tip: When you have recorded the audio from your virtual instrument, you can remove the plug-in and delete the Audio and MIDI tracks that you were using for this to release the DSP resources that these use. Alternatively, you may wish to simply keep these tracks hidden, with the plug-in removed, if you anticipate that you would like to make some changes to the virtual instrument part later on.

Editing MIDI data

Editing issues

MIDI data is displayed in the MIDI tracks in the Edit window. You can edit these MIDI tracks while viewing your audio tracks in the same window – a very useful feature when you want to compare the placing of MIDI notes with audio.

At the top of the tracks display in the Edit window the Ruler View lets you display 'ruler' tracks showing Bars/Beats, SMPTE Timecode, Feet/Frames, Mins/Secs or Samples. Ruler tracks are also available to display Tempo events, Meter events and Marker events – and you can turn any combination of these on or off from the Display Menu.

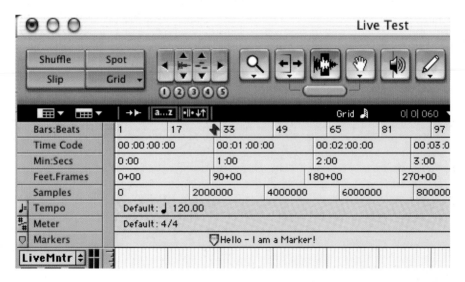

Figure 5.17 Edit window showing all the Ruler 'tracks'.

If you are working on music, you will normally want to display the Bars:Beats ruler so that you can edit according to the bar lines and correct note positions. MIDI data is recorded into Pro Tools with a very high degree of accuracy. The internal 'clock' to which MIDI is resolved has an incredible 960 000 pulses per quarter note (PPQN) resolution.

> Note: When the Time Scale is set to Bars:Beats, the *display* resolution in Pro Tools is 960 PPQN, which provides manageable numbers to work with.

When you are working on music using Bars and Beats, there are several sets of circumstances where you may want to specify tick values. For example, when placing and spotting regions; when setting lengths for regions or MIDI notes; when locating and setting play and record ranges (including pre- and post-roll); when specifying settings in the Quantize and Change Duration windows; and when setting the Grid and Nudge values. It helps to become familiar with the numbers of ticks that correspond to the main normal, dotted and triplet note lengths. For a half note these are 1920, 2880 and 1280 ticks. For a quarter note they are 960, 1440 and 640 ticks. For an eighth note they are 480, 720 and 320 ticks. For a sixteenth note they are 240,

360 and 160 ticks. For a thirtysecond note they are 120, 180 and 80 ticks, and for a sixtyfourth note they are 60, 90 and 40 ticks.

Because MIDI data in Pro Tools is tick-based, if a MIDI region is located at a particular bar and beat location, it will not move from that bar and beat location if you change the tempo of the session – but its sample location will change, which will change its relationship with any audio tracks.

Audio material in Pro Tools, as you might expect, is sample-based. If an audio region is located at a particular sample or SMPTE location, it will not move from this location if the tempo changes in the session – but the audio region's bar and beat location will change.

Note: As the manual explains: because audio material in Pro Tools is sample-based, some amount of sample-rounding may occur with some edits when the Main Time Scale is set to Bars:Beats. This is most evident when you need audio regions to fall cleanly on the beat (as when looping) and you notice that the material is sometimes off by a tick or two. With a few simple precautions, this can be avoided. When selecting audio regions to be copied, duplicated, or repeated, make sure to select the material with the Selector (enable Grid mode for precise selections), or set the selection range by typing in the start and end points in the Event Edit area. Do not select the material with the Grabber (or by double-clicking with the Selector). This ensures that the selection will be precise in terms of bars and beats (and not based on the length of the material in samples).

This question of sample-based versus tick-based behaviour also affects Markers and Memory Locations. Which way should they behave? Fortunately, Pro Tools lets you choose whether any Markers and Selection Memory Locations that you create have an Absolute (sample-based) or Bar|Beat (tick-based) reference. You can select this using the 'Reference' pop-up selector when you create a new memory location.

If you choose Bar|Beat, the Memory Location is tick-based and its bar and beat location remains constant if the tempo is changed – though its relation to audio is scaled, resulting in a new sample location. If you are working on a music session, and you are primarily working with MIDI tracks (and maybe with a few audio loops) you would probably want to use tick-based markers. This would allow you to change the tempo of the session, which you might want to do if you have not completely settled on this yet. If you did change the tempo, all the MIDI data would be fine, and the Markers and Memory Locations would move to match the MIDI. As far as any audio would be concerned, this would stay at the original tempo and would become out of sync with the MIDI data – unless you time-stretched this to match the new tempo.

If you choose Absolute, the Memory Locations are sample-based. So the Bar and Beat positions of any Markers or Memory Locations will shift if the tempo is changed – though their sample locations would remain constant, along with their relationships to any audio material. If you were working to picture, and your Markers and Memory Locations related to SMPTE locations, you would probably use this Absolute reference so that the Markers and Memory Locations, along with any audio, would keep the same relationship with the picture – while the MIDI data would shift as you changed the tempo.

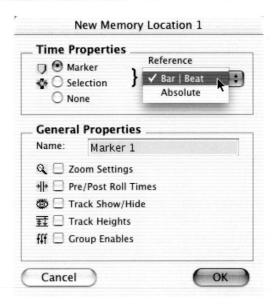

Figure 5.18 Setting the reference for a new Memory Location.

Tip: In the Markers Ruler, Markers that are Bar|Beat appear as yellow chevrons, and Markers that are Absolute appear as yellow diamonds.

Graphic editing

Pro Tools MIDI tracks can be edited graphically in the Edit window. Here you can use the standard Pro Tools Trimmer tool to make notes shorter or longer and use the Grabber tool to move the pitch or position – or 'draw' notes in using the Pencil tool. You can draw in or edit existing velocity, volume, pan, mute, pitchbend, aftertouch and any continuous controller data and the Pencil tool can be set to draw freehand or to automatically draw straight lines, triangles, squares or randomly. With the latest versions of the software, the Pencil tool also lets you draw and trim MIDI note and controller data and the Trim tool can trim MIDI note durations when a MIDI track is set to Velocity view.

It can be very handy at times to insert notes using the pencil tool instead of setting up an external keyboard. Just make sure that the MIDI track is in Notes view and select the Pencil tool at the top of the Edit window. To insert quarter notes on the beat, for example, set the Time Scale to Bars and Beats, then set the Edit mode to Grid and the Grid value to quarter notes. As you move the Pencil tool vertically and horizontally within a MIDI track, the pitch and the bar/beat/clock location is displayed just above the rulers in the Edit window. When you find the note and position you want, you can just click in the track to insert it. It's as simple as that!

To select notes for editing, you can use either the Grabber tool to drag a marquee around the notes, or drag using the Selector tool across a range of notes. Once some notes are selected, you can drag these up or down to change the pitch using the Grabber or Pencil tools – while pressing the Shift key if you want to make sure that you don't inadvertently move the position of the notes in the bar. You can use the Trimmer tool or the Pencil tool to adjust the start and end points of the notes. If you set the MIDI track to Velocity view, you will see the attack velocities of the notes represented by 'stalks'. You can edit these using the Grabber tool.

Sometimes, you may simply need to edit one note. If you select a note using the Grabber or the Pencil, its attributes will be displayed in the Event Edit area to the right of the counters at the top of the Edit window. Here you can type in new values for any of the displayed parameters.

Continuous controller data recorded onto MIDI tracks is displayed as a graph line with a series of editable breakpoints. These breakpoints are stepped to represent individual controller events, in contrast to the standard automation breakpoints, which are interruptions on a continuous line. You can edit pitchbend, aftertouch, mod wheel and other MIDI controller data directly in the Edit window according to which type of data you select using the pop-up selector.

Note: MIDI Controller #7 (Volume) and #10 (Pan) are treated as standard automation data and can be recorded and automated using the volume and pan controls in the Mix window. These can also be recorded from external MIDI devices.

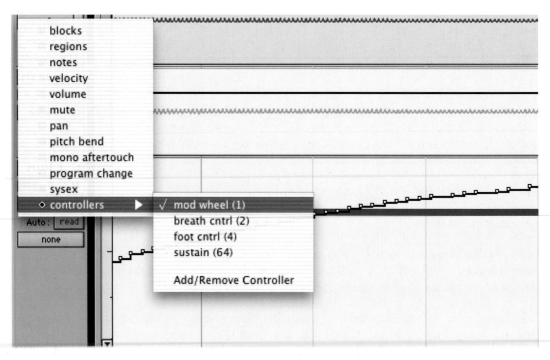

Figure 5.19 Editing MIDI continuous controller data.

Event List editing

You can also view and edit MIDI data using the MIDI Event List, which is available from the Windows menu.

Figure 5.20 Data in the MIDI Event List window.

> Tip: The Pro Tools MIDI Event List editor lets you see your MIDI events listed alpha-numerically. This is often the easiest way to work, especially when editing lists of sound effects spotted to SMPTE time-code locations.

> Note: There is no list editor for audio data – a major omission in my opinion.

You can select, copy, paste, or delete events in the MIDI Event List and any MIDI event, except SysEx, can be inserted using the Insert menu pop-up at the top of the window. The Options pop-up menu has various commands that let you customize the MIDI Event List so that it scrolls the way you want it to and shows what you want it to show while you are editing. You can also define how events are to be inserted. The third pop-up menu lets you choose which track to display in the MIDI Event List window.

The MIDI menu

The MIDI menu has many similar commands to those available in the Digital Performer's Regions menu, such as Quantize; Change Velocity; Change Duration; Transpose; Select Notes and Split Notes – all of which are quite similar to Digital Performer's MIDI Region commands.

You can use the Selector tool in conjunction with the MIDI menu to apply these powerful region commands. Just select a MIDI note or region then choose the operation you wish to apply from the MIDI menu.

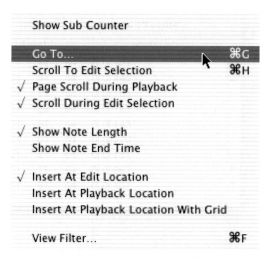

Figure 5.21 MIDI Event List options menu.

One of the most frequently used commands is 'Quantize'. There are comprehensive options provided here, including the very useful 'Swing' factor – which can be set using a % control.

Figure 5.22 The MIDI menu.

Tip: To avoid your music sounding too mechanical, you can use the Strength parameter, which lets you specify a percentage value by which notes will be drawn toward the quantized values. The 100% setting means that notes will be drawn in until they exactly match quantization values, while lower settings will leave the notes some distance away from these.

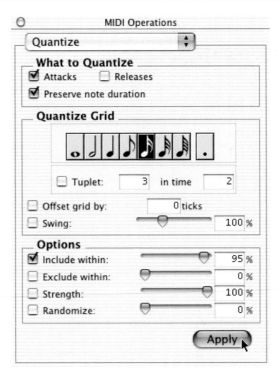

Figure 5.23 MIDI Operations Quantize window.

You can also change the velocities of all the notes you have selected in a variety of useful ways – adding or subtracting a specific value, scaling over the length of your selection, or even introducing a specifiable element of randomization to the values.

Change Velocity, Change Duration and Transpose are all fairly self-explanatory. Select Notes lets you make selections based on criteria such as whether the notes are within a certain range. Split Notes does the same thing as Select Notes but also lets you automatically copy or cut the selected notes when you click Apply.

> Tip: You can use the MIDI Input Filter to filter out any MIDI messages that you don't want to inadvertently record. So, for example, if you are not using the aftertouch messages that are automatically sent by most MIDI keyboards, you can filter these out.

> Note: Remember that it depends on the individual patch settings for a synthesizer sound as to whether aftertouch messages are actually used for any purpose. When these are used, they are often mapped to control vibrato depth or other modulation effects.

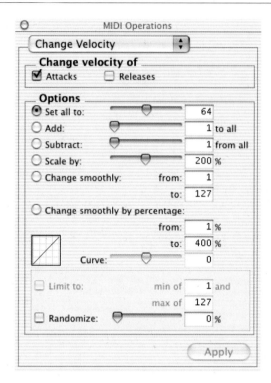

Figure 5.24 The Change Velocity MIDI Operations dialog.

From the MIDI menu you can also enable the Click, set up the Click Options, enable the MIDI Beat Clock, enable MIDI Thru, issue the All Notes Off command to stop stuck notes from sounding, and access the Input Filter and Input Devices setup dialogs.

Three new commands were added to the MIDI menu for version 6 – Restore Performance, Flatten Performance, and Groove Quantize.

The Restore Performance command lets you restore the original performance any time you like – even after the session has been saved or the Undo queue has been cleared. You can select which note attributes to restore – timing (quantization), duration, velocity or pitch – and the selected MIDI notes will revert to the way they were when you originally recorded them (i.e. before any subsequent edits).

The Flatten Performance command 'solidifies' your edits – letting you lock or 'flatten' the current state of selected MIDI notes. Flattening a MIDI Performance also creates a new 'restore to' state for specific note attributes that you can select – including duration, velocity, pitch and timing (quantization). You might use this command, for example, if you made a series of edits that you considered to be 'correct', but you then wanted to be able to try out more edits that may not be correct – while retaining the ability to restore to this 'correct' state.

The third new command, Groove Quantize lets you conform MIDI note locations and durations to an existing groove template instead of to the normal quantization grid. This makes it possible to apply the rhythmic nuances of any audio performance to your MIDI tracks. You can even capture the dynamics from an audio track and apply these to a MIDI track as velocity data.

Digidesign provides Logic-, Cubase- and MPC-style grooves for Pro Tools that let you apply the various quantization and swing factor settings available in these other popular sequencers.

Applying grooves

To apply a 'groove' to your audio, just select some MIDI data then use the Groove Quantize dialog from the MIDI menu to apply the groove of your choice.

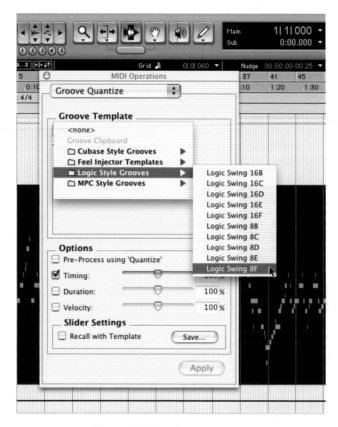

Figure 5.25 Applying grooves.

Drum patterns

When you are developing a musical arrangement, you will often record drums first. For example, you may record bass drum, snare drum and sidestick into one track before adding other tracks containing toms, cymbals, hi-hats, cabassas, or congas. For ease of editing, you may wish to separate out the different drum and percussion instruments onto individual tracks using the Split Notes command. This lets you focus your attention on each part in turn when editing and it makes it easy to play each instrument using a different synthesizer or sampler if you are searching for the perfect bass drum sound or whatever. Simply change each track's MIDI output setting accordingly.

Tip: Sound architect Ernest Cholakis of Numerical Sound (www.numericalsound .com) has provided a default set of 12 Feel Injector Templates for Pro Tools (provided on the Installer CD-ROM). These templates are designed to impart an element of human performance into your MIDI and audio sequences. Cholakis suggests that when applying Feel Injector Templates to MIDI sequences you should use a velocity sensitivity of around 75 to 80 per cent. Apparently, the velocity sensitivity information is based on real-world dynamics, which works well for audio recordings but is a bit too wide for MIDI sequences. The timing sensitivity should be set to 100%. In each Feel Injector Template, a guide tempo is listed. This gives an indication of the original tempo on which the groove was based, so you can take this into account if you like – or you can just go ahead and apply a Feel Injector Template to a MIDI sequence running at any tempo.

For example, let's look at how I worked on a 'cover' version of a Marshall Jefferson 'house' recording called *Give Me Back The Love*. This had a standard four to the bar bass drum pattern, with a typical 'house' snare drum pattern, some congas and hi-hats.

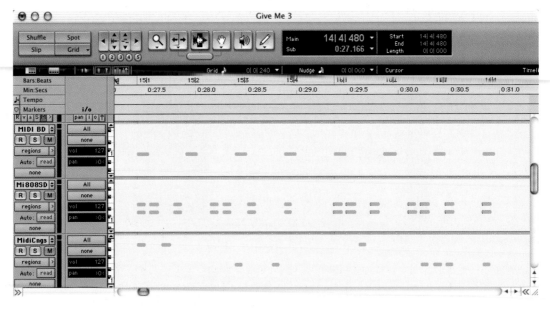

Figure 5.26 Drum parts in regions view.

The basic pattern repeated every two bars, so I started by just recording these two bars. The idea here is that with these two bars perfected, it then becomes trivially easy to paste these wherever in the recording they will be needed.

Tip: When arranging 'blocks' of patterns, it is often best to use the Regions view so that it is easy to see the pattern boundaries.

With the MIDI tracks set to Regions view, you can select the two bars you want to use and use the Edit menu command, 'Separate Regions', to create a two-bar region – not forgetting to name this appropriately.

To build up your arrangement you can use the 'Repeat'… command from the Edit menu. This lays the copies that you specify 'end-to-end' to fill whatever number of bars you need. Of course, you can also use the standard cut and paste commands on a selected pattern or group of patterns and place these wherever in your arrangement you like.

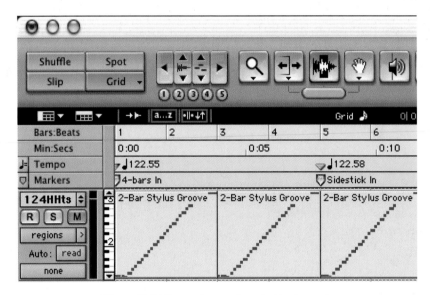

Figure 5.27 Repeating regions in Regions view.

Editing notes

Now let's look at a synthesized brass part I added. Take a look at Figure 5.28. Notice that just one note is incorrectly placed. It is easy to drag this into the correct position using the Grabber tool in Grid mode, with the grid set to 16th notes to suit this 'house' music track.

With a brass part played using a synthesizer, it is always possible that the player did not release all the notes of a chord at the same time – which is exactly what I did here and which sounded messy. To check for this, you need to expand the scale of the display horizontally until you can see clear differences in note lengths. If there are just a few notes that are obviously the wrong length, these can be quickly adjusted using the Trim tool.

If you see that there are a lot of note durations that need to be edited, it can be quicker to use the MIDI Event List editor instead. When you switch to the MIDI Event List, you can see at a glance what is happening with the durations, select any that you want to change, then use the Change Duration dialog to set these the way you want them.

You can set all the durations to a particular value, add to or subtract from the existing values, scale the values, change the values gradually over the selected range of notes, and so forth. See Figure 5.31.

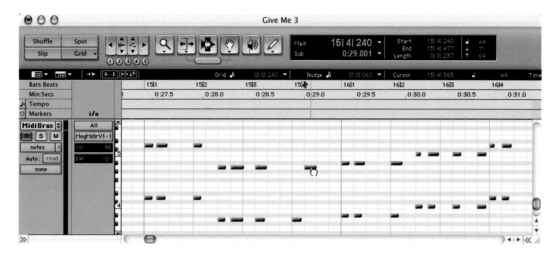

Figure 5.28 Synthesized brass in Notes view.

Tip: If you want the note lengths to finish cleanly on note-length sub-divisions, make sure you are using Grid mode instead of Slip mode.

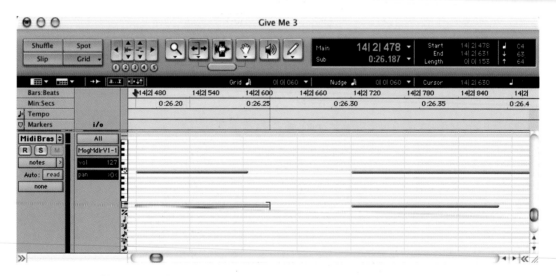

Figure 5.29 Adjusting a note length using the Trim tool in Grid mode.

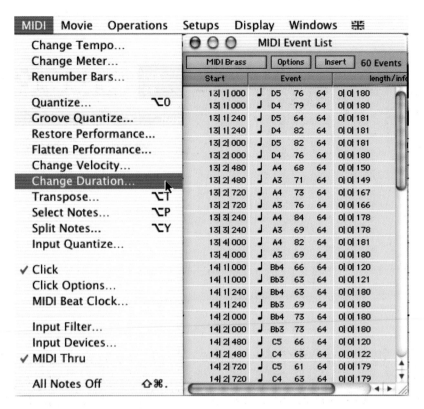

Figure 5.30 MIDI Event List with the MIDI menu in action.

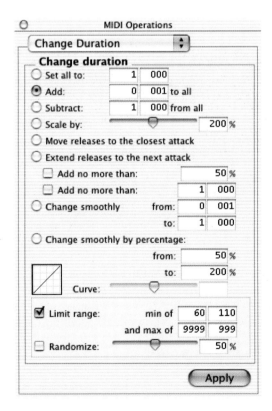

Figure 5.31 Change duration dialog.

Working with patterns

Many people are used to working with short patterns in sequencers like Cubase – finding it very convenient to drag these around in the Arrange window to build up arrangements. Although Pro Tools may appear to be more of a linear recorder, it has good features for working with patterns in a similar way to Cubase.

To work with patterns that you cut and paste to build your arrangement you simply need to change the display from Notes to Regions so you can cut your regions up in whatever way suits you. I find it convenient to cut songs into their conventional sections – Intro, Verse, Chorus, and so forth – which is pretty easy to do using the Markers as a guide.

As is often the case in large software applications, there are several ways to do this in Pro Tools. Let's look at a straightforward approach. First you select a marker in the Memory Locations window. Then Shift-click on the next marker to automatically select all the audio between these markers on whichever track has the insertion cursor active – take a peek at Figure 5.32 to see how this might look.

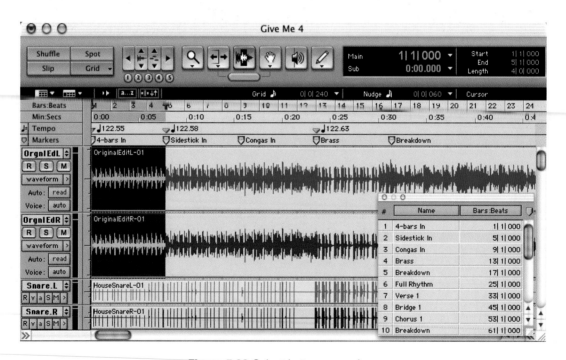

Figure 5.32 Select between markers

Next, extend this selection downwards to encompass all the other tracks, if necessary, by shift-clicking on these in turn – see Figure 5.33.

If you have a lot of tracks in your project, it is probably a good idea to put all the tracks into Small or even Mini size so that you can see them all on-screen at once. You can do this for all tracks by Option-clicking just immediately to the left of the waveform display and selecting appropriately from the pop-up menu that appears – see Figure 5.34.

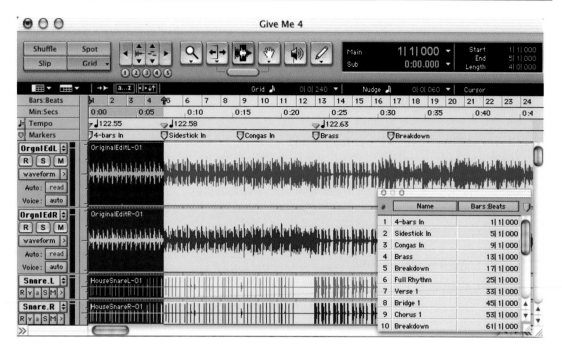

Figure 5.33 Extend selection.

With all the tracks in view, it becomes much easier to extend your selection to include all the tracks in your project – see Figure 5.35.

A quicker way to select all the tracks is to simply click in the Time Ruler at the top of the Edit window while the Selector tool is engaged. This will put a selection point in each track. Now shift-clicking on the next marker will extend the selection across all the tracks at once.

Yet another way to achieve the same result is to engage the Edit Group for 'All' tracks – see Figure 5.36. This exists by default in every Pro Tools session so that you can always carry out edits to all tracks at once when you need to.

With this active, whatever edit (or, in this case, selection) you make to one track will be made to all tracks.

Figure 5.34 Track height selector.

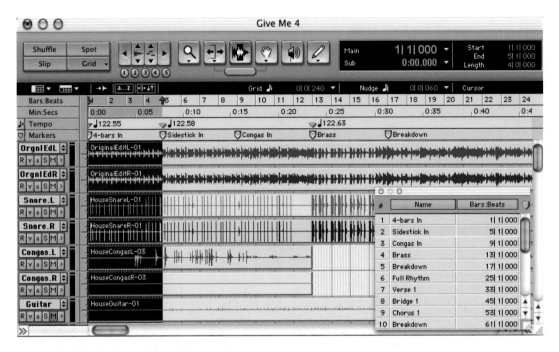

Figure 5.35 Continue to extend selection.

Once you have all your tracks selected, choose the 'Separate Region' command from the Edit menu to split them from any regions occurring before or after the time you have defined. The Region Name dialog appears and you can use this to name the region to correspond with the section of your song or recording – see Figure 5.37.

Separate and name the next region, then repeat the process. Eventually, your edit window will look something like Figure 5.39.

Now you can set a Grid value of, say, a quarter note, and use Grid mode to constrain the regions to this value while moving your regions around.

Figure 5.36 Select Edit Group <All>.

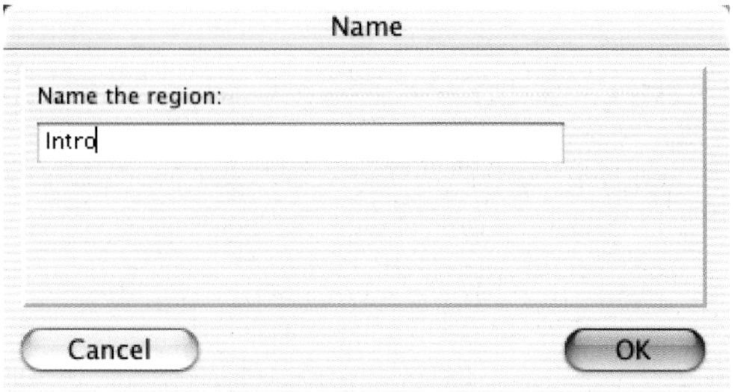

Figure 5.37 Region Name dialog.

Tip: If you still have your first selection made in the Edit window, there is a quick way to shift this to pick up the next section that you want to separate. Just click once on the next Marker, then shift-click on the Marker after that. Your selection will 'flip' over to automatically select the audio between this pair of Markers. See Figure 5.38.

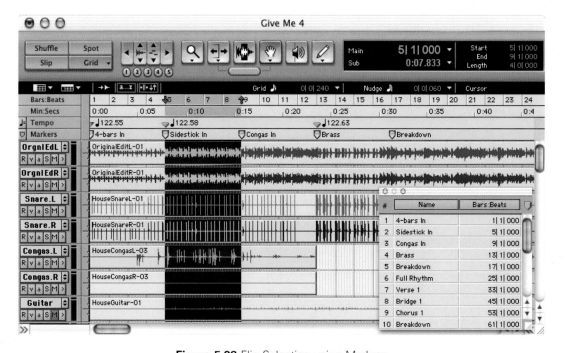

Figure 5.38 Flip Selection using Markers.

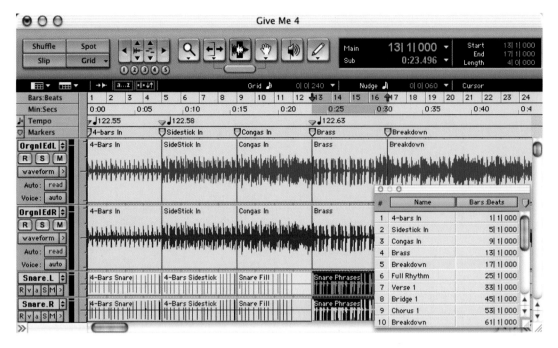

Figure 5.39 Edit Window showing regions in region (block) view.

Click on and hold the down arrow next to the Grid value display above the rulers in the Edit window – see Figure 5.40.

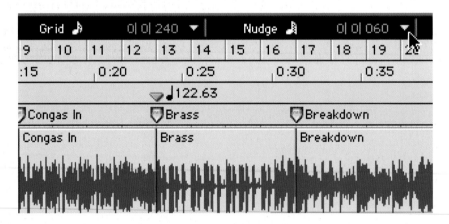

Figure 5.40 Grid resolution pop-up.

A pop-up menu appears with options for different note values when the Grid is set to Bars and Beats.

Figure 5.41 Grid options.

Tip: If you want to see vertical Grid lines in the Edit window, you need to enable the Display Preference for 'Draw Grid in Edit Window'. This is useful when you are dragging regions around in the Edit window using the Grabber tool. Grid lines will appear corresponding to the Grid value you have selected.

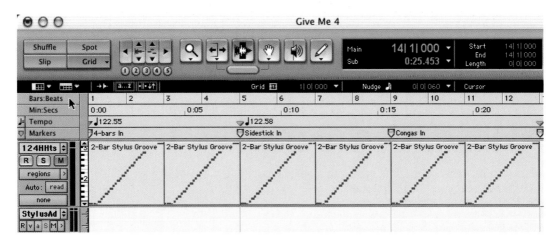

Figure 5.42 Edit window showing cursor about to click on the Ruler name at the left of the Bars:Beats Timebase ruler to make the Grid lines visible.

> Tip: You can quickly enable and disable grid lines by clicking the name of any Timebase Ruler underneath the mode selector buttons at the top left of the Edit window. See Figures 5.42 and 5.43.

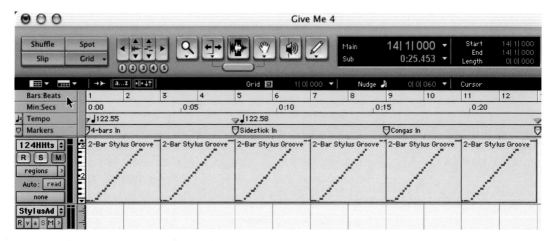

Figure 5.43 Edit window after clicking on the Ruler name at the left of the Bars:Beats Timebase ruler. Note that grid lines have appeared corresponding to each bar line – as set using the Grid pop-up.

> Tip: If you want a region that you are dragging around in the Edit window to automatically butt up against the previous region, you can use Shuffle mode. Be careful to return to Slip or Grid mode as soon as you have made your moves in Shuffle mode though – it is all too easy to accidentally move a region and have Shuffle mode shuffle your regions to somewhere they shouldn't be. And if you don't notice this at the time it happens you may not be able to use even the multiple Undo feature to get back to where you were.

Summary

There can be little argument now – Pro Tools provides a serious set of MIDI capabilities. Yes, Logic is stuffed with MIDI features, Cubase has some neat tricks, and Digital Performer can do some clever things, but Pro Tools will let you do at least 60 per cent of the stuff you can do in these more advanced MIDI environments – including all the most frequently used operations. Also, all the MIDI features in Pro Tools are extremely straightforward and easy to learn how to use. And on busy recording sessions, that simplicity and ease of access definitely counts for a lot!

6 Recording

Introduction

Recording – now that's a big subject. I don't know if you are going to be recording a set of bagpipes, a rock band, a rap track or a symphony orchestra. What I do know is that Pro Tools can handle all of these. And you may even need to work with all of these at some time or other during your career.

Many of you will already be working with Pro Tools while others will be new to the system. I will start out by discussing how digital audio recording works, mentioning various caveats, and highlighting practical issues. Then we will look at areas such as monitoring, setting up clicks, destructive recording, loop recording and the concept of 'voices' in Pro Tools – with tips and notes along the way to draw your attention to particular aspects. I also include a section on how to import audio from a CD – a feature you will use time and time again.

To help get you started with a recording project that incorporates both audio and MIDI, and also involves techniques such as tempo-mapping, I will describe how I went about recording a 'cover version' of a Marshall Jefferson 'house' track called *Give Me Back The Love*. The idea here is to guide you through the process with a mixture of step-by-step instructions, tips and hints.

Digitizing audio

Audio is captured initially in analogue form using a microphone or from a directly connected electric instrument such as a guitar pick-up or synthesizer. Many of today's electronic synthesizers and samplers actually create their sounds digitally, but this is converted to analogue audio before it reaches the standard analogue audio outputs. Some more recent devices are now featuring direct digital outputs, which is obviously the way forward as more and more audio equipment goes digital.

Typically, a microphone is plugged into a microphone pre-amp, often in a mixer, then fed at line level to an A/D converter – which may be in a mixer or in an external unit. Similarly, electric and electronic instruments are connected first to a mixer – either using microphone-level or line-level inputs according to the type of outputs available from the instruments – before being fed from the mixer to an A/D converter.

The A/D converter can be of a stand-alone type, with popular models available from Apogee and many others, in which case a digital output is provided to transfer the digitized audio directly to a digital audio recorder. Digidesign offers a range of interfaces for their Pro Tools systems that include line-level analogue and various digital audio inputs and outputs. Digidesign also offers the Pre – a unit containing eight microphone pre-amplifier channels, the outputs from which can be connected to any suitable A/D converter to feed these into your Pro Tools system.

I am often asked what the difference is between, say, the expensive Prism or Apogee converters, the mid-priced converters used in Digidesign interfaces, and the low-cost converters used in the cheaper audio cards and interfaces. The short answer is that you get what you pay for. And what you are paying for with the more expensive converters is better quality analogue pre-amplifiers and/or amplifiers used in the input and output circuitry, along with higher-quality converters which provide much better dynamic range. The more expensive converters all offer 24-bit 44.1/48 kHz operation and support even higher sampling rates such as 88.2, 96 or even 192 kHz.

So what are the advantages of 24-bit? Well, the converter can sample at smaller increments of the signal level, so when it comes to reproducing the analogue signal more of the original fine detail will be delivered. At higher sample rates such as 96 kHz, frequencies well above 20 kHz will be captured and reproduced, despite the fact that normal humans cannot hear these frequencies. Nevertheless, removing these frequencies from the audio spectrum can have a noticeable effect on the audible frequencies in complex audio waveforms. Such waveforms, typical of most natural sounds, actually contain a mix of frequencies extending well beyond the audible range. Take away these higher frequencies and the 'shape' of the complex waveform will inevitably be changed – albeit by a relatively small amount. It is also said that the designs of the anti-aliasing filters included in all converters are much better in 96 kHz converters – so they will affect the audible sound much less than the typical 'brick-wall' filters used in 16-bit converters to prevent any higher frequencies entering the converters and causing problems. I have regularly worked audio sampled at 96 kHz and each time I am impressed at the clarity of the sound – especially at the higher frequencies.

So what's the downside? Well, if you record using eight more bits of information, your audio files are half as large again compared with 16-bit files. So 1 minute of mono audio which occupies 5 Mb at 16/44 would occupy 7.5 Mb at 24-bit. Similarly, if you double the sample rate you double the file size. So 1 minute of audio at 96 kHz/16-bit would occupy 10 Mb – and at 24-bit would occupy 15 Mb.

Also, you have to consider the distribution medium. If the audio is for CD-release, then this cannot deliver better than 16/44. However, newer formats such as DVD-Audio can handle 96 kHz/24-bit and even higher sampling and bit rates. Even if you plan to release on CD 'today', you may wish to record and archive at higher resolutions for delivery using different formats 'tomorrow'.

Another consideration is the type of audio material you are dealing with. If you are working with audiophile-quality recordings of delicate orchestral music you will want to use the highest sampling and bit rates to capture every nuance of detail and the widest possible dynamic range. On the other hand, for straight-ahead rock or dance music, where the dynamic range is already severely restricted by the use of increasingly massive amounts of compression and where the musical instruments used may not have such a wide dynamic range or offer the same fine detail as orchestral instruments, then 16/44 may well suffice.

Digital audio caveats

It is too easy to assume that digital audio is a perfect system. We all love the advantages – making copies with no loss of quality, random access while editing from hard disk, and so forth – but we sometimes forget that there are imperfections.

The first of these is the sampling rate itself. The theorists, Nyquist, Shannon and others, say that sampling at twice the frequency of the highest frequency you want to digitize will give you an accurate result. Now consider this carefully: humans can hear sounds extending in frequency up to around 20 kHz (although this upper limit normally drops off with age to much lower frequencies). So a sampling frequency of 44.1 kHz should comfortably cope with this range – according to the theory. But stop and think for a moment what this actually means. Suppose you are trying to record a frequency of 20 kHz. In this case you will have just two samples to represent the waveform. If you are sampling a sine wave, the amplitude of a single waveform increases smoothly in a curve from zero level to its maximum, then drops back to zero, then goes to its maximum negative value, then back to zero. If you sample at the maximum positive value and again at the maximum negative value, the best you will get from the DA converter may at least resemble the sine wave. However, if you were to sample at any other points, the waveform would appear more like a sawtooth waveform. And if you happened to sample the waveform at its zero crossing points you would only get back silence!

Even more important as far as audio quality is concerned is the quality of the filters used in the A/D converters to restrict the frequencies put through the system to a maximum of 20 kHz. For the converters to work properly without producing 'alias' frequencies, all frequencies above 20 kHz must be removed by filters in the converters. If this is not done, the higher frequencies will produce spurious 'alias' frequencies which will appear within the audible bandwidth producing a fatiguing, harsher sound. The problem here is getting the design of these filters right without affecting the available sound quality. The first 16-bit/44.1 kHz systems, based on the technologies available in the mid to late 1980s, suffered from having audibly less than perfect A/D and D/A converters. Many people complained about the higher frequencies sounding 'brittle' – or that the recordings lacked 'bottom-end'. Today, converter technologies have been drastically improved to the point where 16-bit 44.1 kHz recordings sound acceptable to the vast majority of listeners – despite the fact that some expert listeners feel that there is still plenty of room for improvement. Of course, it should not be forgotten that A/D and D/A converters have associated analogue circuitry that needs to be designed to the highest standards for best results. This is another reason why some converters sound better than others.

To address some of these concerns, today's 24-bit systems record more detailed information than 16-bit systems and operate at sampling rates of 96 or even 192 kHz. One of the greatest benefits of the higher sampling rate systems is that the filters used in the converters can be kept well out of the way of any remotely audible frequencies.

Also, bear in mind that although humans cannot hear above 20 kHz, the bandwidth of frequencies in many sounds can extend well beyond this – a factor that was recognized and reflected in the design of high-end analogue mixing consoles and other audio equipment from the outset. Removing these inaudible frequencies from a complex waveform inevitably must change that waveform – however slightly.

Practical considerations

The changeover to digital technologies has meant that recording engineers need to change some of the ways they work. For example, with analogue recordings, if recording levels are too high, the recording medium will distort the audio. At the onset of this distortion, the audible effects, counter-intuitively, can actually enhance the sound from a creative point of view – fattening up drum sounds, sweetening guitar sounds and so on. The explanation here is that the even-harmonic distortions produced just happen to be of a type that the brain finds pleasing – at least initially. Of course, if there is too much distortion, the recordings can be ruined, but analogue engineers have learned from experience to judge the levels to take account of these factors.

For this reason, some recording engineers prefer to record at least some instruments, such as drums and guitars, to analogue tape first – to capture these desirable distortions that they use to 'mould' the sound creatively before transferring to digital further down the line. Of course there are still those who prefer to record analogue all the way, even mixing to analogue 1/2-inch stereo format machines. For distribution on CD or other digital media these analogue tapes end up being digitized eventually – while retaining the desirable audio characteristics of the analogue recordings.

With digital recordings, the situation is very different. Once you exceed the maximum level which the system can handle without distortion, the waveforms will be clipped and the resulting sound will contain unpleasant-sounding odd-harmonic distortions which make the recordings sound fatiguing to the listener – and in the worst cases nasty clicks or pops will be audible. Now if you set the record level too low you are not making the best use of the available resolution and your recordings will not capture the dynamic range as accurately as they would if they used all 16 or 24 bits. The answer is to always aim for the highest record level short of exceeding the full scale. (And don't make the mistake of thinking that you can record at lower levels then adjust the gain later to compensate. Once you have recorded an audio waveform at lower resolution you can never increase the resolution – the information simply isn't there on your hard drive.) This can be something of a juggling act to achieve if you are working with very dynamic material, but, with practice, hand-riding the mixer faders or judicious use of compressors can help you achieve the results you are looking for.

> Tip: It is not necessarily wrong to record digitally at lower levels – especially if the audio element you are recording will be kept at a similar low level in the mix. As long as it sounds OK on playback, then it is OK. Just remember that you may have a problem with background noise if you change your mind and raise the level of this audio element in your mix.

Things you should know about before recording with Pro Tools

If you are new to Pro Tools, you do need to know several basic things about the system before you get started with recording: how the monitoring works, how to set up clicks, how to set up meters and tempos, how the rulers work, how the playlists work, how to do punch-ins, and so forth.

Monitoring

Before commencing any recording session you need to decide how you are going to monitor the audio you are about to record. The monitoring latency with TDM systems is extremely low, although there is inevitably a certain amount of latency due to any A/D and D/A conversions necessary to carry audio in and out of the system. Monitoring incoming audio via an external mixer avoids this latency, but is not always the best method to use for a variety of reasons.

Recording engineers, used to working with conventional recording equipment, usually prefer to monitor through Pro Tools as this helps give them confidence that the equipment is working correctly. The philosophy here is that when you are preparing the sound you should hear it the way it is going to be – with the sound of the A/D and D/A converters. Then it is exactly the same when you play it back. That is one of the beauties of digital – unlike with analogue, where you never hear the identical thing coming back. Also, while recording to a particular track or set of tracks, this method allows you to roll back and drop in immediately to replace any recorded material.

Another way to work is to open a new track or tracks for drop-ins. You have plenty of tracks to work with in Pro Tools, so shortage of tracks is not normally an issue. Also, some engineers like the flexibility of being able to simply mute out (or quickly cut out) the section of the original take which is to be replaced, then have the musician or singer play or sing from whatever point prior to the 'drop-in' point that they feel comfortable with. This way they don't have to worry about setting the drop-in and drop-out points accurately.

> Note: This method is not fast enough when you are working with 'live' bands, because you would have to set up new headphone levels, pans and so forth when inserting the new tracks, and this would take far too long.

On a typical multi-track recording session with a band, the engineer will set up one or more separate headphone mixes for the musicians using each channel's Auxiliary Send controls and this will be completely independent of the monitor mix. With up to five separate monitor mixes available you can set these up to suit different musicians, heavily compressing the drums to make them sound great for the drummer, for example.

> Tip: Copy the main mix to each auxiliary send using the Copy To Send command in the Edit menu and then tweak to suit. This is much quicker than setting up each auxiliary send on each channel from scratch.

Pro Tools lets you select between two different ways of input monitoring in the Operations menu. Auto Input Monitoring is the monitoring method which most recording engineers will be familiar with. This works similarly to the way conventional analogue and digital multi-track tape machines work – using auto-switching logic circuitry. With playback stopped, the audio input is monitored through the system. While playing back prior to a drop-in, you will hear any material already recorded on that track. When you drop in, the logic switches to monitor the input signal,

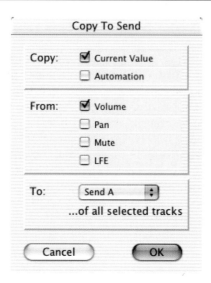

Figure 6.1 Copy to Send window.

automatically switching back to monitor the track when you drop out. In Pro Tools, this switch back to monitoring the track is not instantaneous unless you use the QuickPunch method. With Input Only Monitoring, as its name suggests, you hear the input signal at all times whenever a track is record-enabled. In this case, the Record button in the Transport window turns green to let you know that Input Only Monitoring is active.

> Note: A drop-in is referred to as a punch-in in the Pro Tools manual. The term drop-in is more usual in the UK while the term punch-in is more usual in the USA.

Another issue to be aware of while recording is that a different fader level is in operation than when you are playing back. This arrangement is necessary so that you can adjust one set of levels for monitoring while recording and another set of levels for playback. You can always tell which level is applicable because the volume faders in the Mix window turn red whenever a track is record-enabled.

There is also the issue of how fast Pro Tools can drop in to record. With just a few tracks, this is never an issue, but with a large number of tracks or channels, or if you are playing back large numbers of tracks while recording, you may notice a delay. To get around this, you can use Record Pause mode. After you have clicked on Record in the Transport window, simply Option- or Alt-click on the Play button. Pro Tools then prepares the tracks for recording so that when you hit Play, recording will begin right away.

> Tip: Record Pause mode can also be useful when you are syncing to time code as it will speed up the lock-time when working with large numbers of tracks.

Drum punch-ins

Pro Tools is great for tricky stuff such as drum punch-ins that can be difficult to get right using analogue tape, for instance. In the edit window you can see all the tracks stretching from left to right and it is easy to see on the waveform display where things stop and start – so you can quickly insert the cursor where you want and play from that point. You can also zoom in – all the way to sample level if necessary. To do a drum punch-in you can choose the bar and beat by referring to the timeline – as long as you have recorded your track to a MIDI metronome click. Don't forget – you can always cancel record if you mess up and you will still have the original. Just set up a pre-roll, two measures or whatever, then drop in. You can always record more than you need and then use the Trim Tool to just replace the stuff that really needs replacing. And if you don't like the punch-in at all, you can just delete this and use Heal Separation to get back to the original instantly.

The rulers

At the top of the tracks display in the Edit window the Ruler View lets you display 'ruler' tracks running above the audio and MIDI tracks. Timebase Rulers can be set to Bars:Beats, SMPTE Timecode, Feet.Frames, Mins:Secs or Samples. These determine the format of the Main counter and provide the basis for the Edit window Grid.

Ruler tracks are also available to display Tempo events, Meter events and Marker events – and you can turn any combination of these on or off from the Display Menu.

You can hide or show these using the rightmost of the two pop-up menus immediately above the topmost ruler – but you will always have at least one time ruler showing. Clicking on this pop-up menu reveals the various choices.

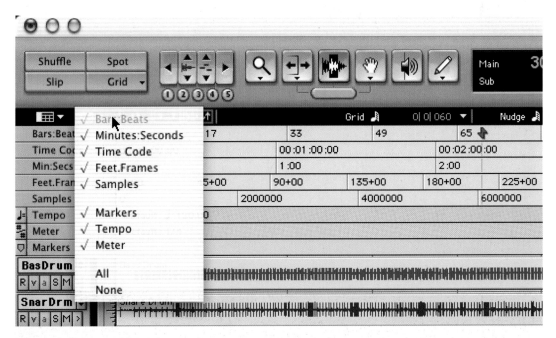

Figure 6.2 Edit window: showing display options for the rulers.

Edit window view options

You will notice that there is another pop-up menu to the left of the pop-up for the ruler views. This lets you choose whether to display the various views that you see in the Mix window – the Comments, I/O, Inserts, Sends and Mic Preamps views.

Figure 6.3 Edit window view options pop-up.

With these views available in the Edit window, it is possible to do most of your work in this window without having to use the Mix window – which can be an advantage if you only have one screen.

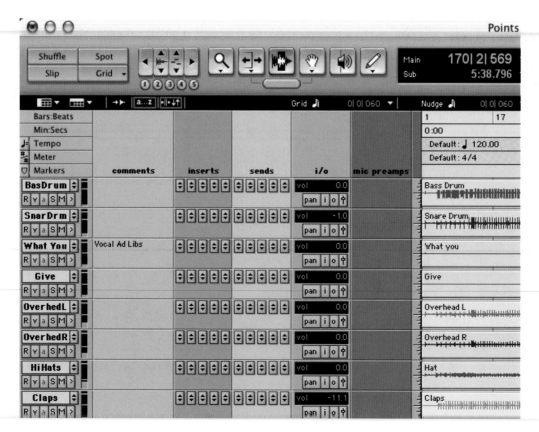

Figure 6.4 Edit window showing additional views.

I find the most useful combination of rulers for working on music productions in Pro Tools is to use bars/beats and minutes/seconds with tempo changes, meter changes and markers. Feet/frames is for some types of film work and timecode is for video post-production – although this can also be useful if you have other audio or video equipment synchronized to Pro Tools.

Meter and tempo

When you open a new session in Pro Tools, the meter defaults to 4/4 – the most commonly used time signature in popular music.

If you intend to record with a click and you want to use a different meter, you will need to set the meter accordingly so that the correct click sounds on the downbeats. You can choose a note value in the Meter Change dialog for the number of clicks that you want to cause to sound in each measure by clicking on the appropriate note value icon. So, for example, if you choose a meter of 6/8, you should click on the 8th note icon so that you will hear a click for every 8th note.

> Note: If you record audio or MIDI without referring to a click, you won't be able to use the editing commands (such as Quantize) properly – because the bar positions in the music won't line up with the barlines in Pro Tools.

> Tip: If you forget or don't realize this, then you should start again and play with a click, or use the Identify Beat and Beat Detective features to create a tempo map for the recording.

You can set the default meter for a session using the Tempo/Meter Change dialog. You can get to this by double-clicking the Meter button in the Transport Window or by selecting Show Tempo/Meter from the Windows menu – or by selecting Change Meter . . . from the MIDI menu.

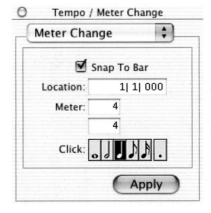

Figure 6.5 The Tempo/Meter Change dialog set to Meter Change.

Tip: To replace the default meter, set the Location to 1|1|000 then enter the Meter you want to use for your session.

If the meter changes within the music at any point, you will need to insert a Meter Change event. These Meter events, which can be inserted at any point along the timeline within a Pro Tools session, are stored in the Meter Track and can be seen in the Meter Ruler.

Setting the default tempo

When you open a new session in Pro Tools, the tempo defaults to 120 BPM. If you want to work with a different tempo, you will need to set the tempo first. If you know the tempo you want to use for the session, go ahead and insert this as a tempo event at the beginning of the Tempo Track. Of course, you can insert tempo events anywhere you like – you might speed up a song in the choruses by one or two beats per minute (BPM) and slow back down in the verses, for example. These tempo events are stored in the Tempo Track and can be seen in the Tempo Ruler.

You can set a default tempo for a session using the Tempo/Meter Change dialog. You can get to this by double-clicking the Meter button in the Transport Window or by selecting Show Tempo/Meter from the Windows menu. If you use either of these methods, you need to make sure that Tempo Change is selected from the pop-up menu at the top of the Tempo/Meter Change window (Meter Change is selected by default). You can also open this dialog directly by selecting 'Change Tempo...' from the MIDI menu.

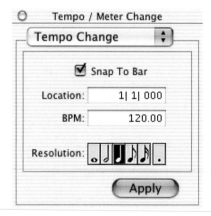

Figure 6.6 The Tempo/Meter Change dialog set to Tempo Change.

To replace the default tempo, set the Location to 1|1|000 then enter the tempo in BPM that you want to use for your session. The Resolution setting lets you choose the note (or beat) value on which to base the beats per minute. Normally this will be 1/4-notes, although other note values are sometimes used instead.

Manual Tempo mode

When the Conductor button in the Transport window is highlighted, tempo is controlled by tempo events in the Conductor track. Un-highlighting this button deactivates the Conductor track and puts Pro Tools into Manual Tempo mode. In Manual Tempo mode you can set the tempo using the Tempo slider in the Transport Window – while pressing Control (Windows) or Command (Macintosh) if you need finer resolution.

> Tip: To base the BPM value on something other than the default quarter note, change the note value in the Beat Value pop-up menu just to the left of the Tap button.

You can also use the Tap button to set a Manual Tempo by clicking this repeatedly (at least four times) at the new tempo. Alternatively, you can click in the Tempo field so that it becomes highlighted and tap in the tempo by playing a note repeatedly at the new tempo on your MIDI keyboard controller.

Recording to a click

I always recommend recording to a click whenever possible, as this means you can start editing right away while referring to actual bar lines that relate to the music you have recorded. Don't forget that you can always construct a tempo map or import an existing one that you have created elsewhere using a MIDI file. But what if there is no time – or the musicians simply don't want to play to a click so they are free to use whatever tempos they want? In this case, you should take some time out as soon as possible to create a tempo map using the Identify Beat command – or using the Beat Detective. I once spent a whole day – and a long night – working with a classical music conductor who insisted on conducting the keyboard player who was playing his score 'live' into a MIDI sequencer – with no click. As soon as the musician had finished, and left, the conductor asked me to edit the score – quite radically in places. I pointed out that we had no barlines to work with unless we created a tempo map – which could take a couple of hours or more on this 15-minute score. He was unwilling to spend this time and insisted that I make the edits without barlines to refer to. The following morning, with his studio budget fully exhausted, we had got maybe 2/3 of the way through the piece, leaving a fairly ragged end section. So that is what went onto the rough mix on DAT that he took away – and I don't think the music ever saw the light of day.

Setting up a click

Before you record MIDI or audio into Pro Tools, you are very likely to want to set up a metronome click to play along to. This ensures that what you record can be aligned with the bars and beats that your Pro Tools session is using.

You can record without playing to a click, but you won't know where the barlines and beats fall in your Pro Tools session – so the MIDI or audio that you record will bear no obvious relation to these. Also, it is often useful to be able to make the click play for one or more bars before the music starts, so that you can prepare yourself to start playing exactly at the beginning. The MIDI

Controls in the Transport window include options to set up and play Countoff bars for this purpose.

To set the click up, choose Click Options from the MIDI menu or double-click either the metronome (Click) button or the Countoff button in the Transport window. In the Click/Countoff Options dialog that appears, you can choose the MIDI port number and channel that will play back the click using the pop-up menu marked 'Output'. Here you can choose an external synthesizer, sampler or drum-machine to produce the click sound.

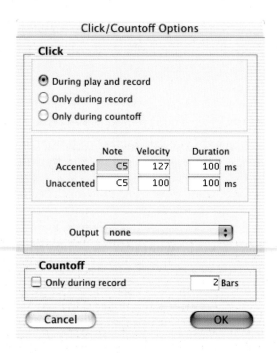

Figure 6.7 Click/Countoff Options.

Pro Tools (versions 6.xx and later) does provide a built-in click generator – the DigiRack Click plug-in. If you want to use the Click plug-in, choose 'None' in the Output pop-up in the Click/Countoff Options dialog. This Click plug-in includes presets with different click sounds, supports accented and unaccented click sounds, and lets you adjust their volumes. One advantage of using the Click plug-in to make the metronome sound is that it is integrated directly into Pro Tools, thus avoiding any possibility of MIDI time delays.

Tip: To hear the click through the Pro Tools mixer, you need to create a mono (or stereo) Auxiliary Input. If you are using the built-in Click plug-in, insert this into this Aux track and you will hear the Click whenever it is set to play. If you are using a virtual instrument plug-in to produce the click, insert this instead. And if you are using an external device to produce the click, you should connect one of its audio outputs to one of your Pro Tools hardware inputs, then select this input as the input to the Aux track.

In the Click/Countoff Options dialog you can select whether the click will be heard 'During play and record', or 'Only during record', or 'Only during countoff'. You can also specify the number of bars to be counted off and choose whether to hear the countoff only when recording – or before normal playback as well. Finally, there are options for Note, Velocity and Duration in this dialog for use with MIDI instrument-based clicks – they do not affect the Click plug-in.

> Note: If you want to record the Click or any other plug-in that you are using to produce the metronome sound as audio, you will need to route the output from the Aux track that you are using to monitor the click sound into the input of an Audio track and then record into this.

Destructive Record

If you prefer to have Pro Tools behave more like a conventional multi-track recorder, where you typically record additional takes over previous takes, you can always enable the Destructive Record mode from the Operations menu. The letter 'D' will appear in the Record button on the Transport to remind you that you are using this mode. When you have recorded the first take, simply leave the track record-enabled, go back and record again. Your new recording will have replaced whatever you had previously recorded into that audio file on disk.

> Tip: Using Destructive Record you can replace part of any previous recording using this technique. Make sure that 'Link Edit and Timeline Selection' is enabled in the operations menu and, using the Selector tool, insert the selection cursor into a region of the track you wish to record to at the point you wish to record from – and put Pro Tools into record. Stop wherever you like, and the new audio will have replaced the original audio in the file from wherever you started recording to wherever you stopped – just as you would expect with a conventional recorder.

> Note: If you insert the cursor at the end of your previous recording, the additional material will be appended at the end of the file – thus extending the length of the track.

Now that we have all got used to the idea of Pro Tools hard disk recording being non-destructive (as long as you always record new takes to new files on disk – which is the default situation), it may seem a little strange for some of you to use the Destructive Recording mode. Experienced recording engineers used to working with conventional multi-track recorders will, of course, be well used to this way of working. Don't forget, this can be a more efficient way to work – especially if you know what you are doing and are confident that you are not too likely to make a mistake. You won't have to take the time and trouble to sort through the alternate takes and delete them from your hard drive afterwards – and if you are running low on hard disk space Destructive Record can be a boon.

Playlists

Playlists provide a simple way to be able to keep recording new takes into the same track without any fuss. You can create a new playlist for the track each time you want to record a new take. This saves you the trouble of inserting or choosing a new track and then having to set the track up with the correct inputs and outputs, foldback for the musicians, inserts, plug-ins – or whatever. Working on the same track you just recorded onto with a new playlist leaves the previous take just as it was – but hidden from view and disabled from playing – so you can simply record another take into this new playlist. Every track lets you create as many of these edit playlists as you like, so if you want to record another take, you just create another new playlist, and so on.

Each track in the Edit window has a pop-up menu next to the track name. Click and hold this to reveal the Playlist menu.

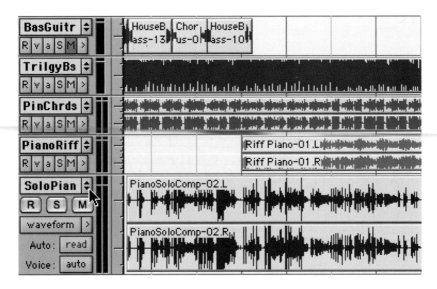

Figure 6.8 About to click on the Playlist pop-up in the Edit window.

You can select 'New   ' to create a new Playlist. The 'Duplicate' command is useful when you start editing your playlists later on, as is the 'Delete Unused . . . .' command. All the playlists you have created for this track are also listed here, and you can even access playlists for other tracks.

When you select 'New . . .', you are presented with a dialog box that lets you type a name for the new Playlist – or you can just OK this and go with the default name if this is fine for you.

Having created a new Playlist, the first thing you will notice is that the track is now empty – there are no regions visible and the previous take does not play back. So you can go ahead and Record-enable the track ready to record the next take.

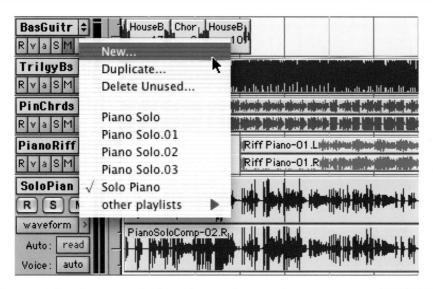

Figure 6.9 Creating a new Playlist to free up the solo piano track to record a new take.

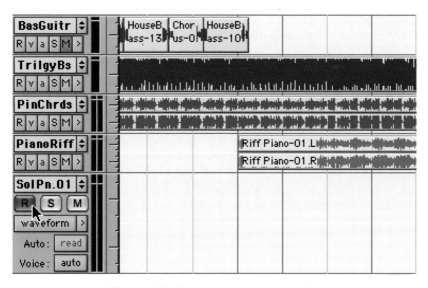

Figure 6.10 Ready to record a new take.

Tip: Playlists are very useful for managing differently edited versions of a particular recording. You can chop up a recorded file one way in one playlist and another way in another playlist and then swap these whenever you like while working on your arrangement.

Punch-in and -out using pre- and post-roll

Don't forget that you can always drop-in on a track by specifying a range to record to first and setting up a pre-roll and post-roll (in the Transport window) as you would do with a conventional multi-track recorder. Playback will start at the pre-roll time and Pro Tools will drop into record at the punch-in point and drop out of record at the punch-out point, stopping at the end of the post-roll time.

The simplest way to set up a range to record to is to select a range in a track's playlist or in a Timebase Ruler at the top of the Edit window using the Selection tool – making sure that the Edit and Timeline selections are linked. As usual, there are other ways to do this – such as typing the start and end times into the Transport window.

Note: One issue to be aware of when using pre/post-roll in PT|24 systems is that two 'voices' are required for each record-enabled track. This becomes particularly relevant when simultaneously recording large numbers of tracks at the same time. For example, if you are recording 32 tracks at once and you want to use pre-post roll on a Pro Tools|24 system you will need to split the tracks between the two DSP engines available to make the additional voices available for pre/post-roll. The first DSP engine handles tracks 1–32 while the second handles tracks 33–64, so in this case you would assign tracks 1–16 to voices 1–16 and assign tracks 17–32 to the first 16 voices of the second DSP engine – i.e. voices 33–48. This way, each engine would have 16 voices free to cope with the pre/post-roll voices needed for all 32 tracks.

Tip: Pre-roll and post-roll do not use additional voices on HD systems, though they may place additional load on the system's PCI bandwidth. The use of pre- and post-roll itself does not require allocation of extra voices in a session. However, if the system is encountering PCI bandwidth limitations, for example, when punching in (switching from playback to recording) on a very large number of tracks simultaneously, it may help to distribute the session's voices across several DSP chips (see the explanation of voice count below).

Voices, channels and tracks

In Pro Tools terminology, 'Tracks' are used to record audio (or MIDI) to your hard drive. These tracks appear in the Edit window and in the Mix window 'channel' strips. Rather confusingly, what you might expect would be referred to as 'channels' or 'channel strips' in the Mix window are referred to as Audio Tracks, MIDI Tracks, Auxiliary Inputs and Master Faders in Pro Tools. Despite this, the volume and pan controls are referred to as Channel Volume and Channel Pan controls in the Pro Tools Mix window. However, other than these two exceptions, Pro Tools uses the terms 'Channel' to refer to an actual physical input or output connection on whichever interface or interfaces you are using. All this is straightforward enough to understand – but it is a little unusual when you are used to a tape recorder having tracks and a mixer having channels. It can be a little 'jarring' mentally to talk about adjusting the Mixer's 'Track' controls rather than the mixer's 'Channel' controls until you get used to the idea. And then you have to make an exception when talking about the Channel Volume and Pan controls...?

Now the concept of 'voices' in Pro Tools deserves rather more explanation. The number of playback and recording 'voices' is the number of unique simultaneous playback and record tracks on your system – a bit like the concept of polyphony in a synthesizer, which dictates the maximum number of simultaneously playable notes. With Pro Tools|HD 1 systems, the software will let you record up to 256 mono tracks – but not all at once. And you can't play all these back at once either – see below. You can only record and play back up to the limit of the available voices within your system, and this depends on the type of cards you are using and the sample rate you are working at.

Note: Pro Tools|HD series systems provide up to 128 Auxiliary tracks (Auxiliary Inputs) while Pro Tools|24 MIX series and Pro Tools|24 systems provide up to 64 Auxiliary Inputs. All TDM-equipped Pro Tools systems also provide a total of 64 internal mix busses, and up to five inserts and five sends per track (depending on the DSP capacity of your system). In addition, Pro Tools TDM systems support up to 256 MIDI channels (Pro Tools 6.x) or up to 128 MIDI channels (Pro Tools 5.1 to 5.3.x).

Tip: The number of tracks you can record to or play back at once also depends on the capabilities of your hard drive system. You will need a SCSI card and a SCSI hard drive system to achieve maximum track counts, especially at higher sample rates. Check the Digidesign website for the latest information on compatible hard drives and achievable track counts.

Pro Tools|24 MIX systems

With Pro Tools|24 MIX systems, you can choose to dedicate one DSP chip (for up to 32 voices at 44.1/48 kHz) or two DSP chips (for up to 64 voices at 44.1/48 kHz) to the Playback Engine.

Pro Tools|HD and HD Accel systems

With Pro Tools|HD and Pro Tools|HD Accel systems, you can choose from a number of different voice count options, which distribute voices across DSP chips to accommodate different system load levels. You can choose to allocate low, medium, or high numbers of voices to each DSP chip used for the Playback Engine. In the Playback Engine dialog, the Number of Voices setting lets you control the number of available voices and how those voices are allocated to DSPs in your system. You will have different choices for voice count depending on the session sample rate and the number of DSP chips dedicated to the system's Playback Engine. For example, the default number of voices on a Pro Tools|HD 1 system is 48 voices, using one DSP (at sample rates of 44.1 kHz or 48 kHz). You can allocate as few as 16 voices or as many as 96 if you use two chips. The maximum number of simultaneous voices that Pro Tools|HD systems can play back depends on the sample rate being used. For example, at 44.1 or

48 kHz, Pro Tools|HD systems can play up to 96 tracks while Pro Tools|HD Accel systems can play up to 192 tracks.

> Note: At 88.2/96 kHz, the total available voice count for each setting is divided by 2; at 176.4/192 kHz, the total available voice count for each setting is divided by 4.

The accompanying screen shots (Figures 6.12, 6.13 and 6.14) show the available voice distributions for a 44.1/48 kHz session on Pro Tools|HD 1, Pro Tools|HD 2, and Pro Tools|HD Accel systems.

With a Pro Tools|HD 1 system, you can choose to allocate 16, 32 or 48 voices per DSP chip, using up to two chips for up to 96 voices at 44.1/48 kHz.

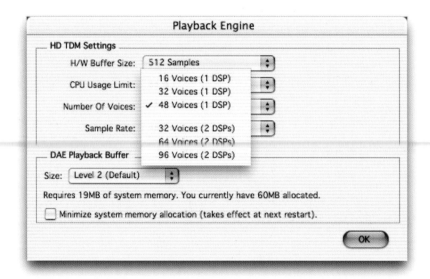

Figure 6.11 Playback Engine dialog showing Pro Tools|HD 1 voice distributions.

With an expanded Pro Tools|HD 2 or HD 3 (1 HD Core and 1 or 2 HD Process cards) system, you can choose to allocate 16, 32, or 48 voices per DSP chip, using up to four chips for up to 128 voices at 44.1/48 kHz.

With a Pro Tools|HD Accel system (with one HD Core and at least one HD Accel card), you can choose to allocate 16, 32 or 48 voices per DSP chip, using up to six chips (for up to 192 voices at 44.1/48 kHz).

> Note: The documentation for MIX systems refers to using one or two 'Playback Engines' – meaning one or two DSP chips. In the context of HD and HD Accel systems it might be more accurate to say that a single Pro Tools Playback Engine uses multiple DSP chips, rather than there being multiple, distinct 'Engines' as such.

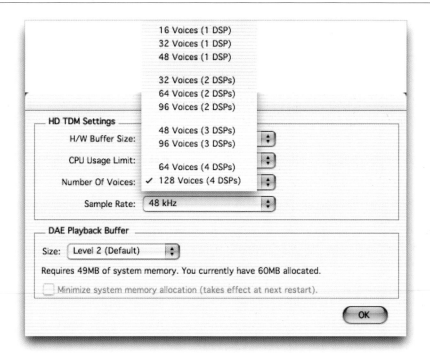

Figure 6.12 Playback Engine dialog showing Pro Tools|HD 2 voice distributions.

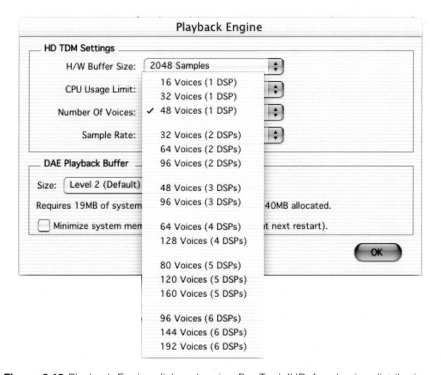

Figure 6.13 Playback Engine dialog showing Pro Tools|HD Accel voice distributions.

The Pro Tools software is not restricted in the same way as the hardware. You can work with up to 256 tracks of audio in Pro Tools, but these are referred to as 'voiceable' tracks as they cannot all be played back at the same time – they have to have a 'voice' first. The 'voice' allocation system lets you assign a particular track to play back using one of the available 'voices'. You can specifically allocate a track to play using a particular voice, or you can use the 'auto' setting. This automatically sets each successive track to the next available voice that is not already in use. Although Pro Tools|HD series systems can open sessions with up to 256 audio tracks (and Pro Tools|24 MIX series or Pro Tools|24 systems can open sessions with up to 128 audio tracks), any audio tracks beyond that system's voiceable track limit will be automatically set to *Voice Off.*

> Note: Mono tracks take up one voice. Stereo and multi-channel tracks take up one voice per channel.

Voiceable tracks

The total number of 'voiceable' tracks is the maximum number of audio tracks that can share the available voices on your system.

So what happens when you run out of voices? The neat thing here is that you can assign more than one track to the same voice and the track with the highest priority will always play back. If there is a part of this track where no audio region is present, which would be the case with a keyboard pad that was only playing in the verses and any empty audio regions had been removed from the choruses, for example, then you will hear any audio regions present on the next track (in order of priority) assigned to that voice play back in the choruses. The track priority depends on the order of the tracks in the Mix or Edit windows – with the leftmost track in the Mix and the topmost track in the Edit windows having the greatest priority.

This system, referred to as 'dynamic voice allocation', makes Pro Tools appear as if it has many more tracks available than allowed by the hardware. All you have to do is experiment with different combinations of track priorities, voice assignments and arrangements of regions within your tracks to be able to play back many more tracks than you can with a conventional recorder.

> Note: The Voice Allocator pop-up can be found next to the record-enable button on the channel strip.

Just to make this even clearer, let me put it another way. When no regions are present within a particular time range on a track, the track with the next highest priority will be able to play back until a region appears again for playback in the first track – in which case the second track will stop playing back and the first will resume playback.

> Note: One thing to watch out for here is that the start time of a region which you want to have 'pop through' in this way must be after the end time of the region on the higher-priority track.

On complex sessions with lots of tracks you may still find yourself running out of voices at times. To help you manage your voice allocations, Pro Tools frees up a voice if you un-assign the track's output and send assignments, or make the track inactive – or when you simply set the Voice Selector to 'Off'. You can also temporarily free up a voice by muting a track during playback, and if you have 'Mute Frees Assigned Voice' enabled in the operations menu the voice will be allocated to the next highest priority virtual track that is assigned to the same voice.

> Note: When you use the 'Mute Frees Assigned Voice' feature, the computer introduces a delay of one or more seconds, depending on your CPU speed, before the mute or un-mute instruction is carried out. If this bothers you, your only option is to turn it off. Also, muting a track using this feature will not make the voice available for use with the QuickPunch feature (described later).

TDM systems also allow you to make tracks inactive by Command-Control-clicking on the track type icon that you will find just above the track name on each mixer channel strip. Simply Command-Control-click again to make the track active once more. To give you visual feedback, mixer channels turn a darker shade of grey and tracks in the Edit window are dimmed when inactive.

Making some of your less-important, or unused, tracks inactive is a great way of freeing up DSP resources and voices for use elsewhere, as all the plug-ins, sends, voices and automation on inactive tracks are disabled.

> Tip: This feature allows you to open Pro Tools sessions on systems with less DSP resources than were available when the sessions were created. Pro Tools will automatically make tracks inactive as necessary to allow sessions to be opened.

QuickPunch

If you like to work even more quickly while doing overdubs, you can use the QuickPunch feature to drop in and out of record on record-enabled tracks up to 100 times during a single pass simply by clicking on the Record button in the Transport window. Normally, when you use Auto Input Monitoring, the switch back to monitoring track material on punch-out is not instantaneous. The big difference here is that QuickPunch provides instantaneous monitor switching on punch-out. With QuickPunch, Pro Tools actually starts recording a new audio file as soon as you start playback, automatically defining and naming regions in that file at each punch-in/out point. An automatic crossfade is inserted for each punch point – to crossfade into the new region and then out again – and you can choose a suitable length for this crossfade in the Editing Preferences.

> Note: If a value other than zero is specified for the QuickPunch Crossfade Length, QuickPunch writes a pre-crossfade at the punch-in point (which occurs up to but not into the punched region boundary), and a post-crossfade at the punch-out point (which occurs after the punched region).

> Tip: You can always edit these later using the Fades window. And if you don't want any crossfades to be inserted, then simply set this length to zero.

> Note: Even if the QuickPunch Crossfade Length is set to zero, Pro Tools always executes a 4 millisecond 'monitor only' crossfade (which is not written to disk) to avoid distracting pops or clicks that might occur as you enter and exit record mode.

How QuickPunch affects the available voice count

As on Pro Tools|24 MIX systems, with Pro Tools|HD systems the QuickPunch feature uses an additional voice for each record-enabled mono track – so the total number of available voices for QuickPunch recording is cut in half.

On a large project, it would not be impossible to find yourself short of voices. Don't forget that you can always turn off the voice assignments to less-important tracks – or make these inactive.

With TDM systems, the recommended procedure is to use the Auto voice assignment setting for each track so that Pro Tools will automatically distribute the voices between the DSP chips. For example, for a 128-voice configured Pro Tools|HD system, auto-voice distributes voices evenly across four sets of voices (1–32, 33–64, 65–96 and 96–128).

However, it can still be a good idea to manually select voices for the most important tracks – especially if you anticipate running out of voices and want to make sure that these will always play.

If you do not use auto voicing, then the track voicing must be set to distribute the tracks across the appropriate number of DSP chips (e.g., with 32 voices, 1–16 on the first chip, 17–32 on the second).

Loop Recording audio

If you are trying to pin down that perfect eight-bar instrumental solo in the middle of your song or the definitive lead vocal on the verse, or whatever, you will definitely value the Loop Recording feature in Pro Tools. To set this up is very straightforward. Just select Loop Record from the Operations menu you will see a loop symbol appear on the Record button in the Transport window. Make sure you have Link Edit and Timeline Selection checked in the Operations menu and then use the Selector tool to drag over the region you want to work with in the Edit window. You can set a pre-roll time if you like, or you can simply select a little extra at the beginning and then trim this back later. Now, when you hit record, Pro Tools will loop around this selection, recording each take as an individual region within one long file. When you have finished recording you can choose the best take at your leisure. All the takes will be placed into the Audio Regions list and numbered sequentially, with the last one left in the track for you. Now if you want to hear any of the other takes, just select the last take with the Grabber tool and Command-drag whichever takes your fancy from the Audio Regions list and it will automatically replace the selected take in the track – very convenient! An even faster way is to Command-click with the Selector tool at the exact start of the loop or punch range. This immediately brings up the Takes List pop-up – making it even easier to select alternate takes.

But what if you want to audition takes from a previous session? These will not be listed here normally, as the start times are likely to be different. The User Time Stamp for each take in Loop record is set to the same start time – at the beginning of the loop – and the Takes List is based on matching start times. That is why it displays all your takes when you bring the List up at the start time of your loop. So if you want to include other takes from a previous session in the Takes List pop-up for a particular location, you can simply set the User Time Stamp for these regions to the same as for these new takes and they will all appear in the Takes List. And if you plan on recording some more takes later on for this same section, you should store your loop record selection as a Memory Location. This way, these takes will also appear in the Takes List pop-up for that location.

You can also restrict what appears in the Takes List according to the Editing Preferences you choose. If you enable 'Take Region Name(s) That Match Track Names' then the list will only include regions that take their name from track/playlist. This can be useful when sorting through many different takes from other sessions, for example. Or maybe you want to restrict the Takes List to regions with exactly the same length as the current selection. In this case, make sure that you have enabled 'Take Region Lengths That Match' in the Editing Preferences.

If you have both of these preferences selected you can even work with multiple tracks to replace all takes on these simultaneously: when you choose a region from the Takes List in one of the tracks, not only will the selected region be replaced in that track, but also the same take numbers will be placed in the other selected tracks.

A third option is provided in the Editing Preferences to make any 'Separate Region' commands you apply to a particular region apply simultaneously to all other related takes – i.e. takes with the same User Time Stamp. You could use this to separate out a particular phrase that you want to compare with different takes, for example.

Note: All the regions in your session with the same User Time Stamp will be affected unless you keep one or both of the other two options selected – in which case the Separate Region command will only apply to regions that also match these criteria.

Tip: It can be easy to forget that you have left this preference selected and end up accidentally separating regions when you don't intend to – so make sure you deselect this each time after using it.

Half-speed recording and playback

A trick which recording engineers have had 'up their sleeves' for years is to record a difficult-to-play musical part at half-speed an octave below. When played back at normal speed, the part plays back at the correct tempo, but pitched up an octave – back to where it should be. I remember being introduced to this technique in the early 1980s while recording to analogue tape. I was struggling to play a tricky clavinet part and the recording engineer just ran the tape at half-speed and told me to play along an octave below. When he played what I had recorded back at normal speed it sounded perfect – much tighter timing-wise than I had actually played it even at half-tempo.

Now you can use this technique when recording with Pro Tools. Press Command-Shift-Space-bar rather than just the Space-bar when you start recording and Pro Tools will play back existing tracks and record incoming audio at half-speed.

If you just want to play back a Pro Tools session at half-speed, all you need to do is hit Shift-Space-bar. This can be very useful when playing along with or transcribing what is being played on existing recordings – which many musicians want to do from time to time.

Recording shortcuts

Command-Shift-N lets you create a new track and you can use the Command-UpArrow or -DownArrow to select the type of track – Audio, MIDI, Aux Input or Master Fader.

One of the most useful shortcuts while recording is to hit Command-Period (i.e. the full stop, '.') to abort the recording without saving the file. You would do this to save time if you realize that you have a useless take and want to avoid having to select this and delete it from your hard drive later.

Markers

To make navigation around your music easier, you should always mark out the sections of the song or piece of music from the outset. To enter a Marker you simply hit the Enter key on your computer keyboard anytime you like – whether Pro Tools is playing back or stopped. A dialog

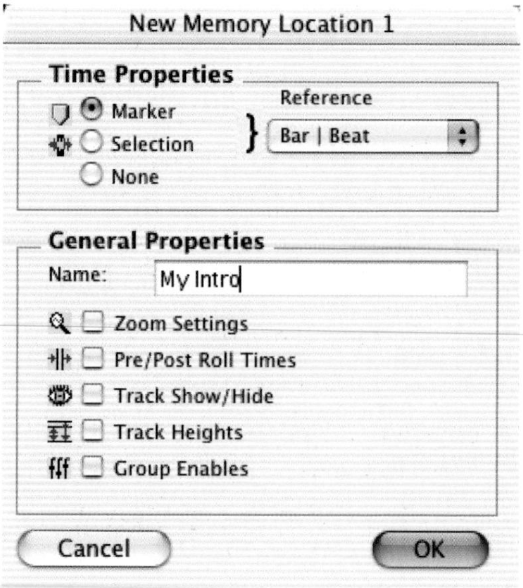

Figure 6.14 New Memory Location dialog.

box will appear to let you name your Marker. This will be stored in the Memory locations window and also made visible in the Markers 'ruler' which runs along the top of the Edit window.

These Memory Locations can also be used to store Zoom Settings, Pre-Post Roll Times, Track Show/Hide, Track Heights and Group Enables.

> Tip: Don't forget that you can always create Memory Locations 'on-the-fly' while you are recording or playing back – and there are up to 200 of these available. Just hit the Enter key and quickly type in a name for the marker before the next one comes up. You can always OK the dialog immediately and go back to change it later if there is not sufficient time between markers to type marker names.

It is a good idea to store the most useful zoom settings using these Memory Locations – as well as having a selection in the five Zoom memories near the top left of the Edit window. The other settings that you can store in the Memory Locations can come in very handy as well – depending on the nature of your session.

Time stamping

Another basic concept worth knowing about before you get started with Pro Tools is the system of time stamping – where every audio region that you record is marked with the original SMPTE time at which the audio was recorded.

When you record audio into Pro Tools it is automatically time-stamped relative to the SMPTE start time of the session. In fact, when *any* region is created in Pro Tools, it will have an original time stamp. This original time stamp is permanently associated with the region and cannot be changed. The benefit of this is that if you subsequently move the region, you can always return it to its original position using the Spot dialog.

Figure 6.15 Spot dialog showing that the audio region was originally recorded at 11:52:15, but is currently located at 11:52:12.

A User Time Stamp, identical to the Original Time Stamp, is also created when you create a region. You can always change this using the Time Stamp Selected command in the Audio Regions list so that you can create a custom time stamp for spotting or re-spotting the region to a time location different from its Original Time Stamp. See Figures 6.16 and 6.17.

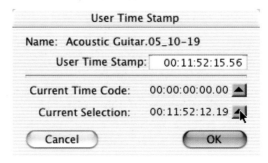

Figure 6.16 User Time Stamp dialog – entering a custom User Time Stamp.

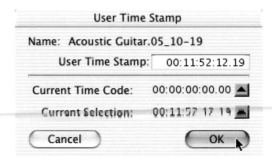

Figure 6.17 User Time Stamp dialog – OK'ing the new User Time Stamp.

If you record some audio and then decide that it really ought to be placed at a different position in time than the position it was originally recorded at, it is worth setting a User Time Stamp up for any such region so that if you subsequently accidentally move the region you can always use the Spot dialog to return it to that position. See Figure 6.18.

Simply click on the arrow to the right of the User Time Stamp (or Original Time Stamp if you want to go there) to enter that location into the Spot dialog's Start field. Click OK and your audio region will 'fly' off to that position. See Figure 6.19.

Importing tracks from CD into Pro Tools

Strangely, Digidesign has never included a dedicated 'Import Audio from CD' facility within Pro Tools. Nevertheless, you can import audio tracks from CD via the CD-ROM drive on the Mac.

Using OS9, this used to be a two-stage process. It first used Apple's standard QuickTime technology to import and save to disk an AIFF QuickTime audio file. Then Pro Tools would open this file, convert it to the dual-mono format for playback within Pro Tools, and save the newly converted files into the current project's Audio folder. Subsequently, you would have to decide what to do with the AIFF QuickTime file – because Pro Tools would leave this on your disk drive. So you would need to do some housekeeping with these files – typically trashing

Spot Dialog
Current Time Code: 00:00:00:00
Time Scale: Time Code
Start: 00:11:51:17
Sync Point: 00:11:51:17
End: 00:11:54:03
Duration: 00:00:02:10
Use Subframes
Original Time Stamp: 00:11:52:15.56
User Time Stamp: 00:11:52:12.19
Cancel
OK

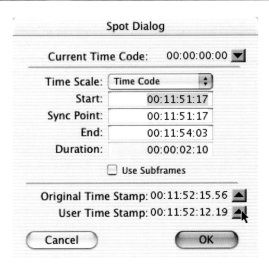

Figure 6.18 Spot dialog – this region was inadvertently moved from its correct position at 11:52:12. Click on the User Time Stamp arrow to set the selected edit field (for the Start time in this example) to this time location.

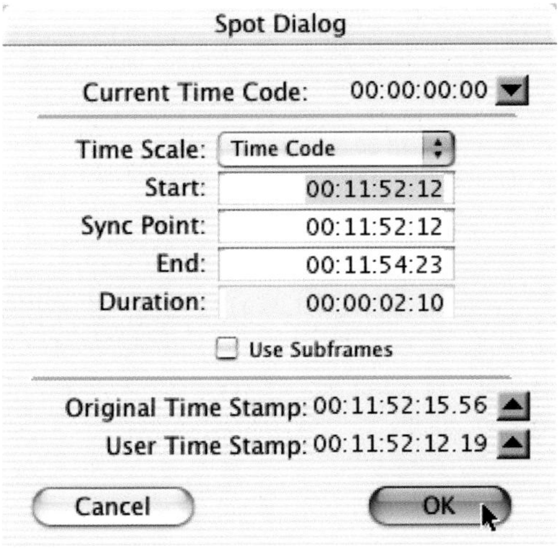

Spot Dialog
Current Time Code: 00:00:00:00
Time Scale: Time Code
Start: 00:11:52:12
Sync Point: 00:11:52:12
End: 00:11:54:23
Duration: 00:00:02:10
Use Subframes
Original Time Stamp: 00:11:52:15.56
User Time Stamp: 00:11:52:12.19
Cancel
OK

Figure 6.19 Spot dialog – OK the restored Start Time for the region.

them unless you had a specific reason to keep them on your hard drive alongside the split stereo versions imported into your Pro Tools projects.

Pro Tools 6.xx running on the Mac with OSX actually offers two ways to get audio into Pro Tools from CD – an updated version of the 'Import Audio From Other Movie…' command; and the new browser drag and drop features. You can still use the 'Import Audio From Other Movie…' command to bring audio from CD into Pro Tools if you wish. This works in a more streamlined way than previously, but still requires several steps.

The best way to import audio is using the new DigiBase drag and drop capabilities provided by the Workspace Browser. The Workspace file 'browser' acts a bit like the Macintosh Finder, providing access to all your disk drives and to the folders and files within these. The DigiBase browsers are more like databases than simple file browsers, however. For example, the Workspace browser is set out rather like a spreadsheet with columns that display meta-data for each item in the list. The bit-rate and sample-depth are displayed, along with lots of other detailed file information, and the Workspace browser provides comprehensive search capabilities.

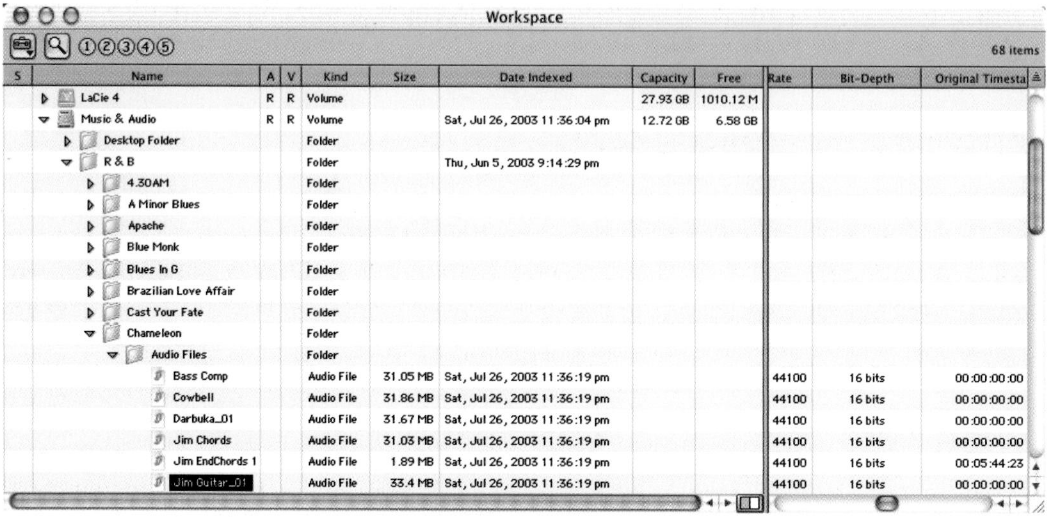

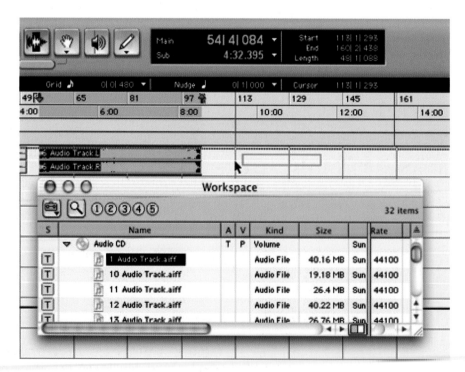

Figure 6.20 The Workspace Browser.

Figure 6.21 Dragging and Dropping a CD track from the Workspace to a Track in the Edit window.

The neat thing here is that you can simply drag and drop any audio file or any CD track that is available in your Workspace onto any audio track in the Edit window of the current Pro Tools session – or into the Audio Regions list.

> Tip: Using this browser drag and drop feature to get your CD tracks into Pro Tools is much quicker than using the 'Import Audio From Other Movie...' command.

Working on a real recording session

Here I will talk about various issues that came up while working on a recording of a 'cover version' of a Marshall Jefferson 'house' track called *Give Me Back The Love*. I transferred the original song into a Pro Tools session and used this as a template while I built up my new arrangement of the song. You can follow this example through using any similar recording that you have conveniently to hand.

Opening a new session

When you launch Pro Tools, which takes a little while, especially if you have extra plug-ins to load, you are presented with the Pro Tools menu bar at the top of the screen. Choose 'New Session' from the File menu to open a new session document.

The New Session dialog box lets you choose the sample rate, audio file type, and I/O Settings for the session. I normally choose 16-bit unless specifically working on an important project that demands the full 24-bit resolution. I also choose 44.1 kHz for most music projects unless I know that the project will end up on DVD or SACD. This way the file sizes are reasonable and the quality is fine for many projects.

> Note: I do plan to use 96 kHz/24-bit the next time that I record a solo jazz guitar album, where I will only be working with one or two tracks at a time. I will also be using 96 kHz/24-bit on various high-budget pop albums and film scoring sessions that I will be working on at the major recording studios here in London. It makes sense here, because the budgets are available to pay for the extra hard disk space and processing times needed for 96 kHz/24-bit operations – and the very highest audio quality is demanded.

So, you choose your options, name your file and save it to a suitable location. Then, when you click 'Save' in the 'New Session' dialog box, the Pro Tools session document opens, to reveal... no tracks?

That's correct! You start out with a 'blank palette' when you open a new project – unless you load a template which you have already prepared or selected from the selection of templates supplied on CD-ROM with the software. For a simple session it is fine to start out this way, as it is very easy to insert just the number of tracks you want. If you prefer to start out with a session containing various tracks set up ready to work with, you can use any of the templates included

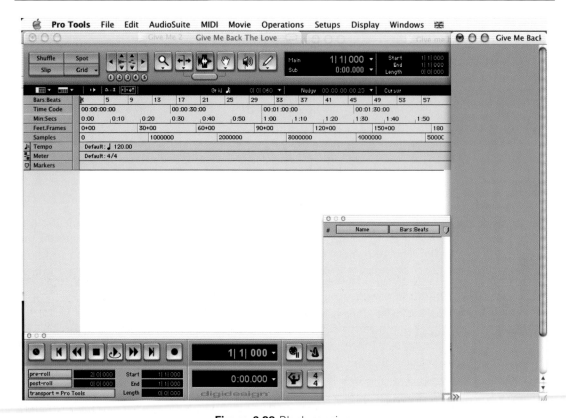

Figure 6.22 Blank session.

on the Pro Tools installation CD-ROM. You can also set up a session exactly the way you like to work and save this as a template which you can use each time you start a new session. Anyway, let's start from our blank session and add tracks as we need them.

Choose 'New Track' from the File menu and you will be presented with a dialog box that lets you select the number and type of tracks you want to insert.

Figure 6.23 New Track dialog.

A pop-up menu lets you choose what type of track you want to create and an editable field is provided to let you type in the number of these tracks you want. You can also choose whether the track or tracks will be Mono, Stereo, or any of the various surround formats covering all the formats from three-channel right up to the 7.1 (eight-channel) format which is used in Sony SDDS Film Sound systems – see Figure 6.24.

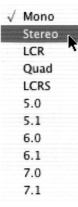

Figure 6.24 Track Format selector.

For now, create one stereo track. You could create two mono tracks and group these together, and this can be a better way to work with stereo tracks if you anticipate that you may need to edit each side of the stereo separately. This is often the case if you are recording 'live' instruments in stereo, such as a Hammond Organ/Leslie Speaker or a grand piano, where clicks or distortions might occur in one channel and not in the other. Here, we will simply be using this stereo track to play back the original recording as a template to work to – so a stereo track is much more convenient.

You could also create a stereo Master Fader at this point, even though this is not absolutely necessary when working with just one stereo track. It is also worth taking a moment or two to arrange everything neatly on your screen (or screens). This lets you work faster than if you have to keep moving the windows around. Of course, you will have to keep adjusting the layout as you add more tracks and so forth. The main point is that you should keep everything as sensibly arranged as possible on-screen so that using the software does not slow you down – particularly when you are trying to be creative.

When working on music productions, I usually display just the Bars:Beats and Min:Secs rulers along with the Tempo, Meter and Markers rulers in the Edit window. Take a look at Figure 6.25 to see how this might look.

When you create a new project from scratch, Pro Tools sets up a folder with the name of your project ready for you to save your session file into. It also creates folders for any Audio, Video or

> Tip: Now is probably a good time for your first 'Save'. You should save very regularly, and especially when you have spent any amount of time making changes to your session that you don't want to have to re-do if the computer crashes or whatever. Hit Command-S or choose 'Save' from the File menu and you will be presented with the 'Save File' dialog box. By default, this will be directed into the Pro Tools project folder for the session you are working on, but if it isn't for any reason, then you should navigate through your disk drives and folders until you find this folder using the pop-up menus at the top of the window.

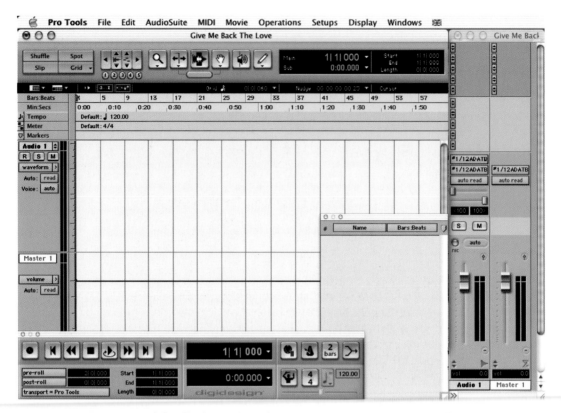

Figure 6.25 A Pro Tools session with one stereo track and a Master Fader.

Fades files that you use during the session and sets up the Save dialog so that it is ready to save these files into these folders by default. If you close your session without creating any of these files, their folders will disappear until you open the session next time. Then, if you save some fades or whatever, next time around you will find the folder for this stays within the project folder to hold the relevant files conveniently alongside the session file.

AutoSave

If you like, you can arrange for Pro Tools to automatically save your sessions. Just tick off the appropriate box in the Operations Preferences window and choose how often you want the session to be saved for you and how many backup versions you want to keep.

> Tip: When you are making a backup copy of your session, to archive or to transfer to another system, you can make sure that all the files you need for this session are included by checking the various buttons in the 'Save Session Copy in...' dialog box. You can also change the audio file type, sample rate and bit depth – which is particularly useful if you are saving a copy to transfer to a different Pro Tools system.

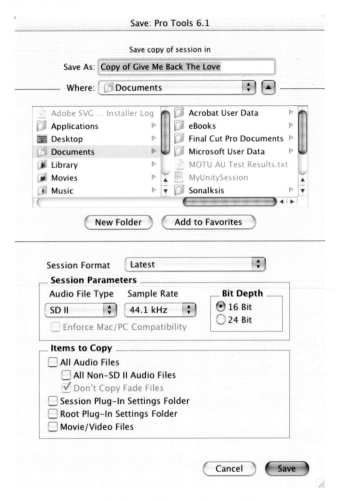

Figure 6.26 Save Session Copy window.

Personally, I never use this option. I prefer to decide myself as to when I will save a file and I use the 'Save Session' as or 'Save Session Copy in . . .' commands from the File menu to specifically save versions of my choosing where and when I want to do this. See Figure 6.26.

Recording audio

My client brought the original recording of the song to be used as a guide on Mini-Disc, along with his own Mini-Disc player. I connected this to my Yamaha 02R mixer, routed it into Pro Tools inputs 1 and 2, and recorded the song directly onto the stereo track I had set up for this purpose in Pro Tools.

To record onto tracks in Pro Tools after routing audio to the appropriate inputs, you just click on each track's 'Record Enable' button in the Edit window, or in the Mix window, then click on the 'Record' button in the Transport Window. Click on the 'Play' arrow to start Pro Tools recording, then play your source material into Pro Tools. As the audio is being recorded, you will see the waveform start to appear within a pink-coloured region that unfurls in the Edit window as the recording proceeds. See Figure 6.27.

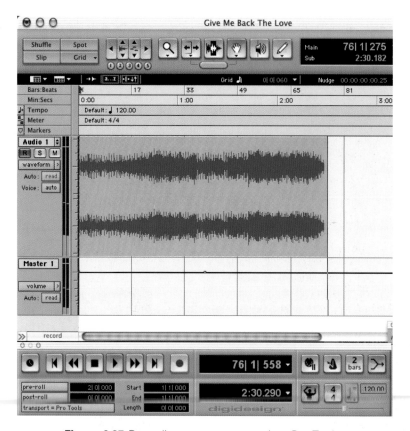

Figure 6.27 Recording a stereo source into Pro Tools.

Tip: If you make a false start in any way, or a problem develops while recording, you can abort the recording immediately by pressing Command-period (Mac), so no audio is saved to disk. If you simply stop by pressing the space-bar or whatever, you will then have the trouble of having to delete the audio region from the Edit window and to delete the audio file from disk – which takes much longer.

When you have made a successful recording, press the square 'Stop' button in the Transport window – or just hit the space-bar. An audio region will appear in the Edit window, and in the Audio Regions list, representing the audio file that will be stored on disk in the Audio Files folder for this project.

Topping and tailing

Very often you will need to immediately remove the 'top' and 'tail' from any recording you make into Pro Tools. If you have recorded material directly from a tape or disk machine, or even from a gramophone turntable if you have vinyl discs, then it is likely that you started recording a little before and finished recording a little after the wanted audio. In which case, you need to remove this so that just the audio you want to hear plays back.

There are two easy ways to do this. The first way is to select the audio you want to keep and then use the 'Trim To Selection' command from the Edit Menu.

Figure 6.28 Using the Trim To Selection command.

The other way is to just use the Trim tool to trim the start and the end of the region, as necessary.

Note: Be aware that you will need to zoom the display resolution to be able to select or trim your wanted audio with the greatest precision. This accuracy will be necessary if you intend to synchronize other material with this audio later on. If you have topped and tailed your audio very accurately at the outset, then, assuming you set the correct tempo to match this audio and set a suitable start position for your audio region, it will play correctly in step with the Bars:Beats display in Pro Tools. Of course, this further assumes that the audio is actually running at a constant tempo – and that you have set this exact tempo in Pro Tools. If you thought the tempo was 120 and it was really 119.98, then the audio and the Bars:Beats will inevitably drift out of sync over time.

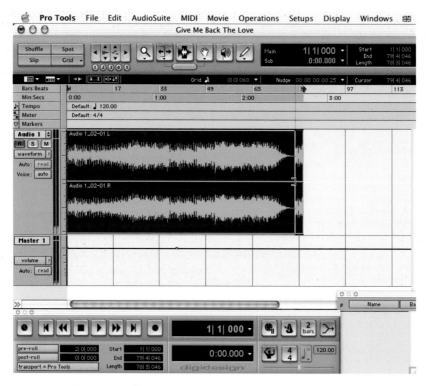

Figure 6.29 Using the Trim tool to crop an audio region.

Naming audio files

Pro Tools automatically names audio files using the name assigned to the track. By default, this will be Audio 1, Audio 2, and so forth. If you want specific names to be allocated automatically to tracks you record, then all you have to do is to make sure that you have typed the names you want into the tracks you are recording onto. Double-click on the Track Name in the Edit or Mix windows and a dialog box will appear that you can use to name the track. Buttons are provided in this dialog box to let you go to the previous or next track and name that before hitting OK to confirm your changes and exiting the dialog box.

> Tip: If you want to change the names of any tracks that you have already recorded, either because they are named Audio 1 or whatever, or because you thought of a better name, you can do this using the 'Rename Selected...' command that you will find in the pop-up menu at the top of the Audio Regions list.

Importing audio

Importing from CD-Audio discs is covered earlier, so let's take a look at how to import existing audio files from any storage disk drive such as a CD-ROM, hard drive or DVD-RAM drive. You can access the 'Import Audio' command from the pop-up menu which appears when you click and hold on the word 'Audio' at the top of the Audio Regions list at the right-hand side of the Edit window.

> Tip: Remember that you can hide or show the MIDI and Audio regions list at the right-hand side of the Edit window by clicking on the small double arrows near the bottom right of the window. If the regions list is hidden, a click on these arrows will instantly reveal it.

You will see the Audio regions list directly above the MIDI Regions list. Click and hold on the word 'Audio' at the top of the Audio Regions List and a pop-up menu will appear containing the Import Audio command. When you select this, a dialog box appears that lets you navigate through your hard disk drives and folders to find the audio files you want to import.

> Tip: Simply double-click on any audio file listed in the top half of this dialog box to add it immediately to the list of regions currently chosen and then hit 'Done' to import this into your session – see Figure 6.30.

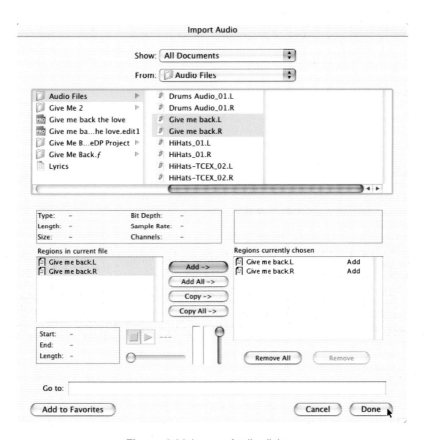

Figure 6.30 Import Audio dialog.

If the audio file you are importing is on CD or if it needs to be converted to the correct format for Pro Tools, the Import Audio command will copy the audio, do the conversion, and save the converted audio into the Audio Files folder for your project, by default. However, you are given the option to choose a different destination folder for your audio before the save is completed.

> Note: If an imported audio file is already correctly formatted for use with Pro Tools and available from a suitable data storage drive, the format will not be converted and the file will not be copied or moved into the Audio Files folder for your project – it will be left wherever you found it. This can be a problem later if you forget and move or delete this file, or remove the data storage drive containing this file.

> Tip: If you do ever need to edit the left or right channels separately, to take out a click from one channel, for example, you can use the 'Split Selected Tracks into Mono' command from the File menu. In the case of a stereo track, two mono tracks will be created and the left and right regions from the channels in the stereo track will be placed into these – so that you can edit them individually – see Figure 6.32. Once you have made your edits, just select these two mono regions, delete them, and carry on working with the stereo track's region, which, of course, references the audio files you have just edited.

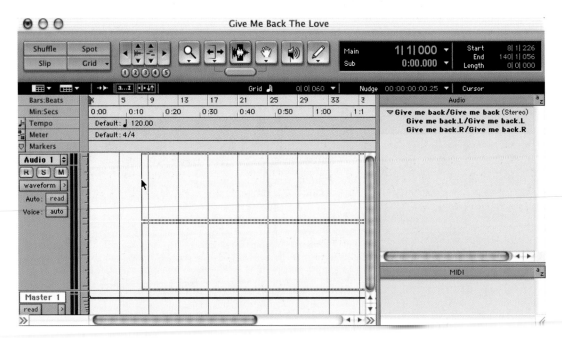

Figure 6.31 Drag and Drop from the Audio Regions list.

When you have imported the song into the Audio Regions list, all you need to do is to drag the audio file out of the list and drop it wherever you like into the stereo audio track – see Figure 6.31.

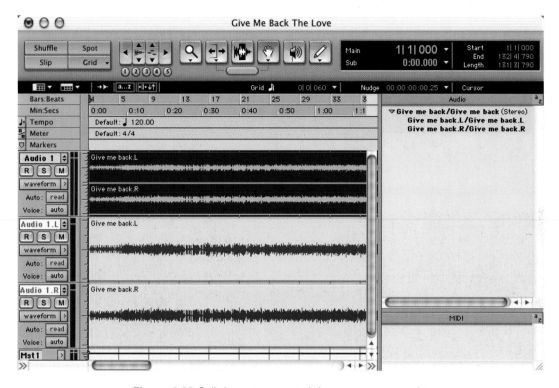

Figure 6.32 Splitting a stereo track into two mono tracks.

A quick edit

If you have started out by recording or importing an existing piece of music to use as a template to work to, as I have done in this example, you may wish to edit this recording at this stage, before recording additional material.

Engage the Selector tool and drag over the part of an audio region that you want to define as a section. Then invoke the 'Separate Region' command from the Edit menu. A 'Name' dialog appears that you can use to type in a suitable name for the region. When you OK this box, a new region, defined by your selection, appears in both the Edit window and in the Regions List with the name you just typed.

If you take the trouble to separate the sections of the song in this way at the outset, it then becomes very easy to re-arrange this recording by simply moving sections around with the Grabber tool in the Edit window.

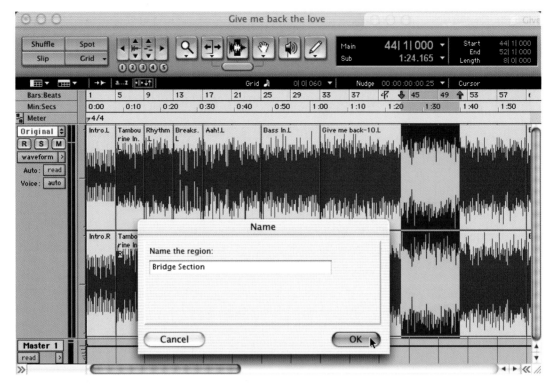

Figure 6.33 Separating and naming regions so that they can be moved around to create new arrangements.

> Tip: If you have already sorted out the tempo and the audio region start time for this session, then you can use Grid mode to move sections around. Otherwise, Slip mode is probably the best choice.

Setting up markers

I normally set up a selection of Markers at this stage so that I can navigate quickly to any point in the session using the Memory Locations window.

To create a marker, just hit the Enter key at any time – whether Pro Tools is stopped or running – and the New Memory Location dialog will appear. Make sure that the Counter is exactly at the time or Bar:Beat location at which you want to place the Marker before you hit Enter – or you will have to drag the Marker forward or backward to the correct place after you have inserted it.

> Tip: If you have already defined the sections of your audio as separate named regions, you can see at a glance where to put your markers. Just type the Bar:Beat number of the start of each region into the counter in turn (I often use the Big Counter for this), hit the return key to update the counter position, and make a Marker at the start of the region.

With the New Memory Location dialog box open for you, check that 'Marker' is selected in the Time Properties area and that nothing else has been inadvertently selected, then type a suitable name for your Marker and OK the dialog box.

Now, when you open the Memory Locations window you will see your Markers, as in Figure 6.35, and you can click on any of these to move the playback position to that location.

New Memory Location 1
Time Properties
Reference
Marker
Selection
None
Bar | Beat
General Properties
Name:
Marker 1
Zoom Settings
Pre/Post Roll Times
Track Show/Hide
Track Heights
Group Enables
Cancel
OK

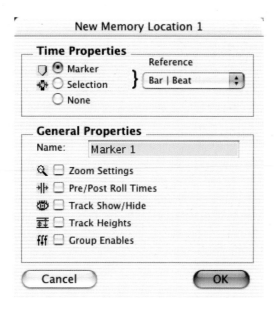

Figure 6.34 Setting up a Marker.

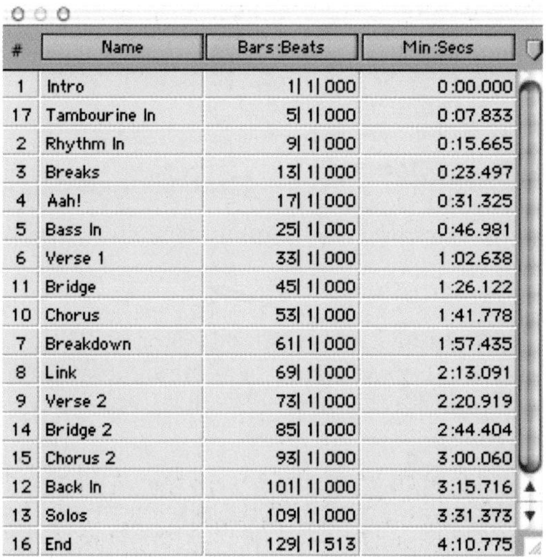

Name Bars:Beats Min:Secs
1 Intro 1| 1| 000 0:00.000
17 Tambourine In 5| 1| 000 0:07.833
2 Rhythm In 9| 1| 000 0:15.665
3 Breaks 13| 1| 000 0:23.497
4 Aah! 17| 1| 000 0:31.325
5 Bass In 25| 1| 000 0:46.981
6 Verse 1 33| 1| 000 1:02.638
11 Bridge 45| 1| 000 1:26.122
10 Chorus 53| 1| 000 1:41.778
7 Breakdown 61| 1| 000 1:57.435
8 Link 69| 1| 000 2:13.091
9 Verse 2 73| 1| 000 2:20.919
14 Bridge 2 85| 1| 000 2:44.404
15 Chorus 2 93| 1| 000 3:00.060
12 Back In 101| 1| 000 3:15.716
13 Solos 109| 1| 000 3:31.373
16 End 129| 1| 513 4:10.775

Figure 6.35 Memory Locations window showing Markers, Main Counter (Bars:Beats) and Sub Counter (Mins:Secs) locations.

Tip: If you don't need to see the Bar/Beat or Timecode locations in the Markers window, you can hide these by deselecting the 'Show Main Counter' and 'Show Sub Counter' options from the pop-up menu which appears when you click on the word 'Name' at the top left of the memory locations window. As you will see, you can have both the Main and Sub Counters displayed here – or just the Markers.

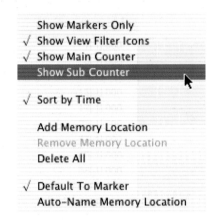

Figure 6.36 Memory Locations window display options.

You can also display Markers in the Edit window along the timeline just above the waveform display. Just click on any marker to move the playback position to that point, and just drag any marker to the left or right if you want to move its position.

Tip: You may need to use the pop-up selector just above the rulers at the left of the display to show the Markers Ruler if it is not already visible.

Using Identify Beat to create a tempo map

Creating a tempo map using the Identify Beat command is a key area that you should become familiar with

Continuing with our example session, I wanted to work with the original Marshall Jefferson song as a kind of 'template' so that I could match any original tempo changes and keep closely to the spirit of the original arrangement. Obviously, I would need to know the correct tempo or tempos to set for my session.

To be able to follow any tempo changes and to allow editing to bars and beats, you have to be able to calculate the tempo of the music at the start and whenever the tempo changes. In other words, you have to create a tempo 'map' that reveals any tempo changes.

To calculate tempos, Pro Tools offers the Identify Beat command. The way this works is that you accurately select a given number of bars, then invoke the Identify Beat command from the

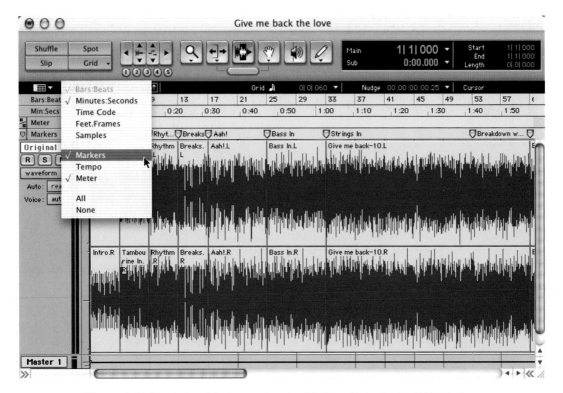

Figure 6.37 Displaying Markers using the Markers Ruler in the Edit window.

Edit menu. This brings up a dialog box where you can type in what you have decided the start and end bar numbers are. If you have made your selection of the waveform with sufficient accuracy, and if you have 'told' Pro Tools the correct number of bars and beats, then all the software has to do is to work out what the BPM would have to be for the tempo of your project to correspond.

So, for example, select exactly four bars of music, making sure that it is exactly four bars (no more and no less) by playing the selection repeatedly in Loop Playback mode until you are satisfied there are no timing glitches around the loop as you listen to it.

> Note: You will need to zoom in closely to 'fine-tune' your edit points and make sure they are on zero waveform crossings for the most accurate results.

When you are satisfied that you have exactly four bars selected, use the Identify Beat command from the Edit menu to bring up the Bar|Beat Markers dialog. If the tempo and position of the region on the timeline are both approximately, but not exactly, correct, then what you will see in this dialog will be approximately, but not exactly, correct also. So the Bar|Beat Markers dialog may look something like Figure 6.38.

Figure 6.38 The Bar|Beat Markers dialog.

Now, at this point you know that you have selected exactly four bars of audio that starts at bar 5 and finishes at the beginning of bar 9 – your own ears tell you that. So all you do is to type 5|1|000 and 9|1|000 into the location fields and hit the OK button – see Figure 6.39.

Figure 6.39 Defining the selected number of bars in the Bar|Beat Markers dialog.

Pro Tools then works out what tempo change is needed to make the session show your selection as four bars exactly between bars 5 and 9 – as in Figure 6.40.

> Tip: Tempo changes can (optionally) be displayed in the Tempo Ruler above the waveform display in the Edit window. Use the rightmost of the pair of pop-up selectors positioned at the left of the Edit window just above the rulers to display or hide the Tempo Ruler – as in Figure 6.41.

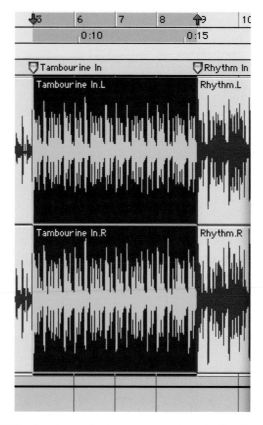

Figure 6.40 The four-bar selection now matches the bar lines correctly.

To create a tempo map for the entire piece of music, you have to repeat this procedure for each section in which the tempo changes.

> Note: If the music that you are tempo mapping has lots of tempo variations, it could easily take you two or three hours or more of intensive work to create an accurate tempo map.

If you are following this example on your computer, your Pro Tools session should now look similar to Figure 6.42.

Inserting Bar|Beat Markers

It is worth mentioning at this point that you can insert Bar|Beat Markers anywhere you like along the Pro Tools timeline by simply inserting the cursor (without making a range selection) at the point you want to change the tempo and using the Identify Beat command. See Figures 6.43 and 6.44.

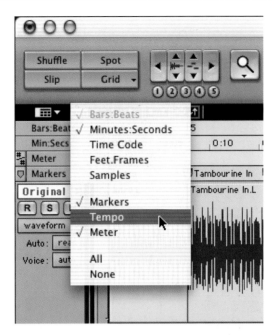

Figure 6.41 Using the pop-up selector to enable the Tempo Ruler.

Figure 6.42 An edited, tempo mapped Pro Tools session showing several separated regions in the Edit window, and Markers for each section displayed in the Markers window and in the Markers Ruler.

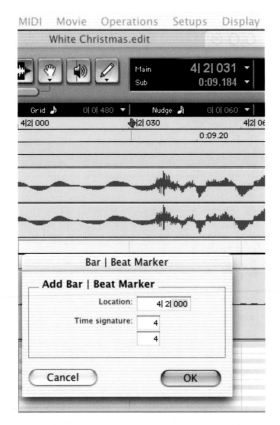

Figure 6.43 A snare 'hit' occurs at bar 4, beat 2, clock 030. You realize that this snare was played correctly, exactly on beat 2, so you insert the cursor at bar 4, beat 2, clock 030 and invoke the Identify Beat command. The Bar|Beat Marker dialog shows the clock location as 030, so you type 000 then hit OK to identify the beat correctly as exactly bar 4, beat 2, clock 000.

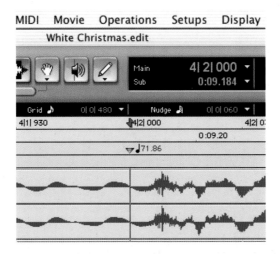

Figure 6.44 After using the Identify Beat command, a tempo change is inserted at this point and the previous tempo is altered so that this point in the audio waveform does correspond correctly with bar 4, beat 2 in the Pro Tools session.

Dragging Bar|Beat Markers

Now, suppose that you have inserted a Bar|Beat Marker at roughly the correct position, but, on closer inspection, you decide it ought to be in a different position close by. It is easy to alter the position of any Bar|Beat Marker by simply dragging the Marker to a new location. New tempos are automatically calculated as you drag the Marker, and any MIDI data is automatically realigned with the new tempo map.

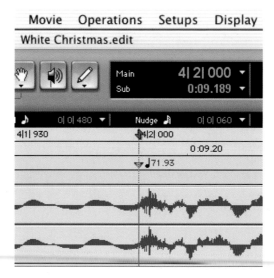

Figure 6.45 To move the position of a Bar|Beat Marker, just point, click and drag the marker to left or right along the timeline.

> Note: When you drag a Bar|Beat Marker, the BPM value immediately to the left is recalculated along with the BPM value at the Marker. BPM values to the right are left unchanged. Also, the bar|beat location stays with the Marker, so, Bar 4, Beat 2 is still Bar 4, Beat 2 after you have moved it – remember that *you* are 'telling' Pro Tools where Bar 4 , Beat 2 is (Pro Tools does not 'know' this; it is just a dumb piece of software after all!).

Editing Bar|Beat Markers

Now if what you really wanted to do was to change the location of the Bar|Beat Marker, there is a dialog box that lets you do this. Just double-click on the Bar|Beat Marker that you want to edit in the Tempo Ruler to bring up the Edit Bar|Beat Marker dialog, enter the new bar, beat and clock location, and click OK.

For example, you may have mistakenly identified a beat at bar 4, beat 2 that should really have been at bar 4, beat 3. Now you know how to quickly correct your error.

Figure 6.46 Edit Bar|Beat Marker dialog.

Why not Beat Detective?

You may be wondering whether to use Beat Detective to work out the tempos instead of using the Identify Beat command, especially if there are a lot of tempo changes. After all – that's what Beat Detective is supposed to do, right? Well, maybe! Beat Detective can work out a tempo map automatically for you, but it works best with fairly straightforward and well-defined rhythmic sections. With short sampled drum loops with very clearly defined beats, Beat Detective works great. Eight or sixteen bar sections of a fairly steady 'live' recording can also work well. But if you have a long piece of music and especially if the beats are not so clearly defined, then your ear will make a much better job of analysing where the beats are than Beat Detective could ever do. Also, if there is only a handful of tempo changes throughout the music, it can be quick enough to use the Identify Beat command – sometimes even quicker than using the Beat Detective.

Pitch shifting to suit a particular vocalist

Now, what if you want a singer to re-record the vocals for a song while singing along to the original version? That is exactly what happened on my remake of the Marshall Jefferson song. The original song was pitched too low for comfort for the new singer, so I shifted the pitch up by a few semitones until he was able to sing the lowest notes OK.

> Note: I used the Digidesign Pitch Shift Audio Suite plug-in for this, as the quality was sufficient to let the singer pitch correctly even though the audio quality was somewhat degraded by this plug-in.

> Tip: The Digidesign Pitch Shift plug-in works reasonably well as long as you don't use extreme settings. For more demanding tasks you will need a third-party tool. I can recommend Speed by Wave Mechanics (www.wavemechanics.com), and Serato's Pitch'nTime (www.serato.com/pnt) – which offers even higher quality and more advanced features. Synchro Arts TimeMod and Waves Time Shifter are also available.

A basic synthesizer tracking session

With everything ready with the tempo track and markers all sorted out and the guide music in the right key, the scene is all set to start recording new drumbeats, bassline, piano, strings, brass and so forth. I set up a MIDI track to record a piano part using a DX7II synthesizer. I also set up a stereo audio track ready to record the audio from the DX7II as soon as I had perfected the MIDI recording. For the bass part I set up a MIDI track to record a bassline using the Spectrasonics Trilogy plug-in. To monitor the Trilogy I had to use a stereo audio track – this was the only option available for this RTAS plug-in. Again, I also set up an audio track ready to record the audio from Trilogy as soon as I had perfected the MIDI recording. For the strings, I used the Arturia simulation of the Yamaha CS80, which is provided in both RTAS and TDM versions. Using the TDM version, I was able to monitor the CS80V using an Aux track – so I did not have to use up two of my available 'voices'. Similarly, for the brass I used Arturia's Moog Modular simulation, which is also available as a TDM plug-in. Take a look at Figure 6.47 to see how this looks.

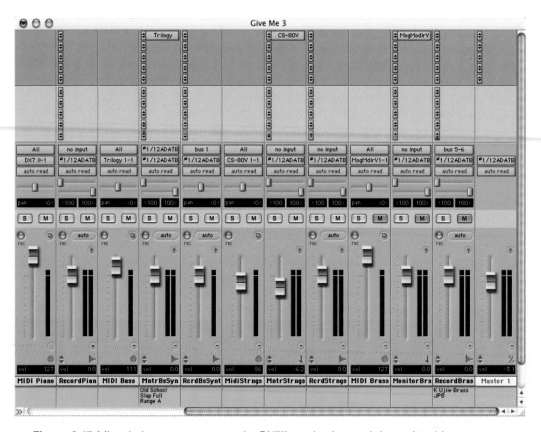

Figure 6.47 Mix window set up to record a DX7II synthesizer and three virtual instruments.

When I had recorded the MIDI tracks for these instruments and had edited these till they were perfect, I recorded them as audio into the waiting tracks that I had set up. Take a look at Figure 6.52 to see how the Edit window looked just after I recorded the audio from the MIDI bass part playing back from the Trilogy virtual instrument.

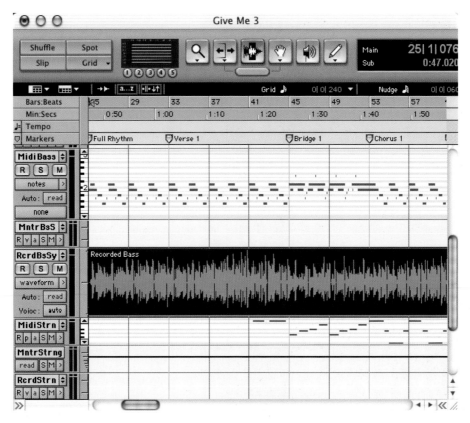

Figure 6.48 Edit window showing a newly recorded bass track with its corresponding MIDI track and audio monitor track above.

When I had finished recording the audio from all my MIDI instruments into Pro Tools, this allowed me to disable all the virtual instrument tracks to reclaim the processing resources.

To make a track inactive, just Command-Control-click (Control-Start-click in Windows) on the Track Type Indicator. The track becomes dimmed with the controls greyed out.

> Tip: You can Shift-click on multiple track names to select multiple tracks and use the Make Selected Tracks Inactive command from the File menu to make two or more tracks inactive at once.

> Note: You cannot make MIDI tracks inactive in this way, but this is not necessary anyway – all you need to do is mute and un-mute the MIDI track as required.

You will probably want to keep the MIDI tracks and any associated Aux or Audio tracks available in your session even after you have recorded these as audio – in case you need to make changes at any time in the future. Of course you should make any Audio or Auxiliary tracks inactive to reduce demand on your DSP resources – and mute the MIDI tracks. You should also

Figure 6.49 Mix window after clicking on the Track Type Indicator to make a track inactive.

strongly consider hiding all these tracks until you next need to use them. This leaves your Mix and Edit windows much less cluttered with tracks. Take a look at Figure 6.50 to see how my Edit window looked at this stage with all the MIDI and associated Audio/Aux tracks hidden – apart from the MIDI Brass track which I was still working on.

> Note: It took me two 10-hour sessions to get to this point – with all the MIDI tracks in place and sounding the way I wanted them to.

Overdubbing piano, guitar, bass and vocals

When you have your basic arrangement in place using MIDI instruments you can start to add 'live' instruments such as guitar, bass guitar, piano – and vocals. It is always the same drill here. First, you connect your microphones to their pre-amplifiers. Then the musician or musicians play the loudest sections from the musical parts they want to record while the recording engineer (that's you in a self-op studio) adjusts the input gain controls on the microphone preamplifiers to avoid clipping during these loudest sections. When everything is ready, you put the track or tracks into record mode and capture the sounds made by the musician(s) onto your hard drive.

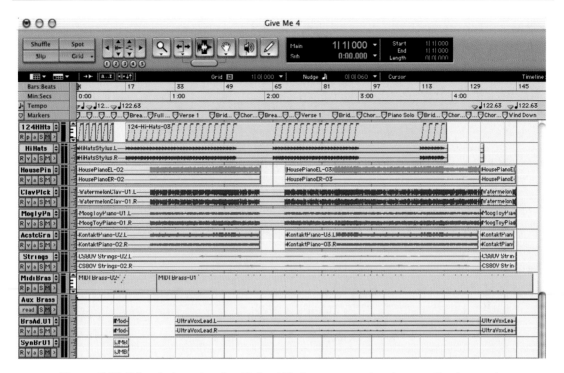

Figure 6.50 Edit window showing Stylus hi-hat grooves and various synthesizer parts.

> Tip: If just one musician is recording all the instruments, as I did here, then you will normally be recording either in mono or in stereo – and rarely with more than two microphones. In this case you should definitely consider using a high-quality two-channel microphone preamplifier such as the Focusrite Red 8.

On the example recording, I put the guitar part down in one take, but had to use a series of punch-ins to record the bass guitar part, as I was not sure what to play in each section until I had tried a few different ideas out. I played piano chords throughout most of the song and added a bass riff on piano in two of the bridge sections and in a chorus section. Then I recorded a piano solo, which took several takes, and I edited the best of these together to form a piano solo 'comp' track. Everything was recorded either in mono or in stereo, some in a single take, some using several takes subsequently edited together, and some tracks – like the bass guitar track shown in Figure 6.51 – were recorded as a series of punch-ins.

I originally recorded the three lead vocals as separate takes into new playlists on a single vocal track – Vocal 1. To discover which sections of vocal from which tracks would fit together best, I created two more vocal tracks and moved the audio regions from the second and third takes into these. Then I created a Vocal Comp track into which I copied and pasted the best sections of the three vocal tracks. Similarly, I recorded two takes of backing vocals into one BV track and then created a second BV track to hold the second take. I was easily able to audition both of these before deciding that the second take was the best – so I did not need a comp track for these.

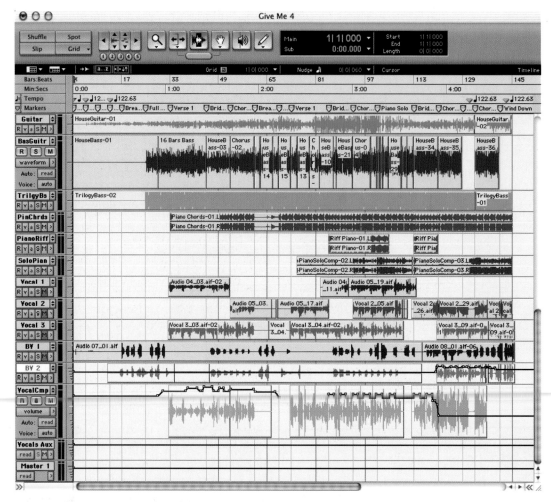

Figure 6.51 Edit window with Guitar, Bass Guitar, Piano and Vocal tracks revealing, for example, how the Bass Guitar was recorded as a series of punch-ins, and showing the volume automation lines with breakpoints along the BV2 and Vocal Comp tracks.

As soon as I had finished recording the vocals, I wanted to be able to play the whole recording back to hear how it was all sounding. The trouble was, the key was still a little low for the vocalist, and the lowest notes that he sang sounded too quiet. It was really easy to sort this problem out by raising the volume automation level wherever the sung notes were inaudible. Take a look at the Vocal Comp track in Figure 6.51 to see how this looks.

Tip: Quite often, you will want to apply compression, EQ and reverb to your vocal tracks and control all the vocals as a group using one fader. To set this up, add a stereo Auxiliary track, insert the plug-ins onto this, and route the outputs from all the vocal tracks via a stereo bus-pair to the input of the Aux track. See Figure 6.52.

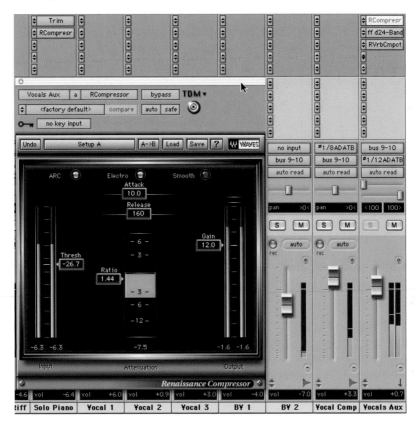

Figure 6.52 Mix window showing all of the vocal tracks routed to an Aux track using bus pair 9–10 so that they can be treated using the Waves Renaissance Compressor, the Focusrite d2 EQ, and the Waves Renaissance Reverb plug-ins.

> Note: It took me about 12 hours to record and edit the 'live' instrument tracks and about 8 hours to record and edit the vocal tracks. All that remained now were some edits and a mix!

Summary

Well, we have covered a lot in this chapter. Beginners will have had the opportunity to see how to get started with a typical audio and MIDI session. Experienced Pro Tools users will probably have brushed up on some techniques and learnt about a few features they have not used before. The only way to really learn how to record with Pro Tools is to put the time in using the system to record in as many situations as possible. It simply takes time to learn all the shortcuts and the best ways of going about things. And there is always another way to do things in Pro Tools! Part of the process is to be aware of what is possible so that when the time comes around you know what you should be looking for. I hope that the examples given here will help.

7 Editing

Introduction

Recording, editing and mixing. The more I think about how I actually use Pro Tools, the more I realize that, for me, these three processes are very often intertwined and overlapping.

Nevertheless, there are plenty of specifics that are concerned with editing. For example, you need to get familiar with the editing tools and with the various transport, navigation and zooming controls to be able to use these to best effect – and you need to know what's available for you to choose from. Similarly, you need to be aware of features such as crossfades, auto-fades, Grid modes, and so forth. There is also a section on Beat Detective for anyone who hasn't had time to work with this powerful feature previously. Reading the first part of this chapter will help you get 'up to speed' with all this stuff.

Now if you are keen to get started on some real edits, there is a section on practical editing techniques further on, so you can skip forward to this if you like. You can always come back to the first part later – that's the beauty of a book!

Edit window features

The programmers working on the software's user-interface have definitely been putting plenty of thought into how to make everything work speedily and efficiently in Pro Tools. If you are upgrading from version 5.x, the first thing you will notice is that the user-interface has a more colourful, three-dimensional 'look' than before, and there are some changes to the way you access some of the controls and settings. The pop-ups that let you select which views and which rulers are displayed in the Edit window, the Tab to Transients, Commands Focus and Timeline Selections buttons, and the grid and nudge values have all been tidied up into a convenient area – just above the rulers display. Everything else is more or less where it was and as it was before, though – so you shouldn't have much trouble finding your way around if you are upgrading.

The Shuffle, Slip, Spot and Grid mode buttons are positioned at the top left of the Edit window – more on these later. To the right of these, a group of zoom buttons lets you zoom audio and MIDI separately – and there are five zoom levels available on buttons underneath so you can always get to your favourite zoom levels quickly. To the right again there are buttons to let you

Figure 7.1 Screenshot shows the Edit Mode and Zoom buttons, the various Tools, the Location Indicators and the Event Edit Area to the right of this, with pop-up selectors for the Views and Rulers, Grid and Nudge values, and a parameter display, underneath.

select the various tools: the Zoom tool, the Trim tool, the Selector tool, the Grabber tool, the Scrubber tool and the Pencil tool.

> Tip: The Selector tool lets you select both horizontally and vertically, so you can select across multiple tracks with a single click and drag.

There are two counters in the Edit window – Main and Sub – so you can see Bars:Beats and Timecode, or Feet:Frames and Minutes:Seconds (or whatever combination you like) at the same time. Also, you can set the Grid to one value and the Nudge to a different value.

The display area underneath the counters shows parameters such as the MIDI note number in a MIDI track, or the exact level in decibels of the volume data, at the far right-hand side.

The current position of the Cursor is displayed just to the left of this. You can keep your eye on this cursor position while finding edit points. It is also useful when zooming the display. You have to zoom in during editing sessions to be able to move in smaller increments, and, by observing how the numbers displayed here change as you move the cursor, you can see whether the slightest cursor movement changes the cursor position by, say, five frames at a time or one frame at a time.

To the left again, you will find pop-up selectors and displays for the Nudge values and the Grid values.

Further to the left there are three buttons and two pop-up selectors discussed elsewhere.

Edit window set-up

After a busy track-laying session you should take time out to tidy up your Pro Tools session. Hide any tracks that you will probably not be using while editing. This could include a click track, some MIDI tracks and associated auxiliary or audio tracks used for monitoring synthesizers or virtual instruments, a demo version of the song used as a template to work to – or whatever. You should also trim the start and end points of your audio tracks to remove any unwanted sounds before the music is supposed to start or after it is supposed to finish. This is always one of the first edits I make at the start of an editing session. You might also want to adjust the track heights according to how important it is for you to see the waveform, volume envelope, or whatever you might want to keep in sight during your editing session.

Tip: If you want to change all the tracks to 'Medium' (or any other size setting) with one move, just hold the Option (Alt) key while you select 'Medium' for any track. The others will all change instantly to this size.

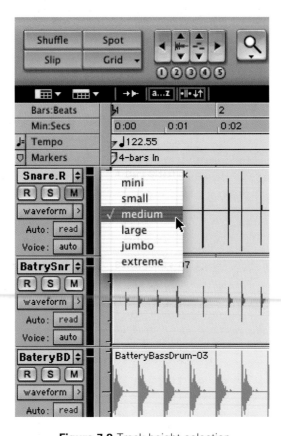

Figure 7.2 Track height selection.

Markers window set-up

It is always a good idea to make sure you have put in a set of Markers for each section of your song, and if you have not got around to this while recording, then you should set up your Markers before you get seriously involved in your editing session. You are almost always going to want to jump to the start points of the different sections within your recording, so section markers are indispensable on all but the simplest sessions. Also, you should make sure that you give the sections meaningful and obvious names – there is no time for 'guessing' games in the middle of busy sessions!

I also recommend setting up various zoom levels in the Markers window at this stage. For example, you can create a 'Normal View' setting in the Markers window which will recall the Zoom Settings, Pre/Post Roll Times, Track Show/Hide settings, Track Heights, and Group Enable settings that you want to use as your 'home' view or normal view.

Figure 7.3 Markers window.

To make a Marker, just hit 'Enter' with all these things set up the way you want them to be. In the 'New Memory Location' dialog box that appears, set the Time Properties to 'None'; type something such as 'Normal View' as the name; then tick the General Properties settings that you want to memorize.

Then you could set up other useful zoom levels, such as all the way out and all the way in, or zooming in or out from the normal zoom in stages. You could even name these zoom levels according to what you will be using them for during your editing session. For example, 'bass edits', 'snare edits' or 'vocal edits'. And when you include Track heights and selections, it is possible to streamline many editing operations by storing frequently used set-ups in these memory locations.

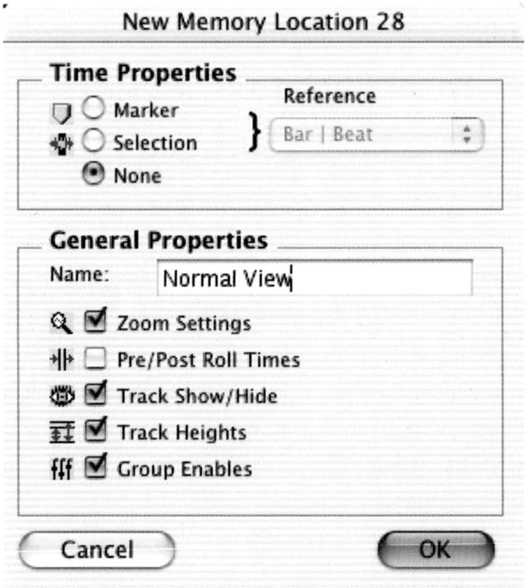

Figure 7.4 New Memory Location dialog.

Scrolling options

The scrolling options in the Edit window include Scroll After Playback, which does what it says – not moving the display until you hit Stop. The Page Scroll option keeps the display stationary until playback reaches the rightmost side of the page, then it quickly changes the display to show the next section (i.e. page) of the waveform that will fit within the Edit window – and so forth.

There are two continuous scroll options. The first of these, 'Continuous Scroll During Playback', does exactly what it says – continuously moving the display during playback to keep up with the playback.

The second option, 'Continuous Scroll with Playhead', shows the 'playhead' – a vertical blue bar indicating where the playback point is at in the Edit window – and moves the waveforms past this.

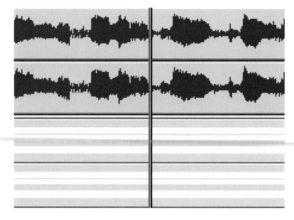

Figure 7.5 Continuous Scroll with Playhead.

With the playhead visible, the waveforms move past a central point and the playback position stays wherever it is when you stop playback.

If you use any of the scroll options without the playhead, the playback position jumps back to the position you started playing from when you stopped playback – unless you enable the 'Timeline Insertion Follows Playback' in the Operation Preferences dialog.

With 'Timeline Insertion Follows Playback' enabled in the Operation Preferences dialog, when you press play, then stop or pause, then press play again, Pro Tools 'picks up where it left off'. In

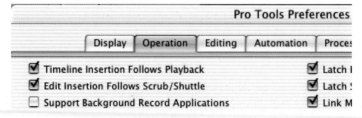

Figure 7.6 Setting the 'Timeline Insertion Follows Playback' preference.

other words, when you stop, the insertion point 'parks itself' at that point – instead of returning to the position it was at when you first pressed play.

Playback cursor locator

The Playback Cursor Locator feature lets you locate the playback cursor when it is off-screen. For example, If you are using the 'No Auto-Scrolling' option, when you stop playing back, the Playback Cursor will be positioned somewhere to the right off the screen, if it has played past the location currently visible in the Edit window. Also, if you manually scroll the screen way off to the right, perhaps to check something visually, then the Playback Cursor will be positioned somewhere to the left off the screen.

To allow you to quickly navigate to wherever the Playback Cursor is positioned on-screen, you can use the Playback Cursor Locator button that appears under certain conditions.

The Playback Cursor Locator will appear at the *right* edge of the Main Timebase Ruler if the playback cursor moves to any position *after* the location visible in the Edit window.

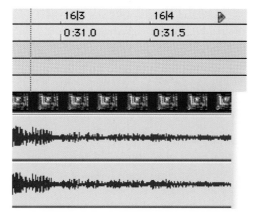

Figure 7.7 The Playback Cursor Locator, the small blue arrowhead, can be seen at the top right in the Main Timebase Ruler (which is displaying bar locations in this example).

The Playback Cursor Locator will appear at the *left* edge of the Main Timebase Ruler if the playback cursor is located *before* the location visible in the Edit window.

Clicking the Playback Cursor Locator immediately moves the Edit window's waveform display to the Playback Cursor's current on-screen location – saving you lots of time compared with any other way of finding this location.

Note: The Playback Cursor Locator is red when a track is record enabled and blue when no tracks are record enabled.

Zooming and navigation

You can always store as many zoom levels as you like in the Memory Locations window, and you can store track heights and selections of visible tracks – recalling your settings by clicking in the Memory Locations window or hitting numerical keys assigned to these.

Also, there are five numbered zoom-level buttons positioned near the top left of the Edit window – just underneath the zoom arrows. Command-click on any of these five buttons to store the current zoom level and simply click on any of these buttons to recall the stored zoom level. I prefer using these ways of zooming, or manually using the zoom arrows, to using the Zoomer tool, as I like the precision.

> Tip: At first I used to end up zooming far too much using this Zoomer tool – ending up wasting too much time finding the zoom level I wanted. Then I discovered a short cut to return to the zoom level I started from – by holding the option key and clicking on the right or left horizontal zoom arrows.

The Universe window provides yet another way to find your way around in a session. If you click to left or right of the highlighted section it will move the display in the Edit window to the left or right (assuming this is not all visible). This is OK, but it should really let you click and hold the highlighted portion and move this to the left or right to wherever you want it. Also, it would be much more useful if it displayed a miniature representation of the waveform so you could see which place you wanted to go to more easily. What is nice is that if you click higher or lower in this window, the Edit window will scroll vertically to show tracks higher or lower than are currently being displayed (assuming that they are not all visible).

Now if you want to jump around a page at a time in the Edit window using keyboard commands rather than the mouse while playback is stopped, you can use the Option (Alt) key in conjunction with the Page Up and Page Down keys instead of clicking in the scroll bar at the bottom of the Edit window. This can be a very useful shortcut – especially if the bottom of the edit window is not currently visible on-screen. By the way, the Page Up and Page Down keys are located directly underneath the F15 key and have up-/down-arrows, each with a couple of short transverse lines, as icons.

A quick way to scroll the display forward a screen 'page' at a time during playback is to hit the down and left arrows at the same time on the computer keyboard. Even more useful, if you have zoomed in the display and are playing back with scrolling off, you can force the window to scroll along with the playback cursor by repeatedly hitting this key combination to keep the

> Tip: If you have made a selection in the Edit window and then you zoom in to fine-tune your edit points, the display will zoom in around the start point of the selection. If you want to jump to the end point, you can use the right arrow key on the computer keyboard to move the display so that the right-hand edge of the selection, i.e. the end point, is in the centre of the screen. To get back to the start point of the selection, just hit the left arrow.

display following the playback cursor. This is just what you need when you are testing an edit or looking for an edit point and you want to watch the waveform just for a short time as it plays back, although most of the time you don't want the display to scroll.

Keyboard commands and short cuts

With TDM systems the keyboard commands work differently according to whether you have chosen the short cuts to work with the Commands, the Audio or MIDI Regions or the Groups list.

You can select which of these areas the commands are focused on by pressing Command-Option-1, -2, -3 or -4 respectively.

With the Commands Focus enabled you can then simply press the 'E', '-', 'R' or 'T' keys without pressing the Control key – making it even simpler to work with these short cuts. And with the Commands Focus enabled, lots of useful keyboard commands come into operation.

For example, 'N' toggles the preference for 'Timeline Insertion follows Playback' on and off. This can be a very useful setting if you are using the Scrub tool to identify where you would like your insertion point to be as soon as you release the tool – having identified the correct edit point. I prefer to work with this option de-selected most of the time, so that the Timeline Insertion stays wherever I last left it – until I want it to jump to wherever the playback last stopped (which I do when I am using the Scrub tool to identify an edit point).

The 'P' key moves the edit selection to the track above, while ';' moves it to the track below. The 'L' and the apostrophe keys move the edit insertion point to the previous or next major waveform peak, respectively.

Other keyboard shortcuts let you really speed up your work. Track Toggle lets Audio tracks toggle between the Waveform and Volume view while MIDI tracks toggle between Notes and Regions views – in other words, between the two most often-used views for Audio and MIDI tracks. Just click in the track you want to toggle, shift-clicking to select multiple tracks if required, then press the Control and Minus ('−') keys simultaneously.

The 'Zoom Toggle Track Height' preference setting lets you specify a default track height that will apply when you are using the Control–Minus keyboard command to toggle audio tracks between Waveform and Volume view, or to toggle MIDI tracks between Notes and Regions view.

Similarly, you can toggle between two views of a selection in the Edit window – adjusting the zoom level and track height automatically. Select one or more tracks and press Control–E. The selection will zoom to fill the Edit window and the tracks containing the selection are set to a track height of Large – with MIDI tracks automatically set to Notes view.

Put the Edit cursor into a track and you can change the track height by holding the Control key and using the Up/Down arrow keys. And you can temporarily suspend Grid mode and switch to Slip mode by holding down the Command key while trimming audio or MIDI regions.

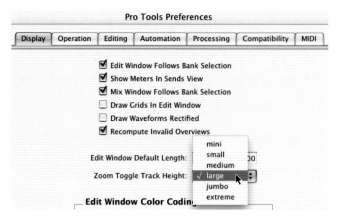

Figure 7.8 Setting the 'Zoom Toggle Track Height' preference.

Control–R and Control–T let you zoom out and in respectively – acting as keyboard commands for the left and right horizontal zoom arrows.

The Edit tools

Zoom tool

The Zoom tool lets you zoom in and out around a particular area within a track using either its Normal Zoom mode or the Single Zoom mode. The Single Zoom mode automatically re-selects the tool that you were using previously after it zooms the display. To zoom horizontally, you just drag with the Zoomer along the track. You can zoom both horizontally and vertically at the same time by pressing the Command key (Control key on the PC) while you drag.

> Tip: If you press and hold the Command and Control keys (Control and Alt on the PC) then move your mouse cursor into the Ruler area in the Edit window, you can zoom in without having to select the Zoomer tool first. If you just click once at any position in the ruler, the display zooms in one level, and if you drag across a range in the Ruler, the display will zoom in around this range. You can also toggle the display to simultaneously increase the height of the track to Large (or whatever track height you have set in the display preferences) and to zoom the display to encompass the selection by clicking Control–E on the Mac (Start–E on the PC) or just by pressing E if the Commands Focus is enabled.

Trim tool

With the Trim tool, you can Scrub and Trim at the same time using the Scrub Trim selection option. Scrub right up to a bass drum, for example, let go of the mouse, and the region will instantly be trimmed up to the start of that bass drum. You can also use the Time Compression and Expansion (TCE) Trim selection option to instantly trim the length of a region and have the audio within this region automatically time-compressed or expanded. You can even replace the default RTAS plug-in which carries out this time compression or expansion by changing this in the Preferences window to use Wave Mechanics Speed or Serato Pitch'n'Time, Synchro Arts Time Mod or Waves Time Shifter – any of which will give you significantly better results than the standard Digidesign plug-in.

Selector tool

The Selector tool lets you select ranges for editing. I use this as my default tool when I am not using any of the other tools. With the Selector Tool engaged, the mouse cursor turns into an insertion cursor and, if you have the Edit and Timeline selections linked, you can click anywhere in a track or ruler to move the playback counter to that point.

Grabber tool

The Grabber tool can use either the normal Time selection, or an Object selection, or a Separation selection. Object-based selection lets you select any regions on any tracks without these having to be next to each other – so you can move just about any combination of regions around in the Edit window. Even more usefully, you can select any area within a region using the Selector tool and then instantly cut this selection out using the Separation Grabber so you can move it to another location or track. You can select from within a single region, or across adjacent regions within the same or multiple tracks. And if you want to leave the original selection where it is, just press and hold the Option (Alt) key while using the Separation grabber.

> Tip: Don't forget that you need to *select your audio first* using the Selector tool, *then* change to the Selection Grabber tool to work with this selection.

Smart tool

The Smart tool, positioned just underneath the Selector tool, links the Trimmer, Selector and Grabber tools. With these three tools linked, you can choose whether the Selector, Grabber or Trimmer tool is operational depending on where you position the cursor in the Waveform display. Hold the cursor above the zero line and it becomes the Selector; hold it below the zero line and it becomes the Grabber; hold it close to either end of a region and it becomes the Trimmer. You can also use the Escape key on the Mac keyboard to toggle through these three options, plus the Zoomer, Scrub and Pencil tool options as well. Or you can use Command-1, -2, -3, -4, -5 or -6 to select these tools specifically – using Command-7 for the Smart tool. Similarly, you can cycle through the edit modes using the key with the tilde (~) symbol or you can choose the edit mode using Option-1, -2, -3 or -4.

Scrub tool

The Scrub tool lets you 'scrub' up to two tracks of audio in the Edit window. Scrubbing emulates the technique of moving the tape relatively slowly back and forth against the playback head on a tape recorder in order to locate edit points. When you have selected the Scrub tool, the mouse cursor changes to a loudspeaker icon. When you drag this back and forth along an audio track, you will hear the audio playing slowly in the forward and reverse directions. To make the edit cursor automatically move to the position where you stop scrubbing, i.e. to the edit point you have located, make sure that 'Edit Insertion Follows Scrub/Shuttle' is enabled in the Operation Preference window. To scrub two adjacent tracks at once, position the mouse cursor (the loudspeaker icon) between the two adjacent tracks and drag. Note that this works with mono and stereo tracks – but not with surround tracks. If you press the Option (Alt) key while scrubbing, this puts you in Shuttle mode – which plays the audio at several times normal speed. You can also control Shuttling using the Numeric Keypad on your computer keyboard – see the manual for details.

Pencil tool

The Pencil tool has Freehand, Line, Triangle, Square and Random versions. For example, you can draw in MIDI controller data using the Freehand tool, or use the Triangle tool to quickly create pan or plug-in automation. The most common use with audio is to repair pops or clicks in an audio file. To make the pencil tool active for waveform repair, you need to zoom in to sample level first – otherwise the pencil cursor is represented by a dotted outline and will not work. Note that using the Pencil tool to redraw a waveform changes the audio file, so you may wish to copy the file first and work on the copy in case you make a mistake.

The Transport window

This provides the usual controls for Play, Record, Stop and Rewind, along with buttons and controls to select the Transport Master, to set and enable the Pre-Roll and Post-Roll times and to set the Start, End, and Length for Timeline Selection.

In the centre, between the Transport and MIDI controls, you can optionally display the main and sub counters.

> Note: Making the main and sub Location counters visible in the Transport window lets you see where you are when you are working just in the Mix window (which has no counters). (These counters are always displayed in the Edit window.)

At the far right there is a group of MIDI-related buttons and controls to let you set the tempo or engage the conductor track, engage and configure the metronome, engage and configure count-in bars, and so forth.

Figure 7.9 The Transport window.

Using the options available in the Display menu, the counters and MIDI controls can be hidden if you don't need these. Similarly, the lower part of the Transport window can be hidden using the Display menu – or by clicking on the size button near the top left of the Transport window.

Crossfades

In many editing situations you will want to create crossfades between adjacent regions to smooth the transitions and avoid abrupt changes in the sound – or to create special effects. Pro Tools lets you define the length, position and shape of your crossfades and saves these as files in a folder marked 'Fade Files' inside each session folder. These files are then read from disk during playback.

To create a crossfade, choose the Selector tool and drag across the end of the first region and the beginning of the second region to set the length of your crossfade. Then you can use the Fades dialog to select, view, and manipulate the volume curves assigned to the fade-out and fade-in portions of the crossfades.

> Note: A Dither option is provided which you should use when mixing or fading low-level audio signals – for example, when fading in or fading out silence, or crossfading between regions with low amplitudes. You won't need to use dither if you are working with regions of high amplitude and the fades will be processed more quickly without this option.

You can also use the crossfade feature to create fade-ins and fade-outs at the beginnings and ends of regions. Detailed instructions on using the Crossfade features are available in the Pro Tools Reference Guide supplied with the system.

There is also a very useful Batch Mode that lets you select across several regions and use the Create Fades command from the Edit menu to create crossfades for all the regions within your selection. This gives you options to Create New Fades, Create New Fade-Ins and -Outs, Adjust Existing Fades, or any combination of these. You can also choose where to place the fades – Pre-Splice (up to the transition), Centred (across the transition), or Post-Splice (after the transition).

Auto fade-in and fade-out

Be aware that on Pro Tools TDM systems there is an automatic fade-in/out option that saves you the trouble of editing to zero-crossings or creating numerous rendered fades in order to eliminate clicks or pops during playback. You can use this to have Pro Tools automatically apply real-time fade-ins/outs to all region boundaries in the session. What is important to note here is that these fade-ins/outs are performed during playback and do not appear in the Edit window, and are not written to disk. You can set the length of the automatic fade-ins/outs in the Operation Preferences window. Here you can enter a value between 0 and 10 ms for the Auto Region Fade In/Out Length. Once this is set, this Auto Fade value is saved with the session, and is automatically applied to all free-standing region boundaries until you change it.

Using autofades saves you the trouble of editing on zero-crossings or having to create numerous rendered fades in order to eliminate any clicks or pops that would otherwise appear at the edit points.

However, since these autofades are not written to disk, these clicks or pops still exist in the underlying sound file. Why might this be a problem? Well, if you subsequently used the 'Duplicate' AudioSuite plug-in or the 'Export Selected as Sound Files' command (from the Audio Regions List) to duplicate multiple regions as a continuous file, these anomalies would still appear – and that could definitely be a problem. One way round this is to render these real-time autofades to disk, using the 'Bounce to Disk' command in the File menu.

A tip straight from the manual is also worth quoting directly here: 'This automatic fade-in/out option also has an effect on virtual track switching in a session. Whenever a lower-priority virtual track "pops thru" a silence in a higher-priority track on the same voice, a fade-in and fade-out is applied to the transition. This feature is especially useful in post-production situations such as dialog tracking. For example, you could assign both a dialog track and a "room tone" track with matching background to the same voice. You could then set the Auto-Fade option to a moderate length (4 ms or so) so that whenever a silence occurred in the dialog, playback would switch smoothly to and from the background track without clicks or pops'.

Editing accuracy

It is easy enough to make rough edits in Pro Tools – just chopping up the waveform by eye while zoomed out – and you can get lucky working this way if you are working with a very straight-forward recording of some sort. However, if you want to achieve the most accurate edits, you normally need to zoom in until you can see exactly where you are placing your cuts or making your selection start and end points.

> Tip: If you want to roll backwards or forwards a short way, you can do this by using the 1 and 2 keys on the computer's numeric keypad – as long as you have the keypad preference set up in the Operation Preferences dialog to 'Transport'. Setting the preference here to 'Shuttle' mode lets you wind back or forward, slow, medium or fast, while hearing the playback – using pairs of keys 1/3, 4/6, 7/9 from the keypad. This is particularly useful during intensive editing sessions.

> Note: Because audio material in Pro Tools is sample based, some amount of sample rounding may occur with some edits when the Main Time Scale is set to Bars:Beats. This is most evident when you need audio regions to fall cleanly on the beat (as when looping) and notice that the material is sometimes a tick or two off. With a few simple precautions, this can be avoided. When selecting audio regions to be copied, duplicated, or repeated, make sure to select the material with the Selector (enable Grid mode for precise selections), or set the selection range by typing in the start and end points in the Event Edit area. Do not select the material with the Grabber (or by double-clicking with the Selector). This ensures that the selection will be precise in terms of bars and beats (and not based on the length of the material in samples).

Tab to transients

To find good edit points quickly, just place the cursor in the waveform display a little before the expected transient and click on the Tab To Transients button up near the top left of the Edit window. The insertion point will then jump to the next transient peak that it detects. The idea of this is to avoid having to expand the waveform to its maximum levels in order to identify edit points with absolute accuracy.

Tab to transients is a great feature that really speeds up editing when you are working with, say, a bass drum track. Here you can expect a clear-cut situation with a fast transient waveform peak at the start of each bass drum hit. To find good edit points, first make sure the Tab To Transients feature is activated by clicking on the small icon at the top left of the Edit window, just underneath the Slip mode button, then insert the waveform cursor into any track and hit the Tab button on the computer keyboard. The insertion cursor will 'tab' through the waveform display for the selected track, finding a suitable spot.

Let's take a peek at how this looks. Open the edit window and examine, say, a bass drum track. Put the cursor in the display near the start of the recording. See how this might look in Figure 7.10.

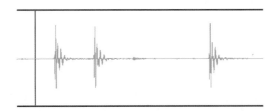

Figure 7.10 Waveform display with cursor inserted near the start of the region, which contains three bass drum beats.

To jump to a transient, Hit Tab. The cursor will jump to the beginning of the first transient – which occurs at the start of the first bass drum beat. See how this might look in Figure 7.11.

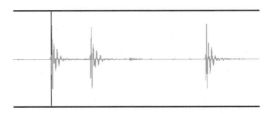

Figure 7.11 Waveform display showing the cursor at the beginning of the first transient – i.e. at the first bass drum beat.

To jump to the next transient, Hit Tab again. The cursor will 'find' the beginning of the second bass drum beat. See how this might look in Figure 7.12.

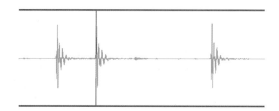

Figure 7.12 Waveform display showing the cursor at the beginning of the second transient – i.e. at the second bass drum beat.

To automatically select the region between the second and third bass drum beats, Hit Shift + Tab. See how this might look in Figure 7.13.

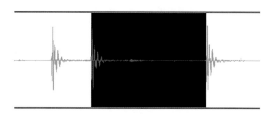

Figure 7.13 Waveform display showing region between the second and third bass drum beats selected.

Now you could copy this selection, or delete it or move it – or whatever.

To cut the track at this point, with the blinking cursor inserted where you want your cut, just hit Command + E.

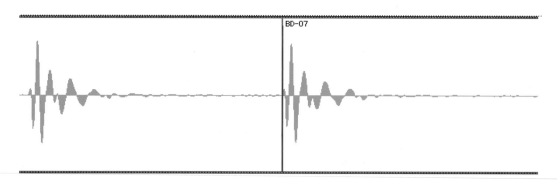

Figure 7.14 Waveform display showing two bass drum beats with a cut separating them into two regions.

> Tip: If you want to cut all the tracks at this point, the quickest way is to click in the ruler at the top of the window so that the insertion blinking cursor extends all the way through every track.

Editing regions

Pro Tools has excellent region editing features. You can slide regions freely within a track or onto another track in Slip mode using the Grabber tool. This allows you to arrange regions with spaces between them or move regions so that they overlap or completely cover other regions. If you want to constrain the movement and placement of regions to bars or beats or to particular time boundaries, you can use Grid mode. This is particularly useful when you are working with Bars and Beats. Shuffle mode, on the other hand, will automatically close up any gaps when you cut out a region – which you will often want to do.

You can select any region and quickly repeat this using the Repeat command in the Edit menu. This is great for working with drum loops, for instance, where you might place a four-bar loop and want to repeat this to fill a verse or whatever. Another typical editing operation is to replace a loop or sample that you have used initially with a better choice later on as the session develops. Similarly, when spotting sound effects you may decide to replace some of these at a later stage. Simply select a region in the Edit window and Command-drag a replacement region from the Regions list. The Replace Region dialog appears and you can use this to fit the replacement region to the original in a variety of ways. See Figure 7.15.

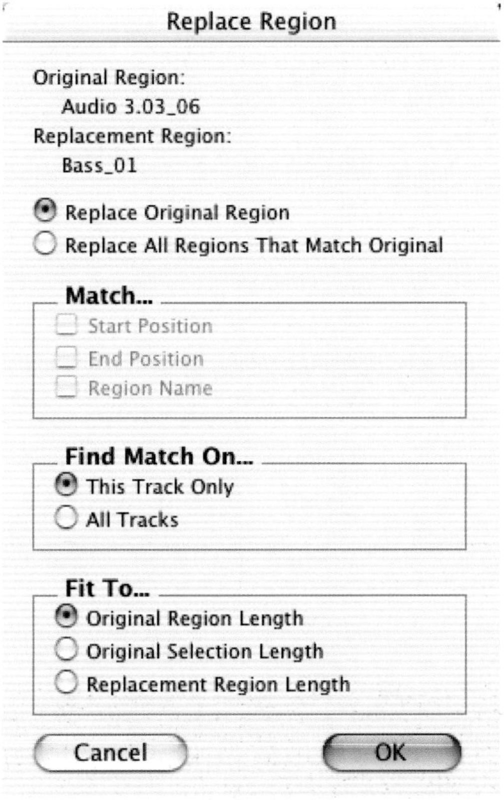

Figure 7.15 Replace Region window.

Grid modes

In Absolute Grid mode, as you move any region to left or right along the timeline, its start point will snap to the nearest grid boundary as soon as you let go. Setting up a grid that lets you move regions around a bar at a time or a beat at a time is very useful when editing pattern-based music, for example. And if you are working to picture, you can move a frame at a time or in increments of Feet and Frames.

Pro Tools also has a very powerful Relative Grid mode that lets you edit audio and MIDI regions that are not aligned with Grid boundaries as though they were. For example, if a region's start point falls between beats and the Grid is set to quarter notes, dragging the region in Relative Grid mode will preserve the region's position relative to the nearest beat.

To choose the Grid mode, click and hold the Grid button to open the mode selector then move to and release the mouse on the one you want to select.

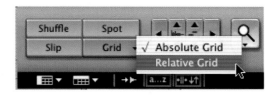

Figure 7.16 Selecting the Grid mode using the Grid pop-up.

You can choose the actual Grid size using the Grid Value pop-up – which is located just above the rulers in the Edit window. The Grid can be based on Bars|Beats, Time Code or other time formats.

Figure 7.17 The Grid Value pop-up.

> Tip: There is a useful Grid mode feature that allows you to place events freely along the timeline, as in Slip mode, but with one very important difference: if you place any event close to a Region Start, End or Sync point, or close to any Marker or Edit selection, the event will snap to that nearby location.

Time Compression/Expansion edits

The way that the Pro Tools software can automatically make use of the Time Compression/ Expansion (TCE) plug-in while carrying out various editing actions is quite unique. For instance, if you want to quickly fit an audio region within a time range that you have selected in the Edit window, you can Command-Option-drag any region from the Regions list and the Time Compression/Expansion plug-in will compress or expand the audio region to fit within that selection.

The 'Compress/Expand Edit To Play' command works in a similar way. This lets you force a region selection that you have made in the Edit window to fit a different Timeline selection. In this case you need to unlink the Edit and Timeline selections, select the audio you want to

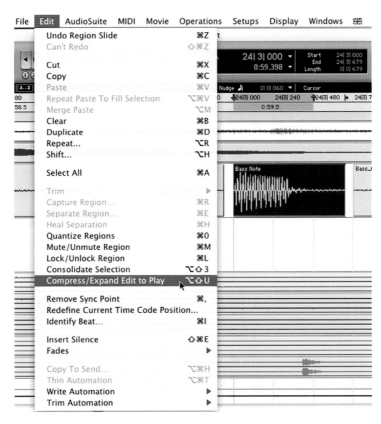

Figure 7.18 Compress/Expand Region to Fit: 1. Select region to be shortened (or lengthened) then use the 'Compress/Expand Edit To Play' command.

compress or expand, select the time range in the Timebase Ruler that you want to fit the audio to, and then invoke the 'Compress/Expand Edit To Play' command from the Edit Menu – see Figure 7.18.

The Edit selection will then be compressed or expanded to match the length of the Timeline selection. Notice in the screenshot, Figure 7.19, that the length of the bass note has been shortened to match the timeline selection in the rulers above the waveform displays.

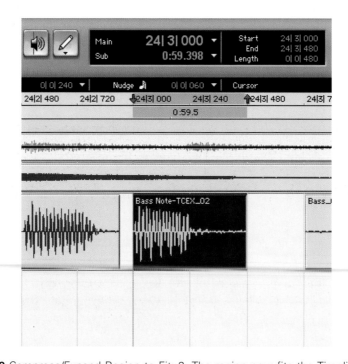

Figure 7.19 Compress/Expand Region to Fit: 2. The region now fits the Timeline selection.

You can also combine Time Compression/Expansion with the Trimmer tool while editing regions. Click and hold the mouse cursor on the Trimmer Tool and a pop-up selector will appear to let you select the alternative Trimmer Scrub or Trimmer TCE modes.

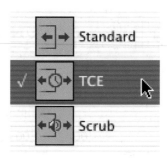

Figure 7.20 Selecting the Trimmer TCE tool.

Using the Trimmer TCE tool, you can easily and conveniently match the length of one audio region to another, or to the length of a video scene, or whatever. As you drag the region's start or end point the TCE plug-in automatically does its work, creating a new file of the correct length and replacing the region you are editing with this new region when you let go of the Trimmer tool.

In Grid mode you can use this tool to make the tempo of a region fit the tempo of another region – or of the session. If the session was at 130 BPM and you added a drum loop running at a different tempo – say 120 or 140 BPM – the new loop would either be too long or too short to fit into a bar at 120 BPM. Using the Trimmer TCE tool you can just drag the start or end point by a suitable Grid value such as 1/16 note increments to make it fit. In Slip mode you can basically do the same thing, but 'freehand' as it were. You would need to expand your waveform display to its maximum to see the start and end points accurately which would make this a slower method to use.

> Tip: You can also use the Trimmer TCE tool in Spot mode. In this case, clicking on a region with the Trimmer opens up the Spot dialog which lets you specify where you want the region to start and end, i.e. how long it should be.

Linked selections

Normally, whenever you make a selection in a track, the same selection is made in the time rulers at the top of the Edit window – and this is the selection that plays when you commence playback. Look at Figure 7.21 to see how this typically looks.

Figure 7.21 Edit window showing linked Track and Timeline selections.

You can also unlink the Edit window and Timeline selections using the Linked Selections button that you will find just underneath zoom presets 3 and 4 near the top left of the Edit window. In Figure 7.22 you will see the timeline selection between the 'Band In' and 'Section 2' markers while the Edit window selection is between the 'Bridge Section' and 'Head Again' markers.

Figure 7.22 Edit window showing different, unlinked Track and Timeline selections

Normally, you will work with the Edit and Timeline selections linked, but you will find it useful to be able to unlink these from time to time. For example, you could select an Edit region that is shorter than the Timeline selection and contained within it. Say you want to move this edit region until it sounds good against the rest of the tracks. In this case, you want to loop around a larger timeline selection so that when you move the Edit region a little forwards or backwards, it remains within the timeline selection and keeps looping. This way, you can hear when you have positioned the Edit region correctly within the Timeline selection that you are looping.

Beat Detective

As the documentation explains, Beat Detective analyses and corrects timing in performances that have strong transient points, such as drums, bass and rhythm guitar. It allows the user to define a tempo map from a performance or to conform the performance to a tempo map by separating it into regions and aligning it to the beats. The idea here is that Beat Detective identifies the individual beats in your audio selection by looking for the peaks in the waveform. You can adjust the settings until you have identified most of these and then edit manually to fine tune the choices. The points identified are referred to as 'beat triggers' in Pro Tools and these can be converted to Bar|Beat Markers. Once the beats have been identified correctly, Beat Detective can extract the tempo from the audio – creating a tempo map. Other audio regions and MIDI tracks can then be quantized to these markers. So, for example, you can use Beat Detective to extract tempo from audio that was recorded without listening to a click – even if the audio contains varying tempos, or material that is swung – and you can then quantize other audio regions or MIDI tracks to this 'groove'.

You can also do the opposite of this: if your session already has the right tempo, you can use Beat Detective to 'conform' any audio with a different tempo (or with varying tempos) to the session's tempo. You can choose to keep a percentage of the original feel if you like, and you can increase or decrease the amount of swing in the conformed material. You can also conform regions which you have previously separated using Beat Detective to a session's tempo map.

First you define a selection of audio material on a single mono or multi-channel track, or across multiple tracks. Then adjust the Detection parameters so that vertical beat triggers appear in the Edit window, based on the peak transients detected in the selection. For example, with a 'boom-chick' bass drum and snare drum beat you would see a vertical line in the display immediately before each bass drum and snare drum beat. You should examine these triggers visually to make sure that there are none in the wrong places for any reason. When you are satisfied that the triggers look OK, you can generate Bar|Beat markers based on these beat triggers to form a tempo map from the selection which you can use for your session.

One of the most useful applications is aligning loops with different tempos or feels. If one loop has a subtly different feel or groove you can use Beat Detective to impose that groove onto another loop. This is great for remixes where you often need to extract tempo from the original drum tracks, or even from the original stereo mix. New audio or MIDI tracks can then be matched timing-wise to the original material, or the original material can be matched to the new tracks.

Beat Detective lets you separate and automatically create new regions, representing beats or sub-beats, based on the beat triggers. You can then conform these new regions to the session's existing tempo map or to a groove template.

After conforming regions, gaps may be left between these, so an Edit Smoothing feature is provided which can fill the gaps between regions – automatically trimming them and inserting cross-fades as required. This can save you an awful lot of detailed editing work that would otherwise be necessary to avoid pops and clicks at the region boundaries that would normally need to be trimmed and cross-faded. It also has the advantage of preserving the ambience throughout the track to keep this constant despite the edits.

Another feature is the Collection mode. This lets you 'collect' beat triggers from different tracks, such as the bass drum, snare drum and overheads on a kit, to arrive at the 'best fit'.

> Note: Beat Detective works best with rhythmic tracks such as drums, bass, and guitars – so don't expect great results with more ambient material, strings, vocals or such-like. And if you have a rhythm track that is too far out of time, you won't get good results with this either.

Beat Detective basic operation

The Beat Detective window is divided into three sections. The first lets you select the Mode, the next lets you define and capture the Selection, while the third section changes according to the Mode selected. It contains the Detection parameters in the first three modes and swaps these for the Conform parameters and the Smoothing parameters in the last two modes. There are five different modes altogether, as explained below.

Bar|Beat Marker Generation automatically generates Bar|Beat Markers corresponding to transients detected in the audio selection.

Groove Template Extraction extracts rhythmic and dynamic information from the audio and puts this information onto the Groove Clipboard or lets you save it as a DigiGroove template.

Region Separation automatically separates and creates new regions based on transients detected in the audio selection.

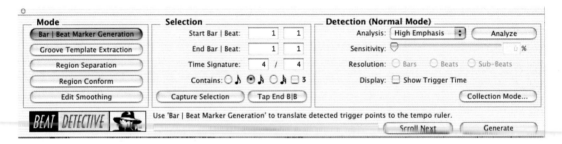

Figure 7.23 Beat Detective window.

Region Conform conforms all separated regions within the selection to the current tempo map. You can preserve some of the original feel of the material with the Strength and Exclude Within option, or impose an amount of swing with the Swing option. Beat Detective can also conform audio regions to groove templates – including its own DigiGroove templates.

After conforming regions, Edit Smoothing can be used to fill the gaps between the regions by automatically trimming them and inserting crossfades if desired.

> Note: Collection Mode is a sub-mode that lets you collect beat triggers for multiple tracks, each with different Detection parameters.

Practical editing techniques

Fixing a note or chord played late

If you click in the Time Ruler at the top of the Edit window, a vertical line will appear which runs through all of the tracks. If you position this at the start of a drum beat, for example, you can easily see if the bass guitar and other instruments have been played 'on the beat', or before the beat, or a little late – as is the case with the Guitar chord in this example.

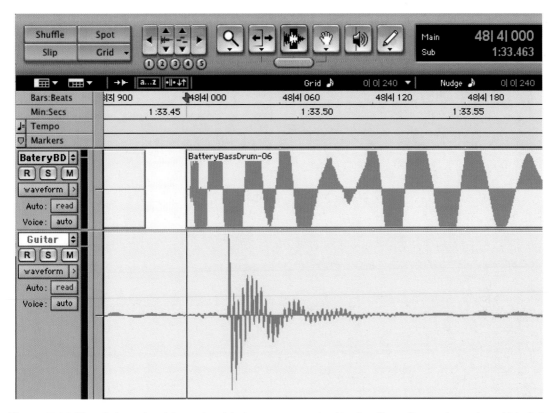

Figure 7.24 The Guitar chord is noticeably late compared with the Bass Drum above it, which falls correctly and exactly on bar 48, beat 4.

To move a guitar chord (that was played late) back into time to coincide with the beat, you first need to select the chord and any decaying sound from this up to where the next chord begins.

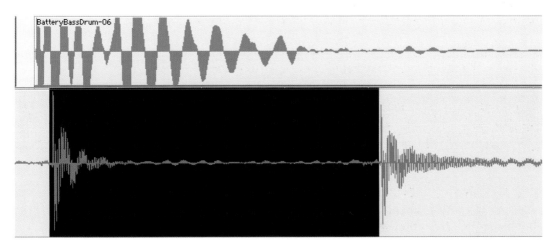

Figure 7.25 Selecting a Guitar chord.

You can then separate this chord from the rest of the audio using the Separate Region command under the Edit menu.

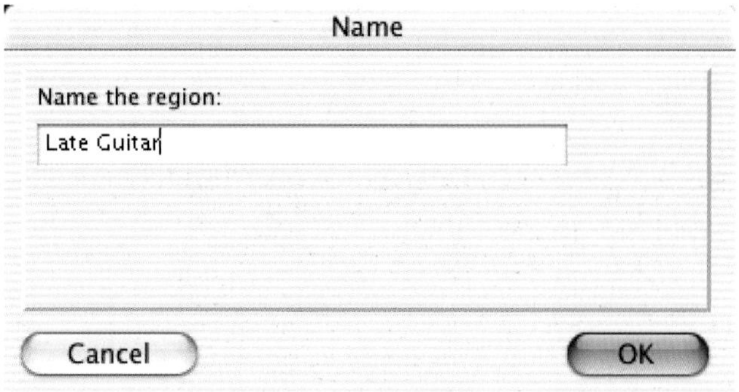

Name

Name the region:

Late Guitar

Cancel   OK

Figure 7.26 Region Name dialog.

Note: You should always try to give your regions as meaningful names as you can – even though you might have to start dreaming up some creative names on an intensive editing session. This helps if you need to look through the regions list for a particular region later.

With the chord now in its own separate region, you can use the Trim tool to make space before this region so you can move the 'late' guitar chord back to where it should have been played – on the beat.

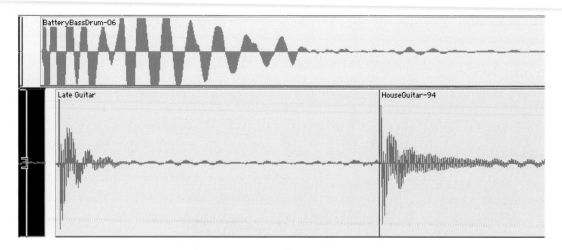

Figure 7.27 Trimming the previous region.

Now you can use the Grabber tool to point at the 'Late Guitar' region, click and hold on this, then drag it backwards to line up with the drums and other instruments.

So should you move every instrument so that they all line up exactly on the beat? This is a question for your own judgement, of course. My view is that with a recording of a band you should simply correct the obvious 'bloopers' where a chord or a note is obviously too early or too late. With good players involved in the recording, you should find that there are not too many

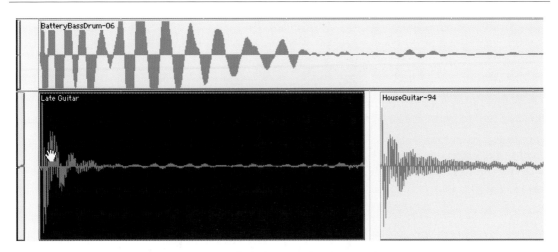

Figure 7.28 Grabbing and moving a region.

edits needed to tighten the recording up very acceptably. If you find that there are mistakes all over the place, I believe that you should spot this at the tracking session and ask the musicians to do more takes until you get an acceptable take – or get better musicians to play the parts. Obviously there are circumstances where this is not possible, and in these cases, you can achieve 'miracles' with edits if you are lucky. But you may end up spending incredible amounts of time moving everything to where you think it should be only to find that it never sounds completely correct – even though you appear to have corrected all the mistakes. And if you are going for a 'natural' sound you can definitely take the 'life' out of a recording by overdoing the edits. Real musicians are not robots or machines, and a little human variation often sounds much better than mechanical precision.

Editing vocals

OK, let's look at editing a vocal track now. You can set the track height to Large or even Jumbo when doing these edits so you can easily see each phrase or word sung. You can quickly turn phrases or words into separate regions – using the Separate Region command – so that you can work with them individually. This makes it really easy to use part of one phrase with part of another.

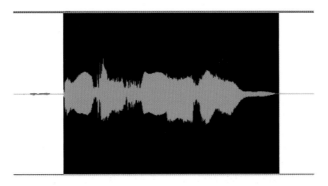

Figure 7.29 Vocal phrase – selection too 'tight'.

Typically, there will be some 'spill' from the singer's headphones, which you will hear in the vocal track in between the vocal phrases. You can easily cut this out in the Edit window.

The trick here is not to edit too close to the vocal phrases or you can end up with unnatural entries. The temptation is to cut out all the 'intake of breath' sounds before the sung phrase.

But often this turns into the first part of the sung phrase so quickly that removing it makes the entry sound somehow unreal – which, of course, it is. The fix? Select some way before the waveform develops to make sure that you have included any important sound components at the beginning of a word or phrase.

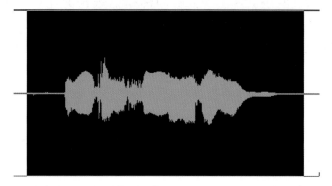

Figure 7.30 Vocal phrase – safe selection.

Similarly, you may fall into the trap of cutting too close to the end of the phrase, so this sounds unnatural. For the greatest accuracy you need to expand the vertical zoom level until you can see any extra quiet sounds near the start or end of the phrase, and it is probably best to solo the vocal and turn the monitoring level up while you do this. This level of editing can be quite time-consuming – especially if there is a lot to do.

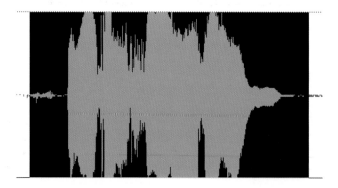

Figure 7.31 Vocal phrase – vertical zoom.

On the other hand, with practice, and with a vocal that is not too convoluted, you can actually edit out vocal phrases very quickly 'on-the-fly', even with all the tracks playing back. While viewing the vocal track in the Edit window at a suitable size and zoom level, you can select a phrase that you have just listened to and decided to cut while the next phrase is playing. Quickly select up to the beginning of the following phrase and hit the backspace key to delete this. You have to be quick at scrolling the display manually, or you have to choose your scroll options carefully for this to be practical – but once you are 'up-to-speed' with a technique like this it can save lots of precious studio time. Don't forget that your edits do not have to be perfect initially, as you can always go back afterwards and fine-tune each edit – at maximum zoom level if necessary.

Tip: Don't forget to use the zoom levels you have set up in your Markers window while selecting edits. And don't be lazy with setting up zoom markers either. As soon as you find yourself regularly using a particular zoom level, then that's the time you should create a zoom marker to let you see this convenient amount of the waveform in the display.

Zero crossing edits

For the tightest edits, and to avoid pops and clicks at the edit points, you should zoom to maximum level, and make sure that you have set your edit point onto a zero waveform crossing.

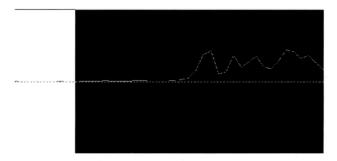

Figure 7.32 Maximum zoom.

The amplitude of the signal is, by definition, at zero as the waveform crosses the zero line on the waveform display – so there can be no sound here. If you cut into a waveform at a point that is not at zero, the value of the signal will jump at the edit and this rapid change in level may be audible as a pop or click. Of course, it may not be audible – depending on many factors. For instance, if you never listen to the track solo, or if the monitoring level is too low, or if the click produced is very quiet, then you are not going to hear it. So – in many practical situations it is simply not going to matter. However, it is certainly good working practice to make clean edits as a general rule.

Tip: To move quickly from the start to the end of your selection you simply hit the left arrow or the right arrow (respectively) on your computer keyboard.

Editing before the downbeat

How do you deal with a phrase that has a couple of notes before the downbeat? This can be fiddly in some software where you have to check out exactly which bar, beat and clock or time location the first pick-up note starts at and then paste the phrase elsewhere – starting this length of time before the downbeat again. With Pro Tools you can simply select the audio from the downbeat to the end of the phrase, either move or copy and paste this to the downbeat of the new location, then use the Trim tool to reveal the pick-up notes before the downbeat.

First you separate the region containing the whole phrase that starts before the downbeat.

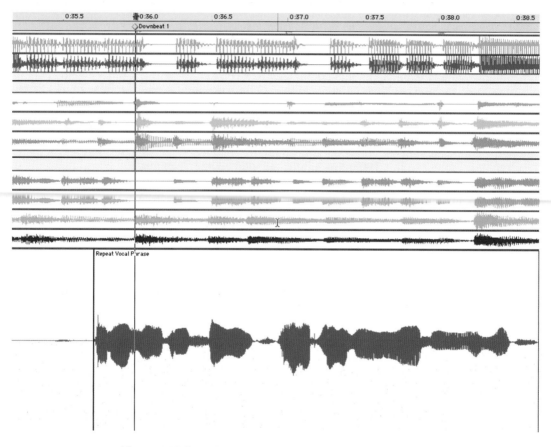

Figure 7.33 Complete phrase starting before the downbeat.

Then you use the Trim tool to trim the start of this region to the downbeat.

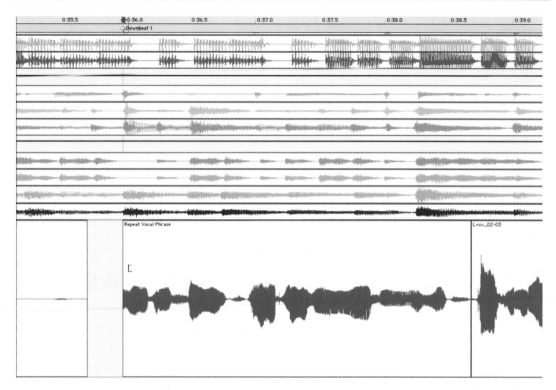

Figure 7.34 Phrase trimmed to start on the downbeat.

Next you might zoom out so you can see the place to which you want to move this region. Then use the Grabber tool to drag the region – while pressing the Option (Alt) key if you want to copy rather than move it.

Having placed the region roughly where you want it, you can zoom in and line it up exactly with the downbeat at which you wish to place it.

> Tip: You can simply select Cut or Copy from the Edit menu instead. Then, if you switch to Grid mode, you can quickly place it using Paste at the downbeat of your new location without fussing around in the Edit window to find the exact edit point.

Finally, you use the Trim tool to reveal the pick-up phrase before the downbeat.

Once you understand this technique, you will realize that you can do this with several tracks at a time by simply grouping them together. Remember that whatever edits you make to one track of a group will automatically be applied to all tracks in the group. This technique is particularly useful for editing drums, for example.

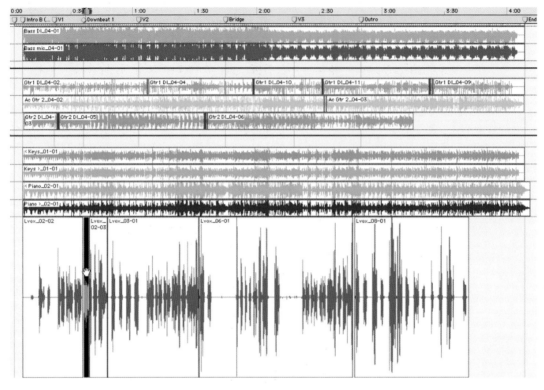

Figure 7.35 Zoom out to view the whole session.

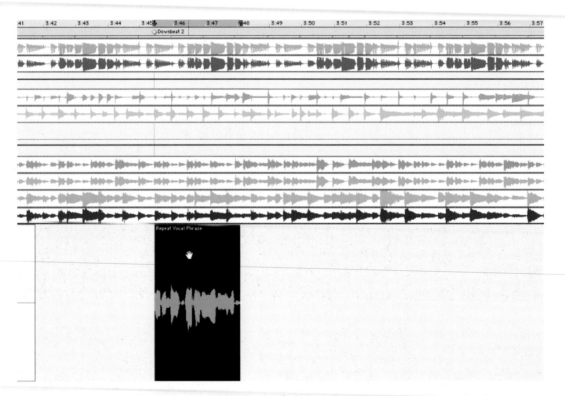

Figure 7.36 Line up moved or copied phrase onto a new downbeat.

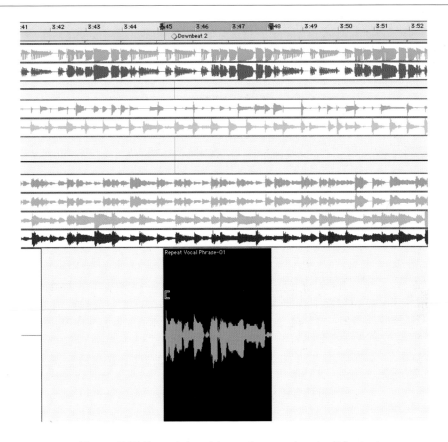

Figure 7.37 Reveal the pick-up phrase using the Trim tool.

Arranging sections

Cutting out, duplicating or re-arranging sections is extremely easy in Pro Tools. A quick way to do this is to click in the Time Ruler using the Selector tool to insert the selection cursor into all the tracks. Then simply click on a Marker placed at the beginning of the section you want to work with and Shift-click on a Marker placed at the end of the section to highlight everything between these two Markers. Now you can use the Cut, Copy and Paste commands to deal with this section as you please.

> Tip: To cut a section and have the gap automatically (and accurately) close up, go into Shuffle mode, do the edit, and the audio after the section you have cut out will instantly butt up against the previous section.

> Note: Don't forget to put Pro Tools back into Slip mode immediately after using Shuffle mode, as it is all too easy to forget that this automatic action will take place – and this could mess up your edits later on during the session.

The 'Repeat Paste to Fill Selection' command in the Edit menu is worth a special mention. With a section already copied into the computer's clipboard, simply select from one location to another in the time ruler across all the tracks and choose this command from the Edit menu. The audio will be pasted to fit, cropping off the end of any regions that would otherwise extend beyond your new selection, and the crossfade window will appear so you can set up to automatically fade each section into the next.

Using playlists

Playlists provide an incredibly powerful way for you to make different 'trial' versions of your edits. An edit playlist in a Pro Tools track is effectively a 'snapshot' of the way the regions are arranged in the track that can be saved, named and recalled at any time. You can create just about as many playlists as you like, and these can be called up from any track – as long as they are not in use by another track. This is a very convenient and flexible way to work that will suit many people.

For example, you may wish to try out various edits on a bass guitar track, cutting out any wrong notes, or moving chords onto the correct beat. The best way to do this would be to duplicate the default edit playlist that contains the unedited region representing the complete bass guitar recording and edit this duplicate.

Figure 7.38 Arrow cursor pointing to playlist pop-up menu selector.

Click on the small arrow to the right of the track name in the Edit window and a pop-up menu will appear.

This pop-up menu contains the list of playlists available for the track along with commands to create a new playlist, duplicate an existing playlist, or delete unused playlists.

By editing a copy of the original playlist you can be sure that you can always return the track to its previous state if anything goes wrong with the edits – or if you simply decide that you like things the way they were previously.

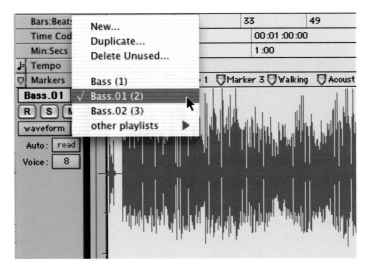

Figure 7.39 Playlist pop-up menu with original bass guitar currently selected for display.

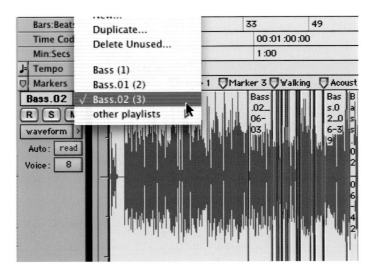

Figure 7.40 Playlist pop-up menu with edited bass guitar currently selected for display.

Each track also has its own set of dedicated automation playlists. With audio tracks, these store data for volume, pan, mute and plug-in parameters. With MIDI tracks, automation playlists simply store mute information. Continuous controller events, program changes and SysEx events are stored in MIDI regions and these regions are already contained within the edit playlists – so there is no need for the automation playlists to also contain this data.

As with many Pro Tools features, you don't have to use playlists. You may prefer the simplicity of having one track containing edited regions which is always visible in the Edit window (unless you have hidden it using the Track Hide feature) and which is always active (unless you have muted it, turned the voice assignment off, or whatever).

Note: Don't forget that Pro Tools 6.1 can keep track of up to 32 of the last undoable operations (Pro Tools 6.0 and lower only support 16 levels), so you can often get back to a previous state this way. There is also a 'Revert To Saved' command under the File menu which lets you return to the last saved version of your session – and you can always save a copy of a session before you start making edits so that you have this available as a fallback as well. So, there are many ways to 'skin this cat'.

Spot mode and time stamping

Spot mode was originally designed for working to picture, where you often need to 'spot' a sound effect or a music cue to a particular SMPTE time code location. The way this works is that you select Spot mode and then click on any region. The Spot dialog comes up and you can either type in the location you want, or capture an incoming Time Code address and spot the region to this, or use the region's time stamp locations for spotting.

There are actually two time stamps that are saved with every region. When you originally record a region it is permanently time-stamped relative to the SMPTE start time specified for the session. Each region can also have a User Time Stamp that can be altered whenever you like using the Time Stamp Selected command in the Regions List pop-up menu. If you have not specifically set a User Time Stamp, the Original Time Stamp location will be set here as the default.

Spot mode is extremely useful in general audio editing within Pro Tools – and particularly if you move a region out of place accidentally. Using Spot Mode you can always return a region to where it was originally recorded, and, as long as you remember to set a User Time Stamp if you re-arrange regions to locations other than where they were first recorded, you can always return regions to these locations.

For example, if you moved a bass guitar region by accident, this would be an ideal way to put it back exactly to where it came from.

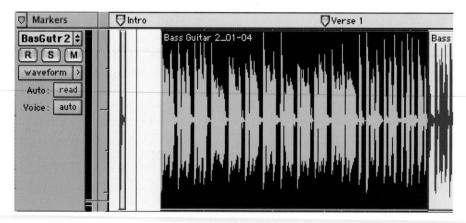

Figure 7.41 Track display with Bass Guitar 2 region accidentally moved from its original position.

Put the software into Spot mode by clicking on the appropriate mode icon at the top left of the Edit window, select the Grabber tool, and click on the region to bring up the Spot dialog.

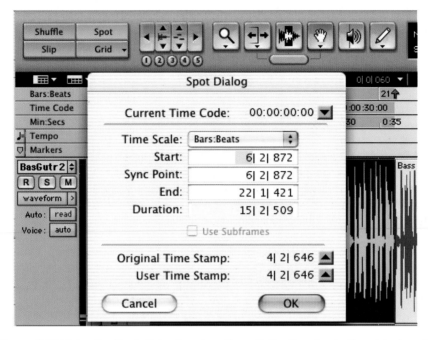

Figure 7.42 Using the Spot dialog to return Bass Guitar 2 region to Its original position.

To quickly enter the original location, simply click on the up arrow next to Original Time Stamp to put this value into the currently selected field – in this case for the Start Time.

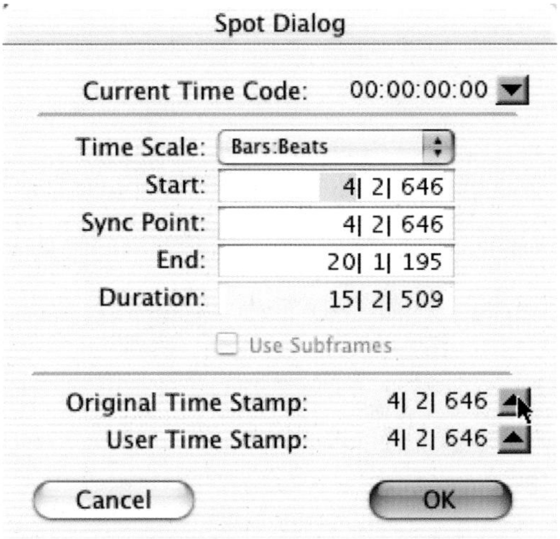

Figure 7.43 Spot dialog detail.

When you 'OK' this dialog, the region will be moved back to the location where it was originally recorded.

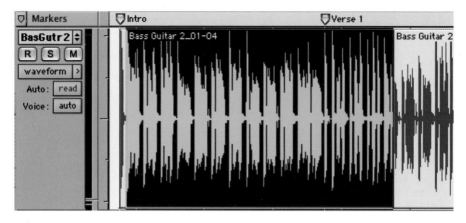

Figure 7.44 Track display with Bass Guitar 2 region back in its original position.

> Tip: You can also spot to the region's Sync Point or End Time – which can be very useful when spotting sound effects to picture.

Summary

The editing capabilities provided by the Pro Tools software let you go far beyond the kind of edits that are possible on tape-based systems. Once you have mastered a selection of the keyboard commands and become used to the various ways of working it is also one of the speediest systems to work with for editing audio. I much prefer editing audio in Pro Tools rather than in any of the MIDI + Audio sequencers, for example. The key to successful edits as far as operating the software is concerned is to use the zoom tools intelligently to carefully make accurate selections – using the Tab to Transients and Auto Fade facilities where appropriate. Get used to trimming up and re-arranging sections at the outset, and if you accidentally move a region out of place and don't know exactly where it should be placed, get used to using the Spot mode to get yourself out of trouble. Get used to the Smart edit tool. Learn as many keyboard commands as you can. Learn how to use techniques like Beat Detective's Edit Smoothing to 'take the sweat out of' laborious editing sessions. Take the time to set up your Markers window properly – and you will be repaid handsomely in time saved later. Also, take time out to tidy up and re-arrange your windows on-screen at different stages throughout your project to make sure everything is as accessible as possible. Once you are up-to-speed with all this stuff you will never want to go back to tape edits – or less capable hard disk-based systems – ever again!

8 Mixing

Introduction

Mixing is the final stage of your multi-track session where you balance all the levels, apply effects and make any last minute edits – or even record any additional material that you feel is needed. Many engineers prefer to build their mixes from the outset of the project – balancing levels, panning and adding effects so that the mix develops continually throughout the process. Others will start afresh after all the recording and basic editing has taken place. Some engineers will particularly value the use of an external control surface during the mixing process as this provides immediate tactile control of fader levels, mutes and so forth. Others will find that using the keyboard commands and mouse control is perfectly acceptable. As with the Recording and Editing chapters, I will introduce various features that you will use during your mixing sessions before presenting practical examples.

Tracks

Audio tracks can be mono, stereo or multi-channel. The outputs from Audio tracks can be routed to Auxiliary Inputs or directly to Master Faders.

Auxiliary Inputs can be used to bring in audio from MIDI devices and other sources or to create sub-mixes of any group of tracks and control these using the Auxiliary Input's fader. They can also be used to apply effects to any group of tracks. You can use plug-in effects on the Auxiliary Input channel or you can send from any of your tracks to an external effects processor and return the signal from the external device via an Auxiliary Input.

Master Faders let you set the master output levels for your output and bus paths. Master Faders have no sends section as they are intended to be the final destinations for signals being mixed or bussed within your session. However, they do have inserts that you can use to process your final output with dither, compression, EQ, reverb or other effects. These inserts are post-fader, which is the best choice when applying processes such as dither and ensures that no signal is left when you fade to zero at the end of your mix. Keep in mind that Master Faders may not only be used for the main mix. You can also create headphone and cue mixes, stems, effects sends, or whatever outputs you need.

You can insert up to five software or hardware inserts on each Audio track or Auxiliary Input to route signals directly to the effect and bring the return signal back into the channel. These inserts are pre-fader so that signals passing through the channel can be routed elsewhere using the sends – with the effects intact and without the channel faders influencing the level. If you insert plug-ins, the audio is routed into and out of the plug-in internally whereas hardware inserts send the signal out through an output channel on your audio interface and back into a corresponding input channel.

> Note: On TDM systems, unlike with LE systems, RTAS plug-ins cannot be inserted onto Auxiliary Inputs or Master Faders – they can only be used with Audio tracks. HTDM plug-ins do not suffer from this restriction.

> Tip: There are some handy short cuts that you can use to quickly assign multiple tracks to inputs or outputs. If you want to assign multiple tracks to the same input, simply hold the Option (Alt) key while you select the input to one track and all the other tracks in your session will be assigned to this. And if you select several tracks first, then hold both Option (Alt) and Shift while you select an input; all the selected tracks will be assigned to this input. Perhaps more usefully you can assign multiple tracks to inputs in ascending order – according to their availability. Hold the Command (Control) and Shift keys down as you assign the input to your first track and all subsequent tracks will automatically be assigned to the next available input path. Similarly, you can hold the Option (Alt) key while you assign any track output to assign all the tracks to the same output or hold both Option (Alt) and Shift to assign all selected tracks to the same output path.

Sends

Up to five sends can be inserted on any Audio track or Auxiliary Input. You can display and edit these from the Mix or Edit windows or in their own Output windows and you can return these sends to your mix using any Auxiliary Input, or Audio track. Pro Tools TDM systems have 64 busses while Pro Tools LE systems have 32 busses. You can send and return signals using any of these busses, or to external equipment via the hardware interface.

Sends can be used for routing signals internally to create sub-mixes or to add processing. You can name the bus paths using the I/O set-up dialog so you can more easily identify which path to use, and these are available as mono, stereo or multi-channel paths according to your system configuration.

> Note: Pro Tools features true multi-channel sends – unlike some other systems, which have to use workarounds.

> Tip: You can always send signals to your interface outputs to provide headphone cue mixes or to use as sends to external effects processors. Don't forget that you will need to return the audio from any external effects using Auxiliary Inputs as these sends do not return audio to the mix themselves – unlike audio sent from inserts.

Output windows for tracks and sends

Pro Tools provides dedicated Output windows for both track outputs and sends. When you insert a send, the Send Output window appears and you can close or open this by clicking on the send in the mixer channel strip.

Figure 8.1 Send Output window.

To open a track Output window you simply click the Output icon which you will find a little way above the channel meter to the right of the fader on each channel strip. Using this track Output window provides an alternative, clearer display for setting the various track output parameters.

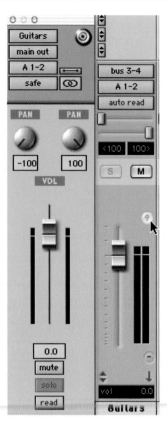

Figure 8.2 Guitars Track Output window with the mixer visible to the right of this showing the cursor about to click on the Output icon to close this Track Output window.

The Track and Send Output windows both have level, pan, mute, solo, and automation controls and the Send Output window has one extra button to let you switch from pre-fader to post-fader operation. Also, the Solo and Automation mode select buttons are separated off in the Send output windows to indicate that they apply to the track – not just to the send.

Tip: Once you have one Output window open, you can use this to adjust the controls for all the sends and track outputs in your session. Click and hold on the track name at the top of the window and a list of all the tracks will appear – see Figure 8.3. Just select each track you want to work with there and the relevant controls will appear in the window. Similarly, if you click and hold on the pop-up selector immediately below the track name selector, this lets you select any of the five sends for viewing in this window – acting as a function selector for the window (see Figure 8.4). So it doesn't matter whether you open a Send Output or a Track Output window – you can view any track or any output or send in this single window by switching using these pop-up selectors.

Figure 8.3 Pop-up track name selector for sends and track outputs.

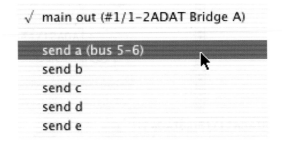

Figure 8.4 Pop-up window function selector for sends and track outputs.

Note: Both windows have a red 'target' button that you can use to keep the window open while you open other output windows. Click on this and it will turn from red to grey – and this window will stay open when you open another window. To keep each new window open you need to click on its target button as well. Having lots of these windows open uses up a lot of screen space, but can be useful at times.

Automation

You can automate virtually every function in Pro Tools. Just choose an automation write mode from the pop-up and play the track. When you move any faders or other controls, your 'moves' will be recorded. It is as simple as that.

When you stop playback, you can view and edit these automation 'moves' graphically in the Edit window. It is very easy to make or edit your automation 'moves' in the Edit window by setting control points or 'breakpoints' manually on the automation line – positioning these to achieve the effects you want. And you can make the changes extremely accurately if you zoom in far enough.

> Note: Moving faders on-screen with a mouse doesn't provide the same tactile feedback that you get with real faders – which is why hardware control surfaces have been developed for Pro Tools. But what level of accuracy can you achieve using a hardware control surface? ProControl's DigiFaders provide 10-bit accuracy, or 1024 steps of resolution. However, Pro Tools interpolates this fader data to provide 24-bit resolution of volume and send automation on playback – so no problems here. On the other hand, most MIDI control surfaces have 8-bit resolution, or 128 steps, although the Mackie HUI has 9-bit resolution, or 512 steps. The Pro Tools software also interpolates this data to a much higher resolution on playback, resulting in fader automation that is smooth enough for most requirements. As well as automating the volume faders, Pro Tools can also automate pan and mute controls for audio tracks and sends, MIDI tracks, and all the parameters in your plug-ins – which is why the Pro Tools automation is one of the most powerful automation systems around.

It is worth spending some time becoming completely familiar with the way Pro Tools handles automation so that you can mix your sessions speedily and confidently. The first thing to understand is the default situation before you have specifically written any automation data. When you first create an Audio track, Auxiliary Input or MIDI track, it defaults to Auto Read mode and puts a single automation breakpoint at the beginning of each automation playlist display. You can move a fader (or any other automation control) and the initial breakpoint will move to this new value and stay there until you move the fader again. If you want to permanently store the initial position of the control you can manually place a second breakpoint after the initial breakpoint at the value you want or you can simply put the track in Auto Write mode and press Play – then press Stop a short time later. If you look at the automation playlist display you will see that this action has inserted a second breakpoint at the time you pressed Stop. Now, if you inadvertently move the control later, it will always return to this value when in Auto Read mode.

The way Pro Tools handles mutes and plug-in bypasses is very neat. Hold the Mute or Bypass button while in any automation writing mode and the Mute or Bypass will be enabled for as long as you hold this down. If you want to clear any of these, just do another automation pass and hold the button again where you want to clear it. To make this even easier, the Mute button, for example, will become highlighted whenever you are passing though a muted section so when you see this you will know where your muted sections are.

The best way to fully understand how the automation works is to write some and look at the automation playlist display to see what breakpoints have been inserted. How do you do this? It's easy. In the Edit window, use the pop-up in the middle of the controls at the left of each track to switch from the waveform display to display the automation type you are interested in – volume, pan or whatever. Here you will see a line with breakpoints where you have written automation. The first thing you will probably want to do is to delete any breakpoints that are obviously in the wrong positions. You can drag the breakpoints to new positions, but if the new position is not close by it is quicker to delete one breakpoint and insert a new one using the Grabber tool. This turns the cursor into a pointing finger when you are in one of these displays. Click on the automation graph line and a new breakpoint will be inserted. To remove a breakpoint you can Option (Alt)-click on it. If you want to remove several it is quicker to change to the Selector tool and drag across the range of breakpoints you want to remove from the graph line to select these – then hit the Delete or Backspace key.

Each breakpoint takes up space in the memory allocated for automation, so thinning data can maximize efficiency and CPU performance. Pro Tools provides two different ways to thin automation data and remove unneeded breakpoints: the Smooth and Thin Data After Pass option and the Thin Automation command. By default, the Smooth and Thin Data After Pass option is selected which usually yields the best performance. You can use the Thin Automation command to selectively thin areas in any track where the automation data is still too dense.

Another way to create automation data is to use the Pencil tool. This lets you create automation events for audio and MIDI tracks by drawing them directly in any automation or MIDI controller playlist. The Pencil tool can be set to draw a series of automation events with the following shapes: Free hand or as a straight Line; or as a Triangle, Square or Random pattern repeating at a rate based on the current Grid value with the amplitude controlled by vertical movement of the Pencil tool. For example, you can use the Triangle pattern to control continuous functions, or the Square pattern to control a switched function such as Mute or Bypass. Since the pencil draws these shapes using the current Grid value, you can use it to perform panning in tempo with a music track.

Let's run through the basic steps to create automation in real-time using the controls in more detail now. First you need to enable the automation type that you want to record – volume, pan, mute, send level, send pan, send mute or plug-in automation. Open the Automation Enable window where you will see buttons for each of these.

Figure 8.5 Automation Enable window.

If you want to automate a plug-in, you will also need to enable the individual plug-in parameters that you want to automate. Open the plug-in window and you will see a button marked Auto.

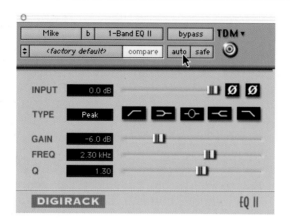

Figure 8.6 A DigiRack plug-in window showing cursor about to click on the Auto button.

Click on this to bring up a dialog box where you can choose from the list of automatable parameters for that plug-in. OK this to make your chosen parameters active.

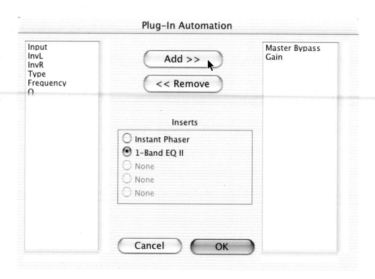

Figure 8.7 Plug-in Automation window.

The next step is to put the appropriate tracks in an automation writing mode – choosing Auto Write, Touch, Latch or a Trim mode. An Automation Safe button is provided in the plug-ins, Track Output and Send Output windows. You can enable this if you want to protect any existing automation data from being overwritten – a wise move to make once you are happy with the way your automation is working.

Figure 8.8 Putting a Track into an Automation mode.

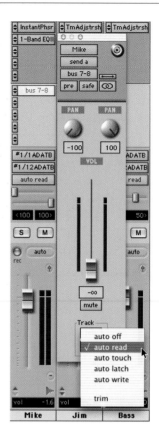

Figure 8.9 Selecting an Automation mode in a send window.

Once you have everything set up correctly, just hit Play and move your controls as you like – then hit Stop. It's as simple as that.

Normally you will choose Auto Write mode for your first pass and use Auto Touch or Auto Latch modes to make further adjustments. Start playing back from wherever you like and just move the control where you want to make your changes. New data will only be written when you actually move the control – the original data will not be altered anywhere else.

Automation playlists

Each Pro Tools track contains a single automation playlist for each automatable control. On audio tracks, these controls include: Volume, Pan, Mute, Send volume, pan and mute, and Plug-in parameters. You can choose which one of these to display in the Edit window by clicking on the Track View Selector.

Here you can choose to display the Waveform or Blocks, Volume, Pan automation or Mute automation, Send automation parameters, or Plug-in automation parameters if you have any sends or plug-ins in use.

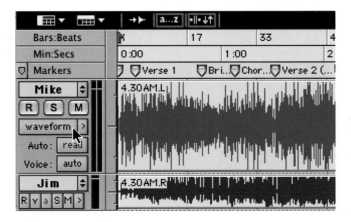

Figure 8.10 Clicking on the Track View Selector.

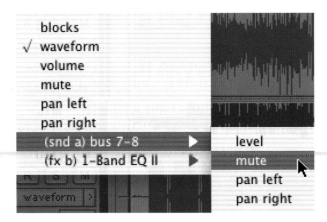

Figure 8.11 Choosing a display mode using the Track View Selector pop-up.

Bear in mind that you won't see any plug-in automation parameters listed here unless you have also enabled these parameters for automation in the plug-in itself. MIDI tracks work slightly differently and only offer automation for Volume, Pan and Mute. The automation data for Audio tracks is held in a separate playlist from the audio data and regions and each edit playlist shares the same automation data. This means you can edit the automation data independently from the audio data if you need to. This automation data is stored in the track automation playlist rather than the audio region playlist. With MIDI tracks, on the other hand, all controller automation data except for Mutes is held in the MIDI region that contains it. This means that each edit playlist on a MIDI track incorporates the controller data – and this can be different for each of these playlists. Mutes are held in the track automation playlist so that you can mute playback of individual MIDI tracks without altering the MIDI controller data.

Automation write modes

The Automation writing modes deserve some further explanation. Auto Write does what it says – it lets you write automation data when you move any control until you stop playback – erasing any previous data up to this point.

Auto Write mode automatically switches to Auto Touch mode when you stop, ready for a second pass to fine-tune your automation pass. In Auto Touch mode, automation is only written when you actually operate any of the controls – and the control will return to its previously automated position when you let go at a rate which you can set using the AutoMatch Time and Touch Timeout settings in the Automation Preferences window.

Auto Latch mode works pretty much the same way, the difference being that the control doesn't return to its previously automated position when you let go of the control – it stays where you left it until you hit Stop. (However, at this point where you hit Stop, the automation value will change instantly to the previous value.) You will find this mode particularly useful for automating pan controls and plug-ins where you will usually move a control to a particular position, let go and leave it there for a while, and then maybe move it to a new position later on. If you complete your automation pass right to the end of the track, the control will stay wherever you last positioned it until the end of the track.

Trim mode

A Trim mode is available for track and send volumes on TDM systems. You can use this mode when you want to keep all your existing 'moves' while moving the overall level up or down – as the fader 'moves' write relative rather than absolute values. When you select Trim mode from the automation selector, this changes the behaviour of the various automation write modes. The track and send volume levels enter Trim mode while the non-trimmable controls enter whichever automation write mode you have selected. These controls behave the same way as in the standard automation modes, apart from when in Trim/Auto Write mode. In this mode, the non-trimmable controls work the same way as in Auto Touch mode – i.e. they don't move back to previously recorded values when you release the controls, until you stop. Also, if they were in standard Auto Write mode it would be all too easy to accidentally overwrite wanted data. The Auto Touch behaviour prevents this.

> Tip: Don't forget to switch out of Auto Write mode when you leave Trim mode or you may still accidentally overwrite your data on these tracks – as the track automation mode is not automatically switched from Auto Write to Auto Touch after an automation pass in Trim mode.

When you enter Auto Write mode while in Trim mode, the faders will automatically position themselves at 0 dB to indicate that no trimming is taking place yet. You can set an initial 'delta' value before you start playback to record your automation trims, then move the fader to trim as required throughout your automation pass. The Auto Touch mode works similarly to the standard Auto Touch mode when using Trim mode, but in relation to the delta value – i.e. when you let go of a fader it will return to the zero delta value until you move it again or stop. The original automation data will be preserved everywhere but where you used the trim control. Again, as with the standard Auto Latch mode, when you are using Trim mode with this the trim delta value stays where you left it when you let go – until playback stops.

So when should you use the different Trim modes? Well, you can use Trim in Auto Read mode to try out your 'moves' before you actually record these if you like. No automation data will be written, but you can quickly switch to an automation write mode after your rehearsal while you still remember what you did. If you know where you want to trim your existing data, it is best to use Auto Write mode as the faders position themselves at 0 dB and stay there until you move them – i.e. the faders do not follow the automation 'moves' previously recorded, although you will hear the effect of this automation while you are playing back your tracks. On the other hand, you might want to be able to see the faders moving so you know where automation exists that you want to trim. In this case, choose Auto Touch or Auto Latch and the faders will follow the existing automation. You will have to chase the faders around as they move, which can be a little tricky at times compared with the Auto Write mode where they stay still until you move them, but you will have the benefit of seeing what is happening with the originally written data.

> Tip: If you want to see how the automation works even more clearly, you can always have the Edit window open and displaying the automation for the tracks you are working with. Here you will see how the automation data you have written changes each time you do a new pass.

> Note: To give you visual feedback to confirm that you are in Trim mode, the volume and send level faders turn yellow and the Automation Mode button is outlined in yellow. This yellow outline flashes when you select any of the automation modes (other than Auto Off) and goes solid while you are moving the fader during playback, to indicate that you are recording trim data.

Snapshot automation

A great way to get started with your mix session is to create a 'snapshot' of your automation data at the start of the session – or even at the beginning of each new section if you already know roughly how things should be. Just put the cursor at the position in the track where you want to write the data, display the automation data you want to edit in the relevant track in the Edit window, and choose 'Write Automation' from the Edit menu. There are two options here, you can simply write data for the automation parameter you are currently working with, or you can write the current settings for all the automation parameters that you have enabled in the Automation Enable window.

If you make a snapshot of your automation values at the start, these values will apply throughout the tracks, as long as you have not inserted any other breakpoints yet, and any newly written values will apply to the rest of the session. This is because when you write your first snapshot, if there is no existing automation data in the track, there will only be one breakpoint – at the beginning of the track. When you write a new value for this, this value will apply throughout the track.

If you have already inserted breakpoints, the automation ramps linearly up or down from the first breakpoint value to the next breakpoint value, and in turn to any subsequent breakpoint values. You can change to new values any time by inserting the cursor at the new location and taking another snapshot using the Write Automation command – thereby inserting new breakpoints at this position.

> Tip: If you make a selection in the Timeline before using the Write Automation command, the automation data will only be applied to this selection – and an extra breakpoint will be placed just before and just after this selection so that data before or after the selection will not be affected. This is useful if you know you want to raise levels just in the choruses, for example.

Mixdown

Once you have your mix sounding the way you want it, you will usually want to record this as a stereo 'master'. You have various options here. The first is to connect your main stereo outputs to a stereo recorder such as a DAT machine, DVD or CD recorder, or even to a 1/2-inch analogue tape recorder. Alternatively you can create a stereo file on disk – either using the Bounce to Disk command in Pro Tools, or by recording the final mix to a pair of tracks in the same Pro Tools session – assuming you have these tracks available.

> Note: You can bounce to disk or record to new tracks anytime during your session if you want to 'print' (i.e. record) effects to disk, so you can free up your DSP to apply more plug-ins, or if you want to create sub-mixes to use as loops or as 'stems' for post-production or whatever.

As the manual explains it, 'Printing effects to disk is the technique of permanently adding real-time effects, such as EQ or reverb, to an audio track by bussing and recording it to new tracks with the effects added. The original audio is preserved, so you can return to the source track at any time. This can be useful when you have a limited number of tracks or effects devices.'

So when should you use Bounce to Disk and when should you record your mix to new tracks in your session? If you have totally automated your mix, or if you have no free tracks available in your session, then Bounce to Disk is the best option. However, some engineers may still prefer to do a manual fade at the end or to tweak faders or other controls on-the-fly during the final mixdown. In this case, recording to new tracks or to an external recorder are the best ways to go.

Later, this session master may be transferred onto a portable medium such as CD, DAT or 8 mm DDP tape to be used as a master for pressing compact discs. Most commercial projects will use a specialist 'mastering' studio to make these transfers.

A note on terminology

When a final mix is recorded in a suitable stereo (or multi-channel) format, this is often referred to as the 'master' mix. Typically, this master mix goes through a further 'mastering' stage – often in what is commonly referred to as a 'mastering' studio under the supervision of a 'mastering' engineer. This stage en route to the pressing plant where CD discs will be replicated for commercial distribution should, perhaps, be more correctly referred to as 'pre-mastering' – as it involves preparing the mixes so they are ready for the 'Mastering' stage at the pressing plant at which the pre-master tapes or discs are used to create the 'Glass Masters' used in the replication process.

Mixing to analogue tape

There are still engineers and producers who prefer to mix to 1/2-inch or even 1/4-inch rather than to any digital format – even when the project has been recorded digitally. Nevertheless, if you do mix to analogue it is wise to simultaneously create a digital master so you can make listening copies without generation loss. Of course, you won't hear the audio 'signature' of the analogue master recording unless you make copies from this – or wait until this analogue master has been transferred onto CD. If you do have access to a high-quality 1/2-inch recorder and if you are using PrismSound, Apogee or similar high-quality converters working at 24-bit resolution, then you are very likely to get a better result than if you master to DAT at 16-bit. But this assumes that when you digitize the 1/2-inch tape during the Pre-Mastering stage prior to Glass Mastering for CD replication you are using the very best converters available. If not, you may lose quality at this stage.

Mixing to DAT

You should be aware that DAT and similar recorders use error-correction schemes to correct any errors that may occur during digital transfer. After all, DAT uses magnetic tape and tape is still subject to drop-outs even if the data is digital. The error correction schemes are good – but cannot always ensure that the data remains entirely faithful to the original data with successive reproductions. So a small amount of generation loss may occur.

Mixing to digital media

Audio files on hard disk or on removable formats such as optical disc or CD-R are not subjected to these error-correction schemes. These files can become corrupted on a hard disk – but you can always check these files to make sure they are not corrupted before relying on them. Once transferred to CD-R, the CD-R can be verified bit-for-bit to make sure that no corruption of the data has taken place during this process. For this reason, using random access digital media for your final master mixes makes a lot of sense, and you can avoid any possibility of losses by making backups onto optical disc, CD-R, DVD-RAM or backup tape drive – ideally making two copies to be kept in different locations. You can then transfer copies at any time onto sequential digital media such as DAT tapes if these are required.

> Note: Bear in mind that Pro Tools|HD and Pro Tools MIX systems are full 24-bit audio recording and mixing environments, supporting record, playback, mixing, and processing of 24-bit audio files. So you can always record to and from other 24-bit recording systems without any bit depth conversion.

Making your own CDs

The ability to make your own CDs is an essential part of any music recording system as the CD is the major delivery medium of our time. If you are using your Pro Tools system commercially or for any other purpose, you should have a CD burner to hand at all times for the sheer convenience of being able to run off a listening copy of the music you have recorded which can be played in the car, on a Walkman CD, in the home – virtually anywhere. Alternatively, you can use Roxio's Toast software to burn listening CDs directly from audio files on disk.

If you intend to do your own 'pre-mastering' on a Mac-based system, Roxio's Jam software is available for OSX. Jam lets you create CD-R discs complete with all the P-Q sub-codes – ready to send to be pressed. But don't be seduced into thinking that you can always save money and get a proper result by doing this part of the job yourself. Unless you have a considerable amount of experience in CD mastering and a great monitoring system to work with, then this job is often much better to be entrusted to a professional mastering engineer who does have the requisite experience and equipment.

Audio compression

Compression is increasingly being used on final masters to allow levels on CD to be increased. Outboard equipment such as the TC Finalizer is specifically designed for this purpose, and several multi-band compression plug-ins from TC, Waves and others are available for Pro Tools TDM systems offering similar facilities. These are very useful tools to use if you know what you are doing and if you are preparing versions of your mixes for broadcast or for other purposes. However, mastering studios prefer that you send final mixes without too much compression applied at the final mix stage. This is because the mastering engineers have much greater experience in this area and use specialized equipment and high-quality monitoring systems which help them to achieve much better results than you will normally be able to achieve yourself. And if you have already applied heavy compression, there is 'nowhere left' for the mastering engineer to go with this.

> Note: If you do use compression on your final mix you will have to keep a very close watch on the overall output level, using the meter on the Master Fader, to make sure you avoid clipping.

Bounce to Disk

You can 'bounce' any Pro Tools session to create a new file (or files) on disk using the Bounce to Disk command from the File Menu.

Basically, you mute everything but the tracks you want to bounce, make sure that all the levels, pans and any effects and automation are the way you want them to be, assign the outputs from all the tracks to the same pair of outputs, then select the Bounce command from the File menu.

Here you can choose the file type as Sound Designer II (SDII), Broadcast WAVE (BWF), AIFF, MPEG-1 Layer 3 (MP3), QuickTime or Sound Resource. File format options include summed mono, multiple mono and interleaved stereo. If you intend to import the files back into Pro Tools you should choose mono or multiple mono files. Interleaved stereo files can be used by other software such as BIAS Peak which you may use to carry out any final edits or processing, or by CD-burning software such as Toast or Jam. Resolutions available include 8-bit for multimedia work, 16-bit for CD distribution and 24-bit for high-quality digital audio systems. All the standard sample rates from 44.1 up to 192 kHz are available with HD systems, including all the pull-up and pull-down rates used for film work.

You can choose whether to convert during or after the bounce. I often choose to convert after the bounce even though this takes longer, as it leaves the processor free to concentrate on one task at a time – bouncing then converting. You can select the option to Import After Bounce if you want to use the new tracks in your session after the bounce – or leave this unchecked if you simply want to create master mixes on disk that you will assemble and check later.

When you hit Bounce, you will be presented with a Save dialog box where you can name your new file(s) and choose where to save these.

> Note: Don't forget to choose your bounce source to match the output assignments on your session. The source selection defaults to outputs 1/2 and you may be using a different output pair in your session. If you do forget, there will be no audio in your bounced file.

The Bounce to Disk command uses all the available voices from your session and all audible tracks will be included in the bounce – including tracks that 'pop through' when other tracks are not using their voices. Muted tracks do not appear in the bounce. All the read-enabled automation is applied along with all plug-ins that are in use and any processing that is being applied via hardware inserts.

> Tip: If you solo any tracks or regions before applying the Bounce to Disk command, only these will appear in the bounce.

Unlike when recording your mix to a Pro Tools track or tracks, the Bounce to Disk method doesn't require the use of these additional tracks – or their voices. Of course, if you import the bounced tracks back into your session afterwards you will need to have tracks/voices available to play these back with.

The Pro Tools software time-stamps the new file (or files) to start at the same point you began your bounce from so you can easily place it at the same location as the original material if you import it back into the session. Although a processing delay is involved relative to the original tracks when you bounce to disk, the DAE compensates for any bus delays due to the bounce so that if you import these files back into the session they can be placed exactly in time with your original tracks with 100 per cent phase accuracy.

Bear in mind that the bounced mix will be exactly the length of any selection you have made in the Timeline or Edit window – which should be linked. If you want to include any reverb trails or other effects which 'hang over' at the end of the track you will need to select additional time to accommodate this. If you don't make any selection, the bounce will be the length of the longest audible track in your session.

> Note: One possible disadvantage of the Bounce to Disk method is that although you will hear the session playing back in real-time during the bounce, you won't be able to make any adjustments to the mix 'on-the-fly'. If you need to do this, you should record to new tracks instead.

Recording to tracks

Recording your mix to new audio tracks is just the same as recording any other input signals into Pro Tools. Obviously, you need to have sufficient free tracks, free voices and bus paths available. The beauty of this technique (compared with Bounce to Disk) is that you can add live input to your mix or adjust volume, pan, mute and other controls during the recording process.

Once you have your mix set up with the levels, pans, plug-in processing and routing all sorted out, you are ready to record your new tracks containing your mix, sub-mix or stem. Simply record-enable the new tracks, click Record in the Transport window, then click Play in the Transport window to begin recording – which will begin from the location of the playback cursor. Recording will continue until you press Stop or punch out of recording – unless you have selected a particular section.

It can help save a little time later if you select exactly the length of audio that you want to record to these new files. Even if you place the playback cursor exactly at the start of the audio you are unlikely to be able to stop recording at exactly the right moment. To make a selection, link the Edit and Timeline selection using the command in the Operations menu and drag the selection cursor over the length you want to encompass.

Don't forget to select some extra time at the end of your selection to accommodate reverb tails, delays or any other effects that may still be sounding after the audio has finished as the new recording will stop automatically immediately at the end of your selection.

And don't be too keen to select too close to the start point either. It is always better to include a bit more than you think you will need in your selection – you may need to edit some 'room tone' that you find there to fill a gap somewhere else, or to analyse some noise that is there so that you are better able to extract this noise from the rest of the session.

When to use the Dither plug-in

Whenever you change the bit depth of digital recordings you need to apply dither to reduce quantization error that can become audible, particularly when fading low-level signals. Dither does this 'trick' by actually introducing very low-level random noise that, counter-intuitively, increases the apparent signal-to-noise ratio. Typically, you will apply dither if you are bouncing a mix to a file on disk that uses a lower bit depth than your session. Two plug-ins are supplied for this purpose with TDM systems – POW-r Dither and Dither.

The POW-r Dither plug-in offers 16-bit and 20-bit resolutions along with three choices for Noise Shaping – Types 1, 2 and 3.

The Dither plug-in has 16-, 18- and 20-bit options to cater for all possible scenarios but only features one standard Noise Shaping type.

> Note: Noise Shaping shifts the noise introduced by the dithering process to frequencies around 4 kHz where this low-level noise is likely to be masked by the audio program material.

If you are mastering from a 24-bit session to a 24-bit digital recorder or to analogue tape via 24-bit D/A converters, there is no need to apply dither. The 20-bit option is provided for compatibility with some digital devices that use this format. On the other hand, if you are mastering to a 16-bit medium – whether this be a file on disk or an external recorder – you should apply dither. You may be seduced into thinking that if the original session is 16-bit you don't need to dither to another 16-bit medium – but you would be wrong. Although 16-bit sessions save their data to 16-bit files, they are actually processed internally while the session is running at higher bit rates – 24-bit for Pro Tool TDM systems and 32-bit floating for Pro Tools LE systems. So it doesn't matter whether you are running a 16-bit or a 24-bit session – you should still dither when mastering to 16-bits.

> Tip: Dither is not automatically applied when you use the Bounce to Disk command – so you need to insert and apply the Dither plug-in on your Master Fader before your bounce if you want to create a dithered file. Bear in mind that if you do not apply dither and you choose to convert to a lower resolution, say from 24-bit to 16-bit, during or after a Bounce to Disk, the resultant file will be converted by truncation – i.e. the low-order bits will simply be 'thrown away' and quantization noise may become audible.

> Note: If you are mastering a 24-bit session to a 24-bit file or recording device, there is no reason to apply dither.

Mixing precision

The Pro Tools mixer is designed to allow the faders to be lowered in level without any loss of resolution. Normally, in a 24-bit system, as you lowered the fader you would be using less than the available 24 bits of resolution. However, the Pro Tools mixer uses registers inside the DSPs on the audio cards to temporarily use 48 bits of precision when mixing signals together. So, even if you lower a fader almost to the bottom of its travel, you still have 24 bits of resolution available. In other words, there is lots of headroom available in the Pro Tools system – which is what allows the faders to be placed in the 'sweet spot' position without clipping, as with a high-quality professional analogue mixing console such as an SSL.

What this means is that as the input signals are summed together onto the mix bus in Pro Tools, these signals can never clip – even with all the channel faders set to the +6 dB maximum gain – because there are more than enough bits and headroom available internally to avoid this.

The situation at the output side of the summing mixer, where audio is sent via a digital output or onto the TDM bus, is different – clipping can occur here.

> Tip: This is why you should always be using a Master Fader to scale the output level of any mix summing point onto a bus or output.

By observing the Master Fader's meters, you can see if any clipping is taking place (the red clip indicators will light up), in which case you simply lower the fader and check your mix again until no clipping takes place. Always check, though, that no clipping is taking place within any plug-ins used on the Master Fader inserts.

> Note: There will be no loss of quality as a result of lowering the Master Fader to adjust the gain at the output stage, because there are 48 bits available within the system. This means that there is no need to trim the individual input faders back to avoid clipping – you can just lower the Master Fader. Engineers used to working with less-capable equipment may find this a little unusual to get used to, but will quickly appreciate the convenience of not having to trim lots of individual faders.

By default, Pro Tools|HD systems use Surround and Stereo Mixer plug-ins that provide 48 dB of headroom.

> Note: By way of comparison, the 24-bit Optimized and the Surround Mixer plug-ins provided with Pro Tools MIX and Pro Tools|24 systems only have 30 dB of headroom, and the 16-bit Optimized Mixer for these systems only has 18 dB of mix headroom but does allow more mixing channels to be used. Note that, unlike with HD-series systems, it is possible to clip the input summing stage of a MIX series mixer plug-in, particularly with the 16-bit Optimized Mixer, although this is less likely with the 24-bit Optimized Mixer.

The Dithered Mixer plug-ins

As an alternative to the Surround and Stereo Mixer plug-ins, the Surround Dithered and Stereo Dithered Mixer plug-ins can be used instead. These add non-correlated dither to any output or bus send.

Applying dither in this way at every output summing point in the mixer avoids any possibility of audible artifacts caused by the truncation of extremely low-level data that occurs when signals pass from the 48-bit 'world' of the TDM mixer to the 24-bit 'world' of a TDM bus connection or a hardware output. Any material that is truncated will lie below these 24 bits, i.e. below −144 dBFs.

To use these alternative Mixer plug-ins, move the Surround and Stereo Mixer plug-ins from the Pro Tools Plug-ins folder to the Plug-ins (Unused) folder and move the Surround Dithered and Stereo Dithered Mixer plug-ins that you will find there into the Plug-ins folder. With OSX systems, these Plug-ins folders can be found in the Library:Application Support:Digidesign folder on your boot drive.

> Note: Adding this uncorrelated dither uses up more DSP power, which reduces by about 15 per cent the number of mix channels that you can use. This is why the non-dithered plug-ins are used as the default.

Mixer automation

Pro Tools provides 24-bit interpolated values between mix breakpoints to provide resolution that is very close to that of analogue controls. So Pro Tools lets you make extremely smooth volume changes, for example. And, unlike some digital systems where the faders suffer from 'zipper'-noise, the DAE 'de-zippers' any 'live input' to the mixer so that fast, real-time fader changes that come in when you move a fader (whether using the mouse or using an external control surface) don't cause audible artifacts as the mixer tries to follow these.

Setting up a mix session

Everybody has their own favourite way of setting up a mix session in Pro Tools. There is no single correct way to do this. Nevertheless, it can help to know how others go about approaching a mix, so let's look at some examples here.

Choosing views

I often set the Mix window to Narrow view so I can see all the tracks on one screen. And I usually add explanatory comments for several of the tracks in the Comments view at the bottom of each channel strip.

Setting up groups

One of the first things I recommend at the start of any mix session is that you set up Groups of faders to let you control the overall levels of any group of instruments such as drums, brass, strings or stereo instruments.

Pro Tools lets you create up to 26 groups and these can be displayed at the bottom left of the Mix window by clicking on the small arrows at the bottom left of the Mix window. Just Shift-click on two or more Mix channels so that the channel names are highlighted in white with blue lettering and hit Command-G on your keyboard. This brings up a window that lets you define a new group based on your channel selection. You can name the group and choose whether it applies to the Edit or Mix window – or both.

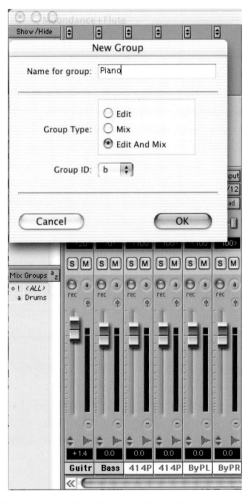

Figure 8.12 Creating a New Group of four piano microphone tracks.

Note: Pro Tools allows you to create groups that apply to both the Mix and Edit windows. However, in some cases you may prefer not to link the Mix and Edit groups. For example, when you are using the Mix window for mixing, you may prefer to work with large, nested groups. However, in the Edit window, you may want to perform editing tasks within a smaller group. If this is the case, you can disable the 'Link Mix and Edit Group Enables' preference so that you can work with different groups in each of the two windows. This is located among the Operation Preferences and lets you link group enabling between the Mix and Edit windows.

There are 26 different locations to store groups – each of which can be identified by a letter of the alphabet. You can then turn these groups on or off simply by pressing the appropriate letter on your computer keyboard. This grouping feature is extremely flexible: you can have groups within groups and you can solo as a group or mute as a group.

When the Group List Key Focus feature is enabled, all you need to do to automatically enable or disable the corresponding group is to type the Group ID letter (any of the lowercase keys, a–z) on your computer's keyboard.

In the Mix window, the Groups List Key Focus is always enabled. But in the Edit window this is disabled by default, so you need to enable the Groups List Key Focus if you want to use it.

Note: There are two ways you can enable the Edit Groups List Key Focus – either using a keyboard command or using the mouse. The keyboard command is Command-Option-4 if you are using a Mac and Control-Alt-4 if you are using Windows. Using the mouse, you just click on the small 'a–z' button in the upper right corner of the Edit Groups List – see Figure 8.13.

Figure 8.13 Edit Groups List just after clicking on the 'a–z' button to enable the Key Focus.

Tip: If you need to adjust one fader within a Group, simply hold the Control key (Mac) or Start key (Windows) on your computer keyboard and you can tweak this on its own. When you let go of the key, you have the Group back in operation again.

Using Automation 'snapshots'

I also like to prepare for a mix session by making sure that everything starts up correctly and finishes cleanly – even if I plan on fading at the end. This way, I don't have to worry about forgetting to mute some unwanted sound that pops up at the start or the end.

Before the start of the music, you might set the faders on some or all of the tracks to zero, then bring them up to the first level you want to use at the start of the music – creating a fade-in. You can set this up very quickly using Automation 'snapshots' in Pro Tools.

Tip: If you want to apply an edit (such as inserting a snapshot or a breakpoint) to all the tracks in your session, you should make sure that the *<All>* Group is selected in the Edit Groups list. Conversely, if you just want to apply the edit to a particular track or selected tracks, make sure that the *<All>* Group is de-selected. Similarly, you can choose which other Edit groups to leave active and which to de-activate.

Because you will be using the Write Automation command, it makes no difference whether the Automation is in Read mode or any of the Write modes – just check that the automation mode in each track is not set to Auto Off. But you do need to make sure that the parameter you want to automate is selected in the Automation Enable window – or no automation data will be written. To automate volume, for example, you would select, say, Auto Read mode for all the tracks and arm the volume automation in the Automation Enable window. It is also a good idea to display the volume (or whichever) automation for the tracks you want to work with, or for all the tracks, in the Edit window so that you can see what is going on as you take your snapshots.

Type the location at which you wish to write your first snapshot into the counter, e.g. Bar 1/Beat1/Clock 000, and hit Return to move to this location. Move all the faders in the Mixer to zero or to whichever value you want to set. To take the 'snapshot', just choose the Write Automation command from the Edit menu and select one of the two options provided. The first will write Automation data for the current parameter, in this case Volume. The second option lets you write automation data for all enabled parameters.

Tip: If you learn the keyboard commands, this makes it very easy to write these 'snapshots'. On the Mac, for example, these are 'Command-forward slash' to 'Write Automation to Current Parameter' and 'Command-Option-forward slash' for 'Write Automation to All Enabled Parameters'.

Once you have written your first snapshot of the automation data at the start of the session, just insert the selection cursor at the next position that you want to set the volume – e.g. at the end of the section you want to fade in – and repeat the process to write your second automation snapshot. The automation curve will ramp up to this second breakpoint value to create the fade-in – and you will see this fade-in represented visually on the automation graph in the Edit window.

Using this 'snapshot' automation technique, you can go through the whole session putting in snapshots of the new values for your automation data at the start of each new section – a great way to get started with your mixing session.

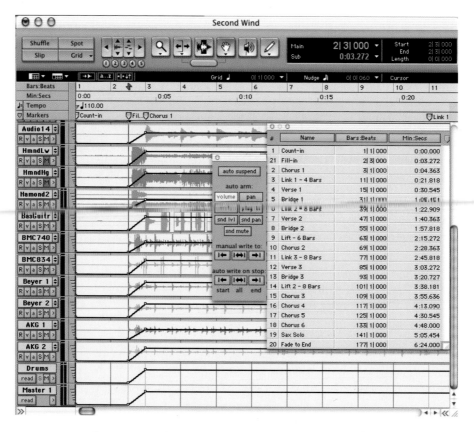

Figure 8.14 Edit window showing automation fade-ins.

Now, having set up the basic levels for each section throughout the session, suppose you want to set the volume of a particular group of instruments to a certain level during part of the session. In this case you should de-activate the <All> edit group if it is activated, and activate (or set up) the Group you are interested in. Choose the Selector tool in the Edit window and select the region you want in the tracks belonging to the group by dragging the cursor in the Edit window to make your selection – or by clicking on a Marker then Shift-clicking on another Marker to select the region between these.

The next step is to set all the faders and other controls to the values you want. If you know the values, you can simply set the controls to these values with playback stopped. If you are not sure what values to set, you may find it helpful to audition your selection.

You can always activate Loop Playback from the Operations menu to loop the selection while you listen and adjust the controls in the Mix window and for any plug-ins.

If you have already set up automation data on these tracks, this will 'fight' you as you play back the track when you try to adjust any automated controls. In this case, you can use the Auto Suspend button in the Automation window to stop this happening. When you have everything set the way you want it to be for this section, use the Write Automation command to take your 'snapshot' of these settings – and don't forget to de-select the Auto Suspend button.

Figure 8.15 Automation Enable window.

Editing and inserting breakpoints manually to create automation

You can always manually insert or edit the automation data in the automation display by inserting breakpoints or dragging these to new values – to create a fade curve, for example. In this case, you simply use the Grabber tool to drag the volume automation breakpoints in each track to the levels you want these to be at. For example, you might want to insert an automation breakpoint at the point where you want the fade-in to finish first – then insert a breakpoint at the start of the session and drag this down to the zero level. (It can be more awkward to do this the other way around, because when you drag the level to zero at the start, the curve lies along the bottom of the display.)

To insert a breakpoint accurately, you will need to zoom in until the display allows you to move the Grabber tool by a suitably small increment, such as one tick or one frame. Then you can watch the Cursor position display above the rulers as you move the Grabber to the right and left along the timeline until it is at the spot where you want to insert the breakpoint.

Tip: If you become adept at editing these breakpoints with the mouse, you may find that you never need to use any of the other automation facilities or automatic fades. Once you know what you are doing with the automation curves you can actually insert breakpoints to control these curves more quickly and more accurately than by recording automation moves in real time – where you often have to trim, edit or re-write the automation several times until you get it the way you want it.

Writing mutes

Perhaps more often, you will want to start the audio exactly at a particular point instead of having a fade-in. In this case you can use Mutes. Make sure that you have armed the Mute automation in the Automation Enable window and display the Mute automation in your tracks. Activate the <All> Edit Group to affect all the tracks, or just activate the Edit Group or groups you are interested in. Of course, if you want to work with individual tracks or selected combinations of tracks, then you may not need to activate any of the Edit Groups.

If you have set up your Memory Locations window to include Markers for each section of your session, you can move the insertion cursor immediately to the first location where you want to hear audio playing by simply clicking on this Marker in the Memory Locations window. You can then use the Memory Locations window to quickly and accurately select between the first Marker at which you want to un-mute and the next Marker at which you want to mute the tracks. This would typically be a Marker right at the end of the session that marks the point where the audio should have finished sounding.

Tip: If you have not set up your Markers, I strongly recommend that you do this right now – and put a marker at the end where the audio should be silent as well as at the beginning where you should first hear sound.

Otherwise, you can type locations into any of the counter windows (the Big Counter, the Transport window or the Edit window counter displays). This is inevitably a slower process, especially with longer sessions, as it is not always obvious where one section starts and another finishes – so you have to spend time checking this out.

When you are ready with your Pro Tools session located at the point where you want to hear the first audio play, click on the Mute button or buttons in the Edit or Mix windows, as necessary, to make sure that the tracks are all un-muted so that you will hear the audio playback from this point. Then use the Write Automation command to write your 'snapshot'.

Now hit the 'Return' key on your computer keyboard to move the playback position to the start of the session so that you can mute the tracks from this point up until where you want to hear them. Go ahead and click on the Mute button or buttons in the Edit or Mix windows again to make sure that these are enabled (i.e. highlighted in colour to indicate that they are muting the audio), and use the Write Automation command to write another 'snapshot'.

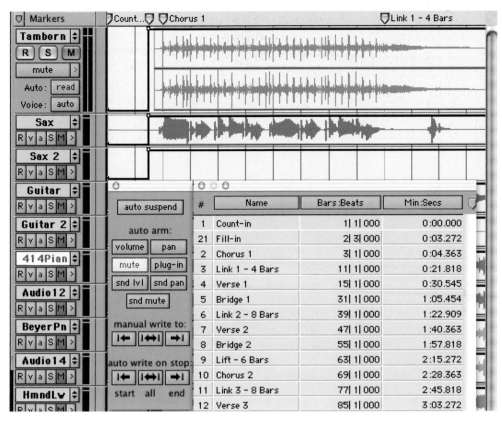

Figure 8.16 Edit window showing Mute Automation enabled between the Count-in Marker and the Fill-in Marker where the music actually starts.

Using this technique, you can go through all your tracks and all the sections in your session at this stage muting out any sections that you don't want to include in your mix.

> Note: If you prefer, you can manually drag the Mute breakpoints to set these the way you want them in the Mute Automation display in the Edit window using the Grabber tool. In this case, you will have to expand the waveform to set the insertion points with accuracy.

Alternatives

As is often the case with Pro Tools, there are several methods you can use to achieve the same end result. To create a fade-in at the start, for example, you could simply select the audio in the Waveform display and use the Fades command to write a fade file to disk. In the Fades dialog, you can select from a range of preset fade curves. With this method, no volume automation is written. Instead an outline of the fade curve is displayed in the Waveform display and the fade is applied during playback from the fade file that is created in the Fade Files folder in your session folder on disk. See Figure 8.17.

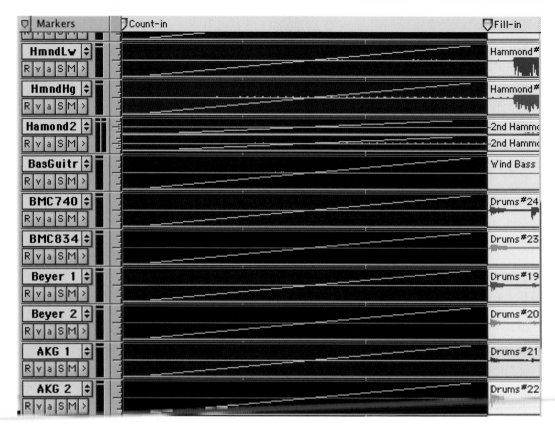

Figure 8.17 Edit window showing Fades.

Also, rather than using Mutes to prevent any audio playback before the actual start of your music, you can always use the Trim tool in the Edit window to make sure that the regions start exactly where you want them to start. I often use this technique during the editing session after the initial recording has been made. See Figure 8.18.

All the above methods can be used during the editing or mixing phases of your project. Of course, you can always use the Mix Automation features in real time during your mix session to create fades and mutes or other mix moves – as you would with a conventional automated mixing console. Post-production editors and people with experience as MIDI programmers will probably be more comfortable with the non-real-time methods outlined above, while recording engineers and producers used to working with conventional mixing consoles may prefer the real-time methods and will probably prefer to use a hardware control surface rather than the mouse.

Setting up Auxiliary routings

You will often want to set up an Auxiliary channel on a mix session so that you can route several audio tracks to this Auxiliary then apply one or more signal processing effects. Using an Auxiliary in this way is more efficient and convenient than inserting plug-ins on each track – it uses much less of your available DSP resources, for example, which is always a good thing.

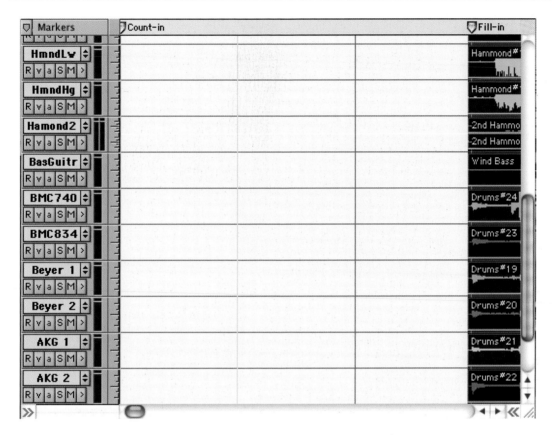

Figure 8.18 Edit window showing trimmed regions.

For example, I often apply a little compression to grouped drum tracks, typically using the Waves Renaissance compressor, which has a brilliantly simple user interface and always 'does what it says on the tin' without introducing unwanted distortions into the sound.

The best way to do this is to add a stereo Auxiliary track and insert the compressor plug-in onto this, then route the outputs of all the drum tracks onto a bus pair that you use as the input to the Auxiliary track. See Figure 8.19.

Tip: Why not label your busses using the I/O Setup window? If you know you will always be using a particular bus pair as inputs to an Aux channel with a Waves compressor inserted across it, you can label this 'WavesCompressor' or some other suitably descriptive name. This will make it easy to select the correct bus when you are routing tracks to this compressor.

Another way to route audio from the drum tracks to the Aux track is to use the individual channel sends. If you send the audio pre-fader, you can leave the drum track outputs set to your main output pair and simply zero the faders or mute the track outputs to remove the uncompressed sound from your mix.

Figure 8.19 Mix window showing six drum tracks routed via bus 3–4 pair to the 'Drums' Auxiliary track, which has a Waves Renaissance compressor and a Time Adjuster plug-in inserted.

If you are using the channel sends, you should get familiar with the Send window. You can use this to set the levels and pans for each track being sent to the Aux channel. Take a look at the screenshot in Figure 8.20. There is a Send window open for each of the drum tracks. Notice how the levels and pans are set up to mirror the way they are set up for playback via the main outputs. You can set up whatever balance and pans work best for you, and you can even automate these controls, as with just about everything else in Pro Tools. Notice also that they are set up as pre-fade sends, and the channels are muted, so no direct sound is fed to the mix.

You can switch the Sends display to show fader and pan sliders, mute and pre-fade buttons for any one of your five sends instead of displaying the five send assignments. This is useful if you need to make any quick adjustments during your mix session. There are two ways to do this. You can choose Sends View Shows from the Display menu and select the Send you want to display, or you can Command-click (Mac) or Control-click (Windows) the Send Selector on the Mixer channel to bring up a pop-up menu that lets you make your selection there instead. If you choose one of the five sends, A to E, the sends section will display controls for just this one send on each mixer channel. This lets you access the controls you may need to use occasionally – without the inconvenience of you having to open the Sends window for each channel. See Figure 8.21.

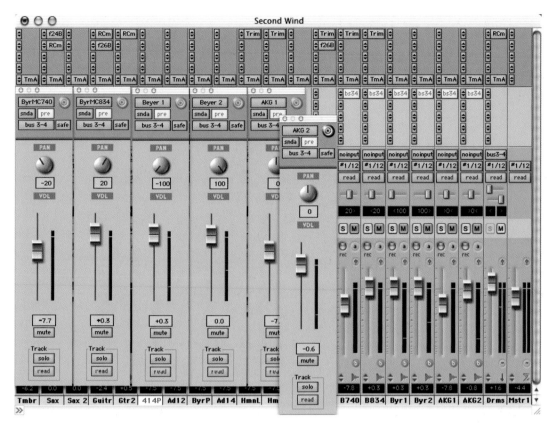

Figure 8.20 Send windows for each of six drum tracks routing audio via bus 3–4 to an Auxiliary track.

Tip: Often you will want to solo tracks when setting up your Auxiliary channels with effects to process these tracks. It can be inconvenient to have to remember to solo the Auxiliary track as well as the individual track or tracks that are being bussed to that Aux channel. A 'solo safe' feature lets you 'warn' the Aux channel to switch into solo as soon as any of the tracks being bussed to this are switched into solo. To enable this solo safe mode, simply Command-click on the Solo button on the Auxiliary channel and this will turn a darker shade of grey. Now when you solo any track being bussed to this, you will hear the output of the Aux channel as well – without having to click on its solo button. I usually leave my Aux channels in this mode throughout my mixing sessions.

Note: You will notice that I have inserted a TimeAdjuster plug-in on every track in my mixing session. This is to make sure that there are no uncompensated delay offsets remaining between tracks after inserting plug-ins on some tracks and not on others. The Waves compressor introduces a delay of 68 samples, for example, so all the other tracks must be delayed to compensate for this. And, as you can imagine, this is particularly important with drum tracks.

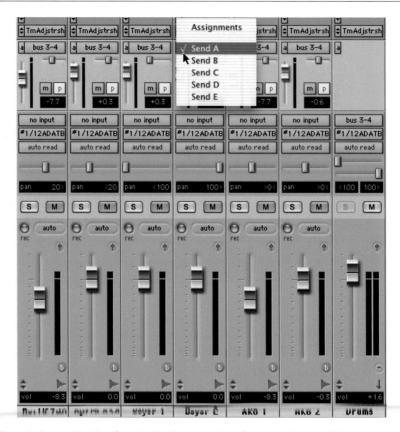

Figure 8.21 Mix window with the Sends display showing fader and pan sliders, mute and pre-fade buttons for the first of the five available sends.

Mixing vocals

The most important part of any song you are mixing is, of course, the lead vocal. This is where you should concentrate your best efforts to make this sound as good as possible in every way.

Let's take a look at an example of how I treated the vocals I recorded for the song *Give Me Back The Love*. In Chapter 6, I showed how I set up the Mix window with all the vocal tracks routed to an Aux track so that they could be treated using the Waves Renaissance Compressor, the Focusrite d2 EQ, and the Waves Renaissance Reverb plug-ins.

I set a trial level for the lead vocal as soon as I had the drums and bass balanced to produce a solid groove and I listened carefully to the vocals as I added more instruments to build the arrangement up. I noticed that the low frequencies were weak and that some of the words were getting lost while others were too loud.

I used the Focusrite EQ to add a little more 'depth' and 'body' to the vocal sound, boosting it by 0.5 dB at 80 Hz and by 1 dB at 159 Hz. Then I gave the vocal a little more 'clarity' or 'presence', boosting it by 0.5 dB at 4 kHz. To top it all off, I gave the vocal a little more 'air' by boosting the higher frequencies around 12 kHz by 1 dB – see Figure 8.22.

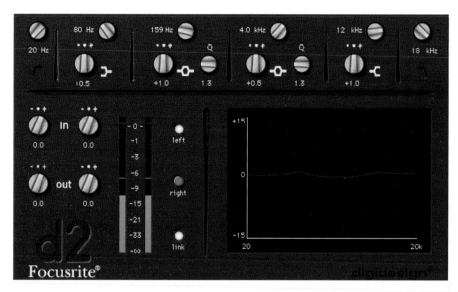

Figure 8.22 Focusrite d2 used to add a little 'body', 'clarity' and 'air' to the vocal.

I used the Waves Renaissance compressor to smooth out level changes a little, but I did not want to change the sound of the voice too much, so I used a low compression ratio of 1.44:1 and a relatively slow attack time with a relatively long release time – see Figure 8.23.

Finally, I added a little reverb to give the vocal a 'bigger' sound. It was recorded in a small room, and I did not want it to sound as though it were in a large room, so I chose the Medium Room

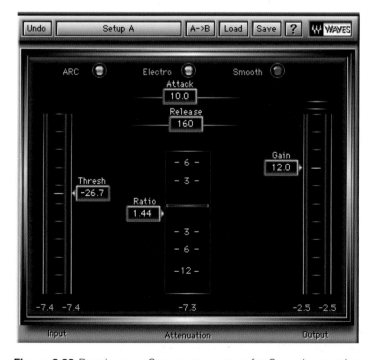

Figure 8.23 Renaissance Compressor set up for Smooth operation.

preset from the Waves Renaissance Reverb and I set the Wet/Dry balance to 80% Dry, 20% Wet.

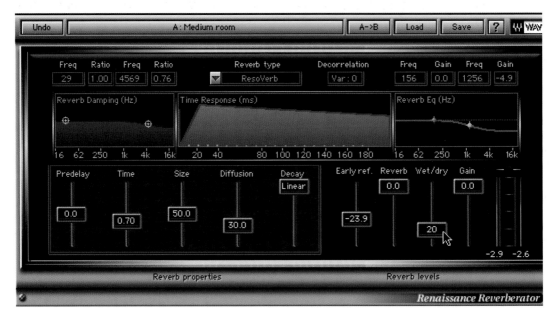

Figure 8.24 Renaissance Reverb showing the Wet/Dry balance being adjusted.

I had balanced all the vocal levels during earlier editing sessions using breakpoints in the Volume Automation displays for each track, and both lead and backing vocals were routed via the same Auxiliary channel that I was using to apply the various effects. This helped me to maintain a consistent vocal sound throughout. And this arrangement also allowed me to conveniently raise or lower the overall level of the vocals using this single Auxiliary channel fader.

Panning the instruments

Panning is the subject of much debate and ultimately comes down to your own creative choices. I have recently been listening to some early stereo mixes of late 1960s pop/soul songs – hit records such as *Goin' Out of my Head* by Little Anthony and The Imperials and *Hey Girl* by Freddie Scott – which typically might have the rhythm section panned hard left, backing vocals, brass and strings panned hard right and lead vocals in the centre. This way of mixing had the advantage that the lead vocal was easy to hear – loud and clear and with no competition from any other instruments occupying this central space. Nevertheless, it is much more usual today to have the bass guitar and bass drum dead centre, the snare sometimes centre or maybe, as with the hi-hat, panned left or right, and with the kit overheads capturing a natural stereo spread of all the drums including the cymbals. Guitars might be panned half left or right with keyboards spread a little further out. One thing to watch out for is the trap of recording several synthesizer parts in stereo – typically with the dry sound panned hard left and a chorused version panned hard right. It is better to make these mono and pan them to particular positions, or spread them over a narrower arc and offset them to one side or the other to allow separate instruments to occupy their own individual spaces as far as possible. It is also worth keeping in mind where the mixes are going to be played. If you are aiming them at the audiophile with a great stereo system listening in the sweet spot, then you can make good use of extremes of panning. On the other hand, if the mixes are intended for a dance club, even if this has some kind of stereo speaker system installed, you can't count on dancers being in any kind of sweet spot for too long – so you should keep the important rhythmic elements of your mix much closer to the centre.

The final mix

Having done all your preparation very thoroughly, the final balancing session can be relatively easy. In the early stages of your mixing session, you will have chosen most of the elements you want to work with and you will have set up the effects you want to use. It can take some time to set up sends to external effects units and Auxiliary tracks with chains of plug-ins to process the drums, the vocals, and the various instruments – especially if you are looking for that 'special' reverb sound or combination of delays, or if you are trying to find the right EQ settings and pan positions to make particular instruments 'sit' properly in your mix, or stand out as features. With a typical pop song, you should allow at least half a day to get everything more or less in place. Then you might take a break for half an hour or so before coming back to do the final balancing of levels and tweaking of effects. Here you should be concentrating on the most important tracks – typically the lead vocal and any featured solo instruments.

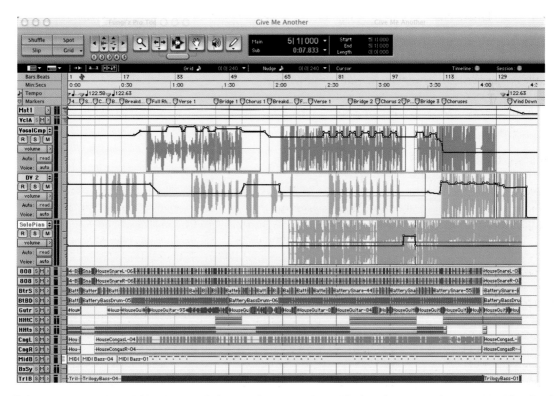

Figure 8.25 Pro Tools Edit window during a mixing session with the three most important tracks, lead vocals, backing vocals, and the piano solo, set to Medium track height with their Volume automation curves visible – ready for the Pro Tools engineer to tweak.

Make sure that all the tracks you are not using are hidden and set the rest to the minimum track height – just leaving the lead vocal, backing vocal and perhaps a solo instrument at the medium track height. It is also a good idea to display the Volume Automation curves in the chosen tracks so that you can see what is going on and so that you can manually adjust these volumes at any stage during the mix session.

If you are using the real-time automation features, you will start out in Auto Write mode. You probably won't get all your moves correct during the first pass, so you can use Auto Touch mode to refine the sections that need changing. As you get closer to what you are looking for, you might use the Trim features to fine-tune the settings even further. For example, if you like the way the guitar rhythm gets softer in the verse and louder in the choruses, but you decide that overall this guitar is too loud, you can use Trim mode to bring the overall level down on the guitar track while retaining the relative automation 'moves'.

I usually end up going into the Edit window to manually edit the automation breakpoints to achieve exactly the right result. Here it is so easy to see exactly where the vocal and instrumental phrases lie and to edit the automation curves to do exactly what you want them to do. And don't forget how easy it is to loop playback of any section that you want to hear over and over again: select a range of time by dragging the mouse through the timeline ruler and choose Loop Playback from the Operations menu – or just hit Command-Shift-l (Mac) on your keyboard – then hit Play.

> Tip: To move all the automation breakpoints in a particular track or section of a track up or down while preserving relative levels, choose the Selector tool first and drag the mouse to highlight the range of interest. Then choose the Trimmer tool, point the mouse anywhere along the line of breakpoints and drag the whole line upwards or downwards – as in Figure 8.26.

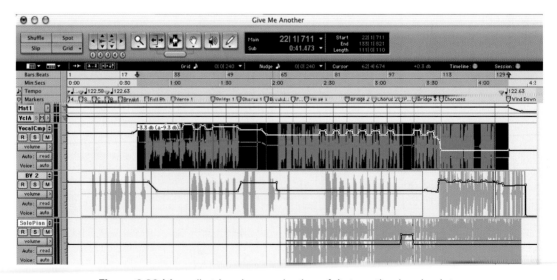

Figure 8.26 Manually trimming a selection of Automation breakpoints.

It took about an hour to complete this track balancing and then I was ready for the final step – mixdown to stereo. I used the Bounce to Disk command for this, as all my mixing was done in Pro Tools with my external mixer simply being used to monitor the stereo outputs.

Summary

Once you have familiarized yourself with how to set up the Mix window and use the automation modes, you will begin to appreciate the greater flexibility of the Pro Tools system compared with most others. The combination of the powerful on-screen editing tools with the sophisticated automation features lets you achieve results that would be totally impossible to get with conventional recording and mixing equipment.

9 Audio Plug-ins

Introduction

Traditionally, recording studios have been built round the mixer, the multi-track recorder and various 'outboard' signal processing units. Digidesign has invested most of its development effort on recording and mixing, rather than signal processing. Nevertheless, realizing the importance of the signal processing equipment, Digidesign had the good sense to create a software 'plug-in' system and then encouraged third-party developers to create software 'modules' which can 'plug-in' to Pro Tools systems to provide this 'outboard' signal processing.

A studio full of conventional outboard gear can set you back thousands of pounds for all the separate boxes – but now you can get all this in software that comes on a CD-ROM or that can be downloaded from a website. The benefits are obvious: you can bring your 'outboard' on-board as an integral part of the system; you don't need any cables to hook it all up; it's all digital; it's cheaper; it's much more flexible; and it's much more easily upgradeable – all to the benefit of the user. And one of the major advantages of digital audio systems is that you don't get the build up of hiss and grunge that you would get with all that analogue gear plus the open effects returns, open inputs, and so forth! The downside is that you need lots of DSP power – so you will need to use the fastest computer you can get hold of and consider adding extra DSP cards to your system.

Plug-in types

Digidesign provides a basic set of plug-ins with Pro Tools systems – the so-called 'DigiRack'. When you buy a Pro Tools system, the DigiRack includes a selection of plug-ins of various types.

There are four types of plug-ins that can be used with Pro Tools systems – TDM, Host TDM (HTDM), Real Time Audio Suite (RTAS) and AudioSuite.

> Note: Some more recent TDM plug-ins will only work on Pro Tools|HD systems. And others will only work on the HD Accel card. If in doubt, check the Digidesign website.

Pro Tools TDM systems, including the HD, MIX and DSP Farm cards, use DSP chips on these cards to process audio using TDM plug-ins. Audio Suite, RTAS and HTDM plug-ins run on the host computer's CPU.

Pro Tools TDM systems can use any of these types – Audio Suite, RTAS, TDM or HTDM. Pro Tools LE systems, on the other hand, can only use Audio Suite and RTAS plug-ins.

All TDM, HTDM and RTAS plug-ins work in real-time – you just insert these into Pro Tools mixer channels to provide immediate effects. The difference is that with HTDM and RTAS plug-ins all the processing is done using the host CPU while TDM plug-ins use DSP chips on the TDM hardware.

With AudioSuite, you select audio in the Edit window first, then choose a plug-in from the AudioSuite menu and tweak the parameters while previewing the effect. When it sounds the way you want it to, you then process the audio via the computer's CPU to produce a new file on disk. Pro Tools lets you choose whether to replace the original file or to use it alongside the original file. The advantage is that once you have carried out this operation, the CPU is left free for other tasks. The disadvantage is that it is a non-real-time process.

To use TDM plug-ins, you simply insert these into Pro Tools mixer channels – as you would with outboard signal processors. You can use RTAS plug-ins in exactly the same ways as the TDM plug-ins – the difference being that the processing is carried out in the computer's CPU rather than on the Digidesign DSP. This is one reason why I always recommend that you use the most powerful computer you can afford, as you will need all the processing power you can get your hands on if you want to make ambitious mixes in Pro Tools.

HTDM plug-ins are a hybrid of TDM and RTAS technologies. They provide all the functionality of standard TDM plug-ins, but, like RTAS plug-ins, all the processing is done on the host instead of using the DSP chips on your TDM hardware. HTDM plug-ins can be used exactly like DSP-based TDM plug-ins, i.e. in conjunction with all types of tracks (audio tracks, aux tracks, and master tracks) within a TDM system.

Note: One thing to watch out for with HTDM plug-ins is that DirectConnect and HTDM share the 32 available audio streams, so a maximum of 16 stereo HTDM plug-ins can be used at the same time, assuming no DirectConnect applications are in use. Also, the audio buffer on HTDM is either 128 or 512 samples, depending on the version of the StreamManager used, while on an RTAS plug-in it is only 32 samples. Thirty-two samples at a sample rate of 44.1 kHz corresponds to around 0.6 ms, 128 samples correspond to 2.8 ms and 512 to 11.6 ms.

Tip: The Convert Plug-In pop-up lets you convert an inserted plug-in from TDM format to an RTAS plug-in of the same type (or vice-versa). It should go without saying that this will only be possible with plug-ins that are available in both TDM and RTAS formats – but I am saying it anyway!

Figure 9.1 The Convert Plug-in pop-up.

Figure 9.2 The Convert Plug-in choices.

Note: One aspect to bear in mind when using RTAS plug-ins is that the more of these you use, the greater will be the impact on the overall performance of your computer system. The available track count, edit density and latency in automation and recording will all be adversely affected at some point – as the CPU has to take care of these as well.

Tip: You should check the System Usage window as you develop your project to see how the available DSP and CPU power is actually being used. This can help you decide how many TDM plug-ins versus RTAS plug-ins to use, for example.

Using real-time plug-ins

There are two ways that you can use the real-time TDM and RTAS plug-ins in Pro Tools.

The simple, obvious way is to insert the plug-in you want to use on the track you want to apply it to using the Inserts section of the Mixer. You can insert up to five plug-ins in this way on any audio tracks, Auxiliary Inputs, or Master Faders and adjust the level of the effect using the plug-in's controls. If you insert more than one plug-in on a track, the track's audio is processed in series by each plug-in, with the signal flowing through the plug-ins in the Mix window from top to bottom.

The other way to use the real-time plug-ins is to use the Sends section of the Mixer to route audio from any track or tracks that you want to process to an Auxiliary Input track and insert the plug-in onto the Auxiliary track – see Figure 9.3. This way, you can process a number of tracks using just one inserted plug-in – so you avoid using up your available DSP resources too quickly. This approach works well with reverb and delay processors, for example, where you might wish

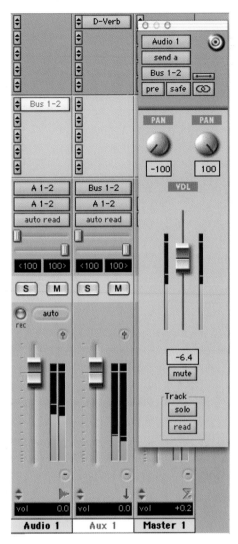

Figure 9.3 Mixer with audio being sent via Bus 1–2 to an Auxiliary Track with D-Verb inserted onto the Auxiliary track. The Send window for the Send to Bus 1–2 is open at the right showing the Send level and pan controls.

Note: Real-time plug-ins are *pre-fader* inserts (except on Master Fader tracks, where they are post-fader), so their input levels are not affected by the track's volume fader. Incoming audio is recorded to disk before it is monitored through the Mixer channel strip, so if you record with a plug-in inserted, although you will hear the effect of the plug-in, the effect will not be recorded to disk. Of course, once you have finished recording, you can always bounce the audio track to disk with the plug-in inserted – creating a new file on disk that includes this effect.

> Tip: If you want to record directly through a plug-in effect, you need to route the incoming audio to an Auxiliary Input, insert the plug-in you want to use on the Auxiliary Input track, then route the output of the Auxiliary track to the input of the audio track you want to record onto.

to apply the same reverb or delay type to several tracks rather than using different settings for each track. Using the Send level for each track you can decide how much of the effect to apply to each individual track, and using the main fader for the Auxiliary track lets you control the overall level of the effect that you will hear in your mix.

Plug-in delays

Don't forget that inserting any kind of plug-in on a track will introduce a delay while the computer carries out the processing for that track. Every plug-in introduces a delay, but the delay varies according to the individual plug-in type. Fortunately, you can get a read-out of what this delay is on each track by Command-clicking on the Volume readout on the Mixing board. Command-clicking causes this display to switch between the readout of the fader setting for the channel, the Peak Value of the signal passing through the channel, and the delay introduced by the plug-in or plug-ins. Digidesign provide a special plug-in called the Time Adjuster to let you add delays to all the other tracks to compensate, but this uses DSP. Another way to compensate is to simply shift the tracks around in the Edit window by the amount of samples needed to compensate for these delays. This uses no additional DSP, and you can always set a nudge value to a suitable number of samples to help you out here.

> Note: You can record through plug-ins as long as the processing delay through these is not sufficient to be off-putting for the musician or musicians. A few milliseconds should not be a problem, but if, say, a guitarist is recording through a plug-in and is playing in the studio control room while listening to the studio monitors, there will also be a delay depending on how far away he is from the monitors – which adds to the plug-in delay. Sound travels fast in air – but not too fast (around 1100 feet/second). I tried this using the Amp Farm plug-in with my Pro Tools TDM system. When I was sitting about 15 feet away from the monitors the delay of around 15 milliseconds was definitely off-putting, while 3 or 4 feet away it was acceptable – although still noticeable.

Making plug-ins inactive

If you need to free up your DSP resources for any reason, such as to try out another plug-in when you have run out of DSP, you can make a plug-in inactive using a Command-Control-click (Macintosh) or a Control-Start-click (Windows) on the Insert button in the Pro Tools Mix or Edit windows.

Alternatively, you can just make the whole track inactive using a Command-Control-click (Macintosh) or a Control-Start-click (Windows) on the Track Type Indicator in the Mix window.

Inactive plug-ins 'remember' their assignments, positions and related automation playlists so that they can be re-instated just as you left them if you make them active again. No audio passes through inactive plug-ins, though, and they do not use up any DSP or TDM resources.

AudioSuite plug-ins

The complete list of AudioSuite plug-ins comprises: one-band and four-band EQ II; Chorus; Compressor II; D-Verb; DC Offset Removal; DeEsser; Mod Delay II; Duplicate; Expander-Gate II; Flanger; Gain; Gate II; Invert; Limiter II; Multi-Tap Delay; Normalize; Ping-Pong Delay; Pitch Shift; Reverse; Signal Generator; and Time Compression/Expansion.

With Pro Tools 6.1 and more recent versions, various Digidesign plug-ins that were previously available as options are now provided with the standard DigiRack set. To help you create a sense of space and depth in your mix you can use the former D-fx AudioSuite plug-ins, which are included as DigiRack Chorus, Flanger, Multi-Tap Delay, and Ping-Pong Delay. D-fx D-Verb and D-Verb, which provide non-real-time and real-time reverb processing, respectively, are included as D-Verb for AudioSuite, RTAS, and TDM. Digidesign's DPP-1 TDM plug-in, which provides real-time retuning and pitch transposition, is included as DigiRack Pitch for TDM.

> Tip: One thing to be aware of when using AudioSuite effects such as delay or reverb is that you need to make a selection longer than the original source material to accommodate the reverb tail or echoes that you will want to hear at the end of the processed region. If you don't do this, the reverb or delay will immediately cut off at the end of the original audio material. This would normally sound wrong, so you need to place the region in a track and select an amount of blank space beyond the end of the region to accommodate the reverb decay or delayed echoes. It is best to select a little more than you think you will need, as you can always trim the end of the newly created region afterwards.

The DigiRack AudioSuite plug-ins would appear to have a major disadvantage in that they do not work in real-time. However, this can actually be an advantage – particularly if you need to use most of your available processing power for other tasks. In this case, it makes best sense to apply the effect using an AudioSuite plug-in to create a new audio file. Then you can dispense with the plug-in processed track and simply play back the new file – complete with the effect you have applied.

> Tip: If you want to process a mono track and obtain a stereo result, first select the desired track or region plus an empty track or region, then select the stereo version of the delay effect. When you process the audio, the result will be two tracks or regions that represent the right and left channels of the processed audio. You should then pan these tracks hard right and hard left in your mix.

Note: If you choose to use a delay effect plug-in in stereo mode, and then select an odd number of Pro Tools tracks for processing, the delay effect will process the selected tracks in pairs, in stereo. However, the last odd (unpaired) track will be processed as mono, using the left channel settings of the stereo delay effect plug-in. If you want the last track to be processed in stereo, you must select an additional track to pair it with an empty one if necessary.

Real-time TDM and RTAS plug-ins

To cope with every situation from mono, to stereo, to multi-channel surround work, real-time plug-ins are available as Mono, Multi-Mono and Multi-Channel plug-ins – depending on the plug-in type and whether the destination is a mono or multi-channel track.

The multi-channel plug-ins are intended for use with stereo and surround formats, although if a Multi-Channel version of a particular plug-in is not available, you can use a Multi-Mono version. This is an excellent system that, by default, links the parameters in the multiple plug-ins so you can adjust, say, a delay parameter and have this apply automatically to all the channels. Otherwise you would have to adjust, say, six sets of plug-in parameters individually. The system is also flexible enough to allow you to unlink the parameters if you want to adjust the plug-ins independently for the different channels. With Multi-Channel plug-ins, the parameters on greater-than-stereo tracks are generally always linked together.

One of the neatest things about the real-time plug-ins is that you can apply automation to all the parameters. The creative possibilities here are simply tremendous. Obvious things such as sophisticated and complex panning moves become extremely easy to create, along with EQ sweeps and so forth. Pro Tools actually creates a separate playlist for each plug-in parameter you are using so you can edit these later if necessary. Before you start making your automation moves you need to click on the Automation button in the plug-in window to bring up a list of Automation parameters that you can choose from. You also need to use the Automation Enable window to write-enable the plug-in you are working with. Once this is all set up, you choose a suitable Automation mode, such as the standard Auto Write to begin with, hit Play to run your tracks, then make your moves – adjusting the plug-in parameters in real-time. Simply hit Stop when you have finished – and that's it, you have recorded your plug-in automation. At this stage, if you are happy with your automation, you can use the Automation Safe button to prevent the automation data being accidentally overwritten. At any time later you can use the Auto Touch or Auto Latch modes to make adjustments to the automation you have recorded. Just make sure you have de-selected the automation Safe button, choose your automation mode and hit Play again. This time, the original automation will apply until you actually change any of the parameters – in which case the new automation data will replace the original.

Note: Real-time plug-ins are configured as pre-fader inserts so input levels to these will not be affected by the volume fader for the track. Occasionally there may be too much level presented to the input of the plug-in – which could lead to clipping. To get around this situation you should insert the plug-in on an auxiliary input so that you can adjust the level being routed to the plug in.

Multi-channel RTAS plug-ins

Multi-channel RTAS plug-ins include: D-Verb; Dither; POW-r Dither; Compressor II; Limiter II; Expander-Gate II; Gate II; one-band and four-band EQ II; Short Delay II; Slap Delay II; Medium Delay II; Long Delay II; and Extra Long Delay II.

D-Verb

The D-Verb plug-in is a basic reverb with most of the controls that you would expect. There are no presets provided, but you can always create your own by adjusting the parameters and saving the settings.

D-Verb is available in TDM, RTAS and AudioSuite formats, and is a very welcome inclusion with the standard set of Pro Tools plug-ins as of version 6.1. It is not the best reverb plug-in that is available, but it does at least provide Pro Tools with a basic reverb capability that was sorely lacking previously.

Dither

In Linear PCM there will almost always be a quantization error, as the quantizer must always choose one level or another to represent the amplitude of the signal and the actual value is unlikely to correspond exactly to any of the quantization steps. This error results in quantization noise or distortion when analogue audio is reconstructed from the digitized version. If the signal

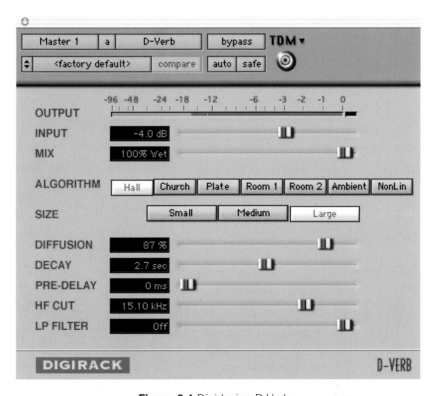

Figure 9.4 Digidesign D-Verb.

amplitudes are large, the error will be random and will resemble white noise. On the other hand, at low signal amplitudes the error becomes correlated with the signal, resulting in potentially audible distortion of the signal. To overcome this, a low-level noise signal known as dither is added to the source signal to randomize the quantization error. The dither signal causes the audio signal to move back and forth between the quantization levels – in other words, to dither. The dynamic range of a practical system using dither with an amplitude equivalent to about one-third of a quantization step will be decreased by about 1.5 dB. Also, when there is no signal present there will be no quantization noise in a digital system – unlike with an analogue system, where noise will always be present and constant.

According to Ken Pohlmann in his book *Principles of Digital Audio*, 'Without dither, a low level signal would be encoded by an A/D converter as a square wave. With dither, the output of the A/D is the signal with noise. Perceptually, the effects of dither are much preferred because noise is more readily tolerated by the ear than distortion.' *An added benefit of dither is that it lets the converter handle amplitudes below the lowest quantization value.* As Pohlmann explains, 'Dither changes the digital nature of the quantization error into a white noise and the ear may then resolve signals with levels well below one quantization level. So, with dither, the resolution of a digitization system is below the least significant bit. By encoding the audio signal with dither to produce modulation of the quantized signal, we may recover that information, even though it might be smaller than the smallest increment of the quantizer.'

Dither is also used when moving audio between digital systems operating at different bit depths. For example, when going from a 24-bit system to a 20-bit or 16-bit system. Simple truncation of the digital word just 'throws away' the lower order bits – including any dither component. So whenever you shorten the word length of digital audio signals, you will need to apply dither again to prevent quantization distortion re-appearing.

You would typically use the Pro Tools Dither plug-in on a stereo Master Fader channel when mixing a 24-bit session down to 16-bit for CD. In this case, you should insert the plug-in post fader – as you will very often be fading the music out at the end.

You should also insert the Dither plug-in (or any dithering plug-in, such as Dither POW-r Dither or Maxim) on a Master Fader track when using Bounce to Disk to create a 16-bit file (or any bit depth lower than 24-bit) from a 24- bit session; otherwise the resulting 16-bit file will be 'truncated' at the destination bit depth.

Three dither options are provided in the Dither plug-in: 20-, 18- and 16-bit. The Dither plug-in also offers a noise-shaping option which further reduces perceived noise in low-level signals by shifting audible noise components into a less audible range.

Now if you are outputting audio to an analogue destination (e.g. mixing to half-inch) from a 24-bit session using your Pro Tools interface's 24-bit D/A converters, for example, then you don't need to use dither. This would simply reduce the signal-to-noise ratio without producing any benefits. You can use the 20-bit setting either to go to a 20-bit digital recorder, such as the Alesis ADAT XT 20, or for output to analogue devices via the Digidesign 882/20 interface. Various digital effects units offer 20-bit digital input and output, so you could also use the 20-bit setting when bussing signals to and from these devices. The 18-bit setting is provided for output to analogue devices via the older 888 or 882 interfaces, which output at 18 bit resolution. When recording to DAT or CD you will, of course, choose 16-bit.

Tip: Bear in mind that the Dither plug-in uses a lot of processing power – so you may prefer to mix 24-bit to a stereo file on disk and convert to 16-bit later using a specialized editor such as BIAS Peak. You will still need to apply dither using this software – the advantage being that you will not be using your available processing power for creative effects processing and mixing at the same time.

POW-r Dither

More recent Pro Tools systems, such as those using version 6.1 software or later, have a more advanced dither plug-in called POW-r Dither. This provides the highest possible audio quality for critical final-stage mixdown and mastering. It should not be used on intermediate stages that may be processed again at a later stage. Also, POW-r Dither should only be used as the last insert in the signal chain – especially when using Type 1 Noise Shaping.

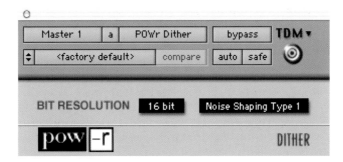

Figure 9.5 POW-r Dither.

You can choose either 16- or 20-bit resolutions using the Bit Resolution pop-up and select one of the three noise shaping types using the Noise Shaping pop-up. Digidesign recommends that you try each noise shaping type and choose the one that adds the least amount of colouration to the audio being processed. Type 1 has the flattest frequency spectrum in the audible range of frequencies, modulating and accumulating the dither noise just below the Nyquist frequency. This is recommended for less stereophonically complex material such as solo instrument recordings. Type 2 has a psycho-acoustically optimized low-order noise-shaping curve and is recommended for material of greater stereophonic complexity. Type 3 has a psycho-acoustically optimized high-order noise-shaping curve and is recommended for full-spectrum, wide-stereo field material.

Note: The POW-r Dither plug-in does not run on third-party software applications, such as Logic, that use DAE.

Dynamics

The DigiRack dynamics processors come in five 'flavours' covering all the commonly used types of dynamics processing including the Compressor, the Limiter, the Gate, the Expander-Gate and the DeEsser.

EQ

The Digidesign EQ comes in two types. With the one-band EQ you have a choice of high-pass, high-shelf, peak/notch, low-pass or low-shelf EQ. If just one part of the frequency spectrum needs tweaking, you will make much better use of your available DSP resources by choosing this plug-in instead of the four-band version.

With the four-band EQ, you get high-shelf, two peak/notches and a low-shelf EQ. Controls are also provided for Input Level, Frequency, Gain, width or 'Q', plus a Bypass button. Clearly this allows much more flexible control of the frequency spectrum, and you can easily notch out troublesome frequencies using high-Q settings or gently roll off higher frequencies using a shelf, for example.

Delays

Delays are used to create effects such as slapback echo, with a single repeat, spacey delay effects with multiple repeats, or chorusing effects. The Gain control lets you attenuate the signal level coming into the delay, allowing you to prevent clipping while the Mix control lets you set the balance between 'wet' (i.e. effected) signal and the 'dry' (i.e. the original) signal at the output. A Low Pass Filter is provided to let you attenuate the higher frequencies of the feedback signal – so the repeats will sound successively duller, as would be the case with a tape delay. The feedback control lets you control the number of repetitions of the delayed signal. You can create the doubling and flanging effects using the modulation controls provided for Depth and Rate. Negative feedback settings can be used to produce a 'tunnel-like' sound for flanging effects.

There are five Mod Delay II plug-ins provided with Pro Tools 6.1 – Short Delay II; Slap Delay II; Medium Delay II; Long Delay II; and Extra Long Delay II.

Each plug-in has a Gain control that you can use to adjust the input level to avoid clipping and a Mix control to set the balance between the delayed (wet) signal and the original (dry) signal

> Tip: If you are using a delay to create a flanging or chorusing effect, you can control the depth of this effect to some extent using the Mix control.

The Low-Pass Filter (LPF) control lets you adjust the cut-off frequency to control the high-frequency content of the feedback signal. The lower the setting, the more the high frequencies are attenuated. The Delay control lets you set the delay time between the original signal and the delayed signal. The Depth controls lets you adjust the depth of the modulation applied to the delayed signal while the Rate control lets you adjust the rate of this modulation. The Feedback control lets you set the amount of feedback applied from the output of the delay back to its input. It also controls the number of repetitions of the delayed signal. Negative feedback settings give a more intense 'tunnel-like' sound to flanging effects.

Short Mod Delay II and Slap Delay II

Short Delay II provides delay times of up to 38 milliseconds at all sample rates – suitable for creating doubling and chorusing effects.

Slap Delay II provides delay times of up to 166 ms at all sample rates – ideal for creating typical slapback echo effects.

Medium Mod Delay II

Medium Delay II provides up to 337 ms of delay at all sample rates and is typically used for chorusing and flanging.

Long and Extra Long Mod Delay II

Long Delay II provides delay times of up to 678 seconds while Extra Long Delay II provides up to 2.726 seconds of delay – at all sample rates.

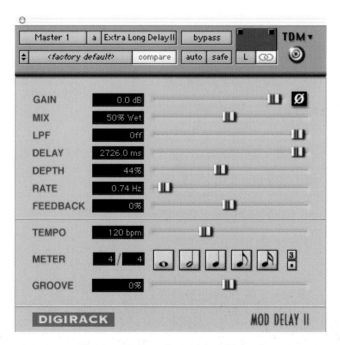

Figure 9.6 Extra Long Mod Delay II.

Tempo, meter, duration and groove

The medium and long delays also have controls for tempo, meter, duration and groove.

You can set the Tempo in BPM for the delay, independently of Pro Tools' tempo. Of course, you may want to set the same tempo that your Pro Tools session is running at.

Tip: If you have not specifically set a tempo for your session you can always use the Identify Beat command or Beat Detective to find the tempo of any audio or MIDI that you have recorded.

You can use the Meter parameters to enter either simple or compound time signatures. To the right of the Meter parameters a row of icons lets you set the Duration of the delay using musical values. Just click on the whole note, half note, quarter note, eighth note or sixteenth note to choose an appropriate delay. Two more buttons are provided to let you dot the note value or turn it into a triplet. The Groove control provides fine adjustment of the delay in percentages of a 1:4 subdivision of the beat. It can be used to add 'swing' by slightly offsetting the delay from the precise beat of the track.

Note: When a specific Duration is selected, moving the Tempo control will affect the Delay setting. Likewise, the range of both controls will be limited to the maximum available delay with the currently selected Duration. To enter very short or long delays it may be necessary to de-select all Duration buttons.

Tip: If you want to create even longer delay times, you can insert any combination of these plug-ins, cascading their outputs one into the next to achieve the delay times you want. In this case, you should avoid using any feedback on the intermediate delays – reserving this for the last delay in the chain to set the number of repeats.

Note: The Mod Delay II plug-ins support all sample rates up to 192 kHz – although there are various exceptions. For example, the TDM versions of the Extra Long Delay mono-to-stereo and stereo plug-in are not supported at 96 kHz. And none of the TDM versions of the Extra Long Delay plug-in are supported at 192 kHz. On the other hand, RTAS versions of the Extra Long Delay plug-in are fully supported at all sample rates.

Mono In, Stereo Out Mod Delays

A common application of delays is to create a stereo effect from a mono track. The Mod delays allow you to do this as Mono in, Stereo out versions of all the Mod delay plug-ins are provided. You can have different delay settings applied to each side of the stereo output if you wish and these can be panned anywhere you like using the stereo pan faders which are automatically created in any track using these stereo plug-ins.

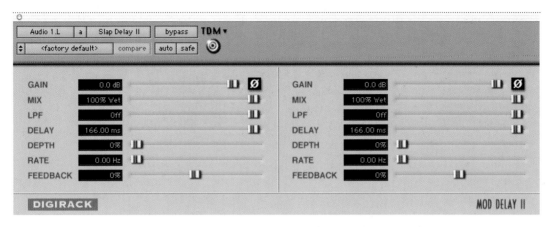

Figure 9.7 Mono In, Stereo Out Slap Delay II.

Multi-channel TDM plug-ins

All the Multi-channel RTAS plug-ins are available as Multi-channel TDM plug-ins. Multi-channel TDM plug-ins also include Pitch and three versions of the TimeAdjuster plug-in – Short, Medium and Long.

The Real-time Pitch Processor

The Real-time Pitch Processor (previously known as the DPP-1) offers up to four octaves of pitch transposition and manipulation at 24-bit resolution. Applications include chorusing, doubling and delay effects, pitch correction and extreme pitch shifting for special effects.

> Note: When the processed signal is transposed in pitch, it still keeps the same overall length as the unprocessed signal – i.e. the pitch is changed while the time is kept constant. This is in contrast to the way audio can be changed in pitch by speeding up or slowing down the speed of a tape recording – in which case the audio lasts a shorter or longer time, depending on the direction of the pitch change.

This Pitch processor is best used on an auxiliary return track to conserve processing power if you want to apply pitch processing to more than one track at a time. You can use it as mono in/out, mono in/stereo out, or stereo in/out, and clicking on any note in the ascending musical staff lets you quickly select a relative pitch transposition value. Alternatively, you can use the Coarse pitch slider which lets you adjust the pitch in semitones over a two octave range.

All-in-all, the Real-time Pitch Processor is a useful pitch-processing tool for the more musically inclined Pro Tools engineer – featuring an effective and intuitive interface.

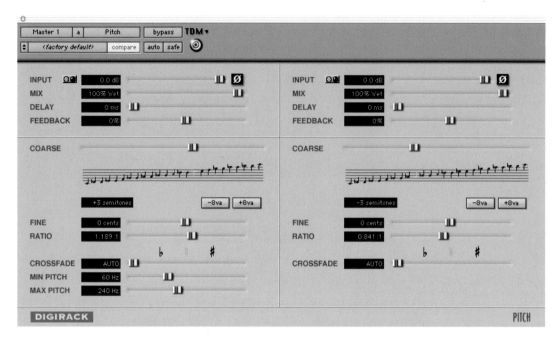

Figure 9.8 Digidesign Pitch Multi-channel TDM plug-in.

TimeAdjuster

The TimeAdjuster plug-in has gain compensation and phase inversion functions like the Trim plug-in, although there is no Mute button. The gain control is different from the one in the Trim plug-in, in this case offering up to ±24 dB of gain adjustment. This is particularly useful if you have recorded an audio signal at a very low level. Without using this plug-in, you might choose to alter the level to a more realistic value using the AudioSuite Gain plug-in, for example. This takes time and means you create a new file. It is much simpler to make this kind of gain adjustment in real-time using the TimeAdjuster. You can also use the TimeAdjuster for general adjustment of phase relationships of audio recorded with multiple microphones – with sample-level control. Nevertheless, the main function of the TimeAdjuster is to insert a delay to match delays existing on other channels – for whatever reason. The main reason delays will occur between tracks is because you have inserted plug-ins. The DSP processing that takes place as the audio passes through a plug-in will always cause the output from the plug-in to be delayed by a number of samples compared to a track without a plug-in inserted. And delays from different plug-in types will be different. The TimeAdjuster lets you apply an exact number of samples of delay to the signal path of a Pro Tools track to compensate for delays caused by specific plug-ins. The default setting for the TimeAdjuster delay is just four samples, which is the minimum delay that insertion of an individual plug-in, such as a DigiRack EQ, onto a Pro Tools mixer channel will cause. The more complex plug-ins will create much larger delays, of course, and if you are using several plug-in inserts, these can add up to significant values.

There are three different TimeAdjuster plug-ins – Short, Medium and Long. The Short delay allows up to 259 samples, medium allows up to 2051 samples, and the Long delay allows up to 8195 samples.

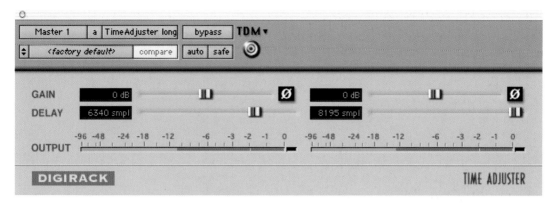

Figure 9.9 TimeAdjuster.

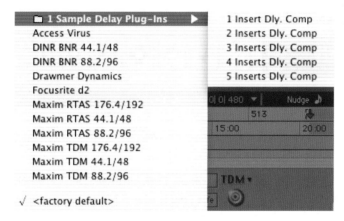

Figure 9.10 TimeAdjuster presets menu.

Using this feature lets you adjust all your mixer channels so they have the same amount of delay – keeping them all perfectly in step. Obviously you need to know what delays are being caused by your plug-ins before you can make the correct TimeAdjuster settings to match these. If you know the delay that will be introduced by each plug-in, you can add these together if you are using more than one plug-in on a track – then insert compensating delays using TimeAdjuster wherever necessary. A list of the delay amounts for DigiRack AudioSuite, RTAS and TDM plug-ins is supplied in the documentation that comes with the software, although third-party plug-ins are not listed there. Of course, this information may be supplied in the manuals that come with the third-party plug-ins – or it may not. Fortunately, you can get a read-out of the delay on any Pro Tools mixer channel by Command-clicking (Control-clicking in Windows) on the Track Level Indicator. This indicator shows the level of the track's volume fader numerically as a default, but switches to show headroom and then channel delay (in samples), respectively, when you Command-click on it.

Now, considering that just a few samples of delay is going to represent less than a millisecond in time, should you be concerned about correcting for these delays if you are only using the basic DigiRack plug-ins? The reality is that often this will not strictly be necessary unless you have several plug-ins inserted on particular tracks or are using plug-ins that cause longer delays

Tip: When working with pairs or groups of microphones that are phase-coherent you may sometimes want to insert a plug-in on just one of the microphone channels. To work out the compensating delay to use with TimeAdjuster on the other channels, you can use the following technique. First place two identical audio regions on two different audio tracks and pan them centre. Then apply the plug-in whose delay you want to calculate to the first track and a TimeAdjuster plug-in to the second (i.e. the target) track. Invert the phase of the target track using the TimeAdjuster plug-in's Phase Invert button, and adjust the plug-in's delay time until the signal disappears. To do this, you can Command-drag (Control-drag in Windows) to fine-tune the delay in one-sample increments – or use the up/down arrow keys – to change the delay one sample at a time until the audio signal disappears. Finally, you deselect the Phase Invert button. The audio disappears when the signals are out of phase at the exact point where you have precisely adjusted and compensated for the delay. This will always happen when you monitor identical signals and invert the polarity of one of them – as the signals will be of opposite phase and cancel each other out. You can always use this method to find the exact delay setting for any plug-in and then use the delay value arrived at to compensate your other channels.

such as some of the dynamics processing. Where it is particularly important is if you have two or more tracks that need to be kept strictly in phase – for example with stereo pairs or multiple microphone set-ups.

The important thing is to be aware that these delays are being created each time you insert a plug-in. So if you want to maintain absolutely accurate relationships between tracks you will have to take the trouble to work out the delays for each track and use TimeAdjuster where required to bring all the tracks back into step with each other.

Interestingly, Digidesign did suggest as long ago as 1995 (in the documentation for their original ADAT interface) that future versions of TDM's mixing environment may have automatic delay compensation. This would mean that any DSP process on a given mixer channel would auto-matically cause an equal amount of delay to be introduced on all other channels, ensuring that the output of the mix would be time and phase synchronous. This technique is used in Steinberg's Nuendo, for example. However, at the time of writing, in the autumn of 2003, this feature has still not been implemented in Pro Tools.

Multi-mono TDM and RTAS plug-ins

All the Multi-channel TDM and RTAS plug-ins are available as Multi-mono TDM and RTAS versions. Multi-mono TDM and RTAS plug-ins also include Click, DeEsser, Signal Generator and Trim plug-ins.

Click

The Click plug-in serves as a built-in click generator for Pro Tools, using Digidesign's DirectMIDI protocol to avoid MIDI timing delays. It offers individual level control of both accented and unaccented click sounds and has a MIDI IN indicator to confirm that it is receiving data. You can choose the sound of the click from a list of presets, such as cowbell and marimba, that often make musicians feel more comfortable than the standard electronic click sounds.

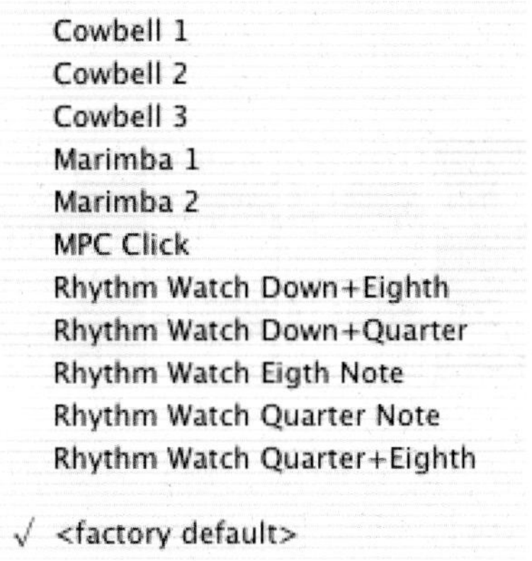

Cowbell 1
Cowbell 2
Cowbell 3
Marimba 1
Marimba 2
MPC Click
Rhythm Watch Down+Eighth
Rhythm Watch Down+Quarter
Rhythm Watch Eigth Note
Rhythm Watch Quarter Note
Rhythm Watch Quarter+Eighth

√ <factory default>

Figure 9.11 Click presets menu.

DeEsser

The DeEsser plug-in reduces sibilants, the 'esses' and 'effs' and other high frequency sounds that you often find in vocals, dialogue, or wind instruments such as flutes. These so-called sibilant frequencies are often much greater in amplitude than the rest of the audio and it is all too easy for these unwanted peaks to cause distortion. The DeEsser uses fast-acting compression to work on just these frequencies to avoid this problem. There are just two simple controls here: a Threshold control to let you set the level above which de-essing takes place and a Frequency control to let you tweak the frequency at which the de-essing takes place. The Key Listen button lets you audition the audio being de-essed to help you choose the correct frequency.

Signal Generator

If you need to generate reference signals to calibrate your interfaces or other studio equipment, the Signal Generator plug-in provides a useful selection. You will hear the default tone as soon as you insert the plug-in on a track and, of course, you can always mute this using the Bypass button. The frequency can be set anywhere between 20 Hz and 20 kHz and levels can be set from –95 dB all the way up to 0 dB – the full-scale setting. The waveform can also be changed from the default Sine wave to a Square wave, Sawtooth or Triangle – or to white or pink noise waveforms.

Trim

Occasionally, you may need to boost or cut the level of an audio signal to let you keep your channel faders, say, around the 0 dB position as a sensible starting point for you to use while mixing down. The Trim plug-in provided on TDM systems lets you add up to 6 dB extra gain – or reduce the gain if necessary. A meter showing the new output level is provided, along with a Phase Invert switch and a fader to let you set the gain boost or cut.

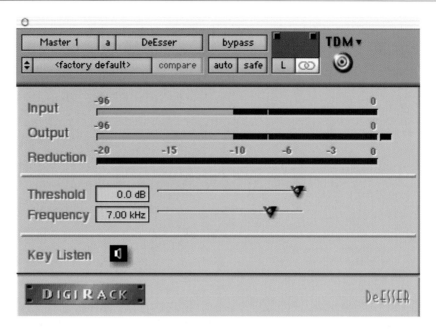

Figure 9.12 DeEsser.

Tip: The Mute button provided in the trim plug-in is particularly useful in surround work – as this will allow you to mute any multi-mono plug-in individually. (The normal Track Mute button on each Pro Tools mixer channel mutes all the channels of a multi-channel track.)

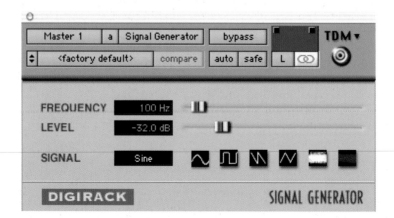

Figure 9.13 Signal Generator.

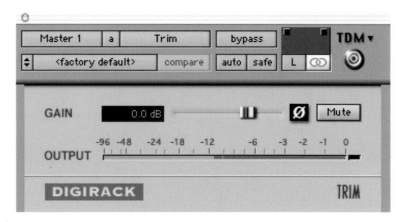

Figure 9.14 Trim.

Using the Trim plug-in

While you are working on your session, you may notice that the clip light on a particular channel has come on at some point during playback of the track. You know that you never moved the fader above the zero position, and you know that the loudest signal playing back from disk ought not to exceed this zero position. The only way this can have happened is that your concentration lapsed while you were recording this track and the level got too high at this point. Now, maybe you can't hear anything wrong, but you don't want to run the risk of overloading the inputs to a plug-in, for example. The Trim plug-in is exactly what you need to sort this problem out.

Take a look at Figure 9.15. You will notice that the red clip light is lit on the channel labelled NL01, while the channel fader is set at the zero position. Lowering the channel fader has no

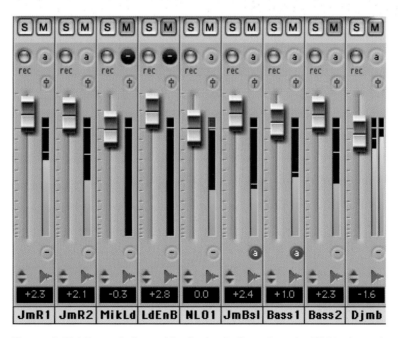

Figure 9.15 Mixer window with clipping indicated on the NL01 channel.

effect on the level of the audio that is playing back from the hard disk other than to control how much or how little of this signal is being fed to the monitors. And the red light tells us that the level being fed into the channel is overloading. Say, for instance, that the audio sounds a bit uneven in level and you want to use a compressor to tame it a little.

Insert the Trim plug-in first, and reduce the level in 0.5 dB steps until the clip light no longer lights in the 'problem' sections. Take a look at Figure 9.16 to see how this is set up. If, after playing through the whole track, the clip light is no longer lit, it is now safe to insert the compressor plug-in – secure in the knowledge that the input to this will not be overloaded at any point.

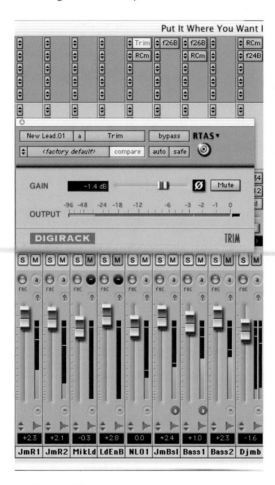

Figure 9.16 Mixer window with the Trim plug-in set to reduce the level at the input to the NL01 mixer channel. Note that there is no longer any clipping on the NL01 channel.

Inserting plug-ins during playback

With Pro Tools 6.0, plug-ins can be inserted or removed 'on-the-fly' during playback on Pro Tools|HD and Pro Tools|24 MIX systems. There are some limitations, of course. You can't insert or remove a plug-in that changes a mono track to a stereo output or replace a TDM plug-in by an RTAS plug-in. And you can't create a side-chain input, or remove a plug-in containing automation during playback – and you have to stop playback if you want to enable any plug-in parameters for automation. You can't insert or remove plug-ins while recording, as this would interrupt the recording process. Nor can you drag to or copy to different locations during playback or recording.

Optional Digidesign plug-ins

Digidesign offers several plug-ins as options, including Reverb One, DIN-R noise reduction, D-Fi for creating low-bandwidth 'grunge' effects, Sound Replacer for sample-replacement, Bruno and Reso for synthesizer effects, the Maxim limiter, and the SurroundScope for multi-channel phase, surround position and level display.

Digidesign distributed third-party plug-ins

Digidesign also distributes various third-party plug-ins including Dolby Surround Tools for LCRS monitoring, Synchro Arts VocAlign Project, VocAlign Pro, Titan and ToolBelt, Drawmer Dynamics for gating, compression and limiting, the Focusrite D2 EQ and D3 Compressor/ Limiter, Aphex Aural Exciter and Big Bottom Pro, Sonic NoNoise, Massenburg Design Works MDW Hi-Res Parametric EQ, and the DUY range of plug-ins, including Valve, Shape, Wide, Max, Tape, Z-Room, DSPider, ReDSPider and SynthSpider.

Plug-in packs

Digidesign regularly offers special deals on promotional bundles or packs of plug-ins to use with the various Pro Tools systems. These bundles include selections of the various Digidesign and Digidesign distributed plug-ins, along with third-party plug-ins from McDSP, Bomb Factory, TC Electronic, Wave Mechanics and others.

More goodies

Digidesign is obviously aware of the growing popularity of software such as Reason and Live, especially among those who like to work with samples and loops. IK Multimedia is also kicking up a storm with Sampletank (one of the best-sounding software samplers around), Amplitube (guitarists love to try different amps and effects), and T-RackS (to give mixes that 'mastered' gloss). So, the old saying goes, 'if you can't beat them, join them' – and that's exactly what Digidesign has done!

In September 2003, Digidesign announced: every new Pro Tools TDM or LE system includes specially adapted versions of Propellerhead Reason, Ableton Live, IK Multimedia AmpliTube, SampleTank SE and T-RackS EQ.

Together, these give you an amazing rack of virtual gear and tools that you can use throughout the production process. With Pro Tools 6.1's support for ReWire, you can tap into Reason's sound libraries or use Live for loop- and sample-based composition. AmpliTube and SampleTank give you even more scope to experiment with sound design and samples, while T-Racks EQ lets you sweeten your tracks.

Propellerhead's Reason Adapted provides a virtual rack of innovative synthesizers, samplers, drum-machines and effects. All the audio outputs from Reason Adapted can be streamed directly into the Pro Tools mixer via ReWire for additional processing within Pro Tools.

Ableton Live Digidesign Edition brings most of Live's key features to Pro Tools users. Live is a flexible sample sequencer that enables you to create, modify and play loops, phrases and songs. Using its unique real-time time-stretching features, multiple samples (originally at different BPM) can all be played back simultaneously at a user-defined tempo. As with Reason, Live's audio outputs can be routed into Pro Tools via ReWire.

SampleTank SE is a powerful sample playback module providing quick access to a host of included samples. Implemented as an RTAS plug-in within the Pro Tools mixer, SampleTank SE makes a great resource for MIDI compositions inside Pro Tools.

AmpliTube LE is an RTAS guitar processing and amp modelling plug-in that gives users easy access to several amplifier and cabinet models as well as classic stomp box and other effects. Use them on guitars and/or experiment with other track treatments.

T-RackS EQ, the EQ module from IK Multimedia's acclaimed T-RackS analogue mastering suite, offers Pro Tools users a powerful, high-quality tube-modelled parametric EQ as an RTAS plug-in for use in mixing and mastering.

'Our new partnerships with Ableton, IK Multimedia and Propellerhead provide Pro Tools users with an incredible rich, varied and powerful palate of tools that allow them to start creating in new ways "right out of the box,"' says Dave Lebolt, Digidesign general manager. 'Now musicians will have a vast new "construction set" of music-making possibilities available to them, regardless of their differing styles or interests. Wherever you want to go creatively, these tools can help take you there, and they're included free in every Pro Tools system, from Mbox to Pro Tools|HD – across the entire Pro Tools product line.'

For information on upgrades to the full versions of Ableton, IK Multimedia and Propellerhead software, visit www.ableton.com, www.ikmultimedia.com and www.propellerheads.se.

There can be no doubt about it – the inclusion of these products with Pro Tools systems will help to take Pro Tools even further ahead of its competitors.

Summary

All the typical signal processing that you would expect to have access to in a digital mixing console is available within Pro Tools – although you have to specifically insert the plug-ins as you need them. The advantage of this is that you have choices about how much signal processing you need to use on any particular session. If you don't need any EQ – don't insert an EQ plug-in. If you don't need compression – it is not inserted by default, so it is not using up any of your precious DSP resources. The disadvantage, of course, is that it is not always there when you need it.

There are many excellent third-party plug-ins available and these are easy to install. Of course, they do go out-of-date and need upgrades much more often than hardware-based signal processors. The strength of the plug-in system is that it gives you access to such an incredibly wide range of signal processing options for your Pro Tools system. And the range (and quality) of plug-ins available for Pro Tools systems is unmatched anywhere else.

10 Virtual Instruments

Introduction

Computer processing speeds have now reached the point where it is becoming increasingly viable to run software simulations of synthesizers, samplers and drum-machines on personal computers within the popular MIDI + Audio DAW's such as Pro Tools, Digital Performer, Logic Pro and Cubase SX – providing an even more highly integrated environment for music recording and production. Just as the wide range of software signal processing plug-ins has been developed – bringing the outboard into the computer environment – now the programmers are bringing in the MIDI gear as well. These software simulations are also known as virtual instruments as they are constructed using computer code rather than real hardware.

> Tip: Laptops (especially top-of-the-range PCs with 2.5 GHz or faster processors, such as the Sony Vaio) loaded with virtual instruments are increasingly being used on-stage and in the studio to run virtual instruments. They have the advantage of being much more compact than the racks of MIDI synthesizer and sampler hardware that they are beginning to replace.

Digidesign

Digidesign distributes the Access Virus Indigo and Waldorf Q synthesizers for TDM systems, along with the Prosoniq Orange Vocoder RTAS plug-in.

Access Virus Indigo TDM Synthesizer plug-in

Using the same DSP algorithms as the original Access Virus synthesizer, the Virus Indigo plug-in sounds exactly like its hardware counterpart. The plug-in has even more parameters than the original Virus that you can tweak to build dense, layered textures with the distinctive Virus analogue-type sound. And if you want results fast, you can use the new 'Easy' page, which provides rapid access to the most-used parameters. Virus Indigo does not disappoint when it comes to presets either – there are more than 1000 of these ready to go right out of the box. And this plug-in is efficient when it comes to DSP usage – allowing you to work with up to eight multi-timbral synthesizers on a single DSP. Virus Indigo supports up to 96 kHz sampling rates

Figure 10.1 Access Virus Indigo Easy page.

and offers up to 20 voices per Pro Tools|HD DSP, with a maximum of 160 voices total at 48 kHz, or 80 voices at 96 kHz.

Offering near-zero latency, the Virus Indigo TDM plug-in doesn't suffer from the CPU bottle-necks and reduced voice counts typical of many software synthesizers. The plug-in also provides many more visible controls and features than the Virus Indigo hardware synthesizer. And you can use a wide variety of control surfaces, such as the Control|24 or ProControl, or a hardware Virus unit or other controllers to manipulate the Virus Indigo's parameters. You can even load patches directly from the Access Virus hardware – so anyone who brings a hardware Virus into your studio can transfer their favourite patches to your system. And you can use Virus Indigo as an effects device – the special 'Input Mode' lets you route complete mixes through the Virus filter section. All-in-all, this is one heck of a useful plug-in to have available on any Pro Tools system.

Waldorf Q TDM

The Waldorf Q TDM is a synthesizer plug-in for Pro Tools|HD systems based on the powerful synthesis architecture of the Waldorf Q synthesizer. Due to its various oscillator and filter models, its ultra-fast envelopes and LFOs and its extensive FM routings, it creates previously unheard sounds and faithfully reproduces classic analogue patches.

Figure 10.2 Waldorf Q synthesizer.

The Waldorf Q TDM plug-in works with Pro Tools|HD and Pro Tools|MIX systems on Mac OS9, Mac OSX, Windows 2000 and XP.

Orange Vocoder

Not exactly a synthesizer, but something that you may want to use with a synthesizer is a 'vocoder'. This allows the sonic characteristics of an input signal to be imprinted onto a synthesised signal – so you get a talking synthesizer effect, for example. Back in 1978, Herbie Hancock had a big hit with a song called *I Thought It Was You* – featuring Herbie 'singing' his synthesizer using a Sennheiser vocoder. Vocoder effects have waxed and waned in popularity over the last 30 years or so, and are currently enjoying something of a comeback.

Digidesign distributes the Prosoniq Orange Vocoder RTAS plug-in. This includes an eight-voice analogue synthesizer section, breakpoint-configurable EQ section, and a filterbank reverb section all in a single plug-in. The presets include Robot Voice, Rotating Robot, Jazz Vocoder, Synthetic Speech, Talking Voices, Ethereal Voices, F Maj Vocoder, Rubber Tongue and Weird Talk – you get the idea.

Figure 10.3 Orange Vocoder.

Spectrasonics

Spectrasonics offer their Stylus, Atmosphere and Trilogy sample replay plug-ins with excellent and comprehensive libraries of sounds. Spectrasonics' founder, Eric Persing, has been programming sounds for popular Roland synthesizers since time began and has been producing sample libraries on CD-ROM throughout the last decade featuring his innovative 'groove control' system. Now Persing has combined the best of his innovations and his creative programming and production talents to produce a kind of hybrid software instrument that plugs in to Pro Tools via RTAS. Versions are also available for Digital Performer, Logic and Cubase.

Note: TDM users must open Spectrasonics virtual instruments using audio tracks. Aux tracks cannot be used because RTAS is not available for these tracks. (However, this is possible for LE and Pro Tools Free users.)

Tip: Logic Platinum users working with TDM hardware can use the Spectrasonics VST plug-ins within Logic Platinum and output the audio from these into the TDM mixer using Emagic's ESB software.

Stylus

Stylus comes with a 3 Gb library of 'groove-control' elements, i.e. audio samples, recorded by Persing. Stylus is not a sampler and you can't load in your Akai or AIFF samples – but if you want a sampler, there are plenty out there to choose from. Instead, Stylus combines really powerful control over its own library of sounds with a really simple interface.

Figure 10.4 Spectrasonics Stylus Vinyl Groove Module.

The sounds load faster than just about anything out there – so you can load up basic beats with lots of variations and swap between these like lightning. The loops are marked with the BPM they were sampled at. There are 700 'grooves' to suit the various dance music genres, including a whole section of 'killer' percussion loops. If the loop you pick is running at the wrong tempo, you just load up the 'groove control' version. This puts all the elements of the loop you have chosen onto your keyboard. Then you select the accompanying MIDI file and drag and drop this onto the sequencer track you want to use. Now you can speed the loop up or slow it down without hearing any artifacts. You can also raise or lower the pitch as well without affecting the tempo. Even better, each slice mapped to each key can have its own parameters – put a filter on one slice, pitch another up or down, and you have a new sound for your groove. You can automate the changes to these parameters using the plug-in and you can even apply random changes every time through the loop. It's dead easy to swap snare sounds or whatever once you have found a groove you like. And there's even a selection of scratched sounds like brass, guitars, or whatever – so you can do the DJ thing.

Atmosphere

Atmosphere features an excellent selection of synthesizer pads, ambient sounds, belltones, Swells, Evolving sounds, Sweeps and so forth that will keep anyone working on ambient or film music absorbed for hours on end. Atmosphere's interface and custom UVI engine features a dual-layer architecture that can create extremely powerful and dynamic sounds.

Figure 10.5 Spectrasonics Atmosphere Dream Synth Module.

Atmosphere features a massive 3.7 Gb core library, which provides much more variety than any hardware instrument. Spectrasonics created this library using everything from processed vocal recordings, to prepared pianos and glass harmonicas, to vintage synthesizers – and even their experiments with hundreds of plug-ins and signal processors. Now you can take advantage of all this detailed preparation and use the sounds in the core library as starting points for your own synthesized sounds. And, for the busy composer, the presets will serve you well.

Trilogy

Trilogy completes the, er, trilogy of plug-ins – providing the 'bottom end' for your recordings with its tremendous selection of bass sounds. Overall, I rate the Total Bass Module extremely highly.

Every patch in Trilogy has two layers that you can tweak individually – editing each independently. You can also mix and match any of the layers in the core library, to combine the sound of a real Minimoog with a Fretless bass, or a Virus with a TB-303, or even to add a Juno sub-oscillator to an Upright Bass!

The highly detailed, chromatically sampled, Acoustic Upright Bass is one of Trilogy's highlights. An incredible variety of tones can be produced from this because the interface allows separate control of the Neumann U-47 Tube Microphone signal and the Direct Pickup signal, which was sampled through a vintage Neve 1083 Console. This acoustic bass, miked using the U47, is the most faithful reproduction of the sound of a double bass that I have heard anywhere – a sound 'to die for'!

There is also a huge selection of Electric Basses including classic four, five and six string models, performed in Fingered, Picked, Muted, Rock and Roll, Slapping, Ballad, Fretless and R&B techniques through rare, custom-made tube pre-amps. Special variations include Harmonics, Glisses, Fuzz, Trills, FX and thousands of Slides. This selection provides plenty of scope when you need electric bass sounds.

Figure 10.6 Spectrasonics Trilogy Total Bass Module.

Synth bass sounds include samples from legendary Analog Bass Synths like the Minimoog, Roland Juno 60, Roland TB-303, Roland SH-101, Oberheim SEM, Moog Taurus, OSCar, Virus, Yamaha CS-80, Arp Odyssey and 2600, Studio Electronics SE-1, Omega and ATC Tone Chameleon, Sequential Circuits Pro One, the mighty Moog Voyager and many others. As with the acoustic bass recordings, these sampled synthesizer sounds are extremely impressive and useful.

Propellerheads

Reason

Propellerhead's Reason is a rack of synthesizers, drum-machines and samplers allied to a simple-to-operate main sequencer. From the point of view of the user on a budget, Reason has just about everything you might want in one package – a simple sequencer with all the sound modules you need to put some impressive pieces of music together.

Sound modules in the rack include the SubTractor Analog Synthesizer, the Malstrom Graintable Synthesizer, the NN19 Digital Sampler and the NN-XT Advanced Sampler. Reason's DR.REX Loop Player lets you play back REX files created in ReCycle. In case you haven't come across this before, Propellerhead's ReCycle software works with sampled loops. By 'slicing' a loop and making separate samples of each beat, ReCycle makes it possible to change the tempo of loops without affecting the pitch and to edit the loop as if it were built up of individual sounds. As a further convenience, if you have Propellerheads' ReBirth drum-machine simulation, you can run this on your computer in sync with Reason and feed the audio from ReBirth directly into Reason using Reason's ReBirth Input Machine module. Reason also has its own drum-machine module called Redrum. This is a sample-based drum-machine with ten drum sound channels into which you can load the factory samples – or your own sounds in AIFF or WAVE format. As with Rebirth, Reason has a built-in Roland-style pattern sequencer, allowing you to create classic

Figure 10.7 Reason Rack Modules.

drum-machine patterns. You can also use Redrum as a sound module – playing it live from an external MIDI controller or from the main Reason sequencer, rather than using its built-in sequencing capability.

Reason's monophonic pattern sequencer, the Matrix, is similar to a vintage analogue sequencer. Just connect this to any of the MIDI devices in Reason and it sends simulated CV (pitch) and Gate CV (note on/off plus velocity) or Curve CV (for general CV parameter control) signals to the device or device parameter. Before MIDI was invented, monophonic analogue synthesizers could be hooked up to a hardware sequencer using patchcords and Reason's Matrix simulates this type of sequencer – even down to the patchcords!

To handle the audio outputs there is a Mixer in the rack, based on the popular Mackie 3204 rackmount model. This has fourteen stereo channels, a basic two-band EQ section, and four effect sends. You get a bunch of effects units as well, including the RV-7 Digital Reverb, the DDL-1 Digital Delay Line, the D-11 Foldback Distortion, the CF-101 Chorus/Flanger, the PH-90 Stereo Phaser, the COMP-01 Compressor/limiter, the PEQ2 Two Band Parametric EQ, and the ECF-42 Envelope Controlled Filter. The latter is a synth-style resonant filter with three different filter modes and you can use a drum-machine or the Matrix sequencer to trigger its envelope to get some truly 'nasty' sounds!

Reason lets you start out with an empty rack and add devices as you need these, although the default song opens with a useful selection of devices already there for you to work with – all hooked up automatically. But what if you want to change the routings? Just press the Tab control on your computer keyboard and the rack 'turns round' to reveal the back panels of the equipment. Here you can see the connections between devices indicated by 'virtual patch cables'. Connections between instrument devices and mixers use red cables; connections to or from effect devices use green cables; and CV connections use yellow cables. Simply make your connections by clicking and dragging from one 'socket' to another on the back panels – just like on the 'real thing'.

You can control Reason from Pro Tools via ReWire and either run its sequences in sync with Pro Tools with the audio routed into the Pro Tools mixer – or you can go the whole way and play Reason's rack of synthesizers using MIDI tracks within Pro Tools. So you can use the Reason sequencers if you like – or you can ignore these and use the much better MIDI sequencing facilities in Pro Tools instead.

Bitheadz

Unity Session

Unity Session is a whole set of plug-ins – eleven in number – that you can run from within a wide range of host environments including Pro Tools. You get software sampling, analogue synthesis, physical modelling, MIDI effects, and audio effects – all from within your favourite sequencer. Stand-alone operation is also provided for, using the Unity Editor, Player and Mixer applications.

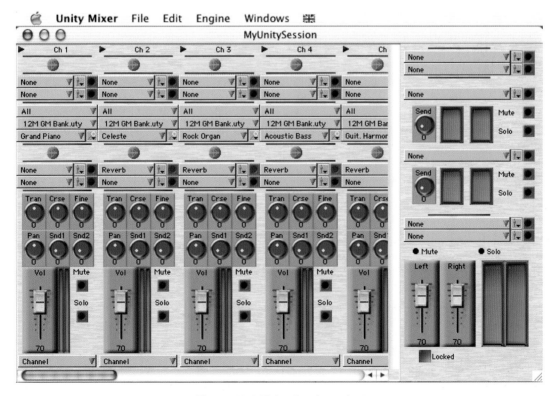

Figure 10.8 Unity Session mixer.

The various samplers and synthesizers are implemented as plug-ins that use the basic Unity 'engine' for playback. These include the original DS-1 sampler and AS-1 analogue synthesizer, plus the new SP-1, SampleCell, GigaSampler, SoundFonts, and DLS sample playback modules, and four physical modelling synthesizers – the CL-1, FL-1, BW-1 and HS-1 which model the clarinet, flute, bowed string and hammered string, respectively. A set of seven MIDI plug-ins lets you process any MIDI that you are sending to a Unity synthesizer to create arpeggiated chords, splits, layers and other effects. A set of 23 Audio plug-ins is also provided to let you add effects such as EQ, delay, or reverb using the 'Unity Editor' and 'Unity Mixer' applications. Unity works with ASIO, ReWire, MAS, VST, DirectIO, RTAS and DirectConnect. These interfaces provide excellent integration with products such as Pro Tools|HD, Digital Performer, Logic Pro, Nuendo and Cubase SX.

Native Instruments

Native Instruments has become one of the leading developers, with a range of software available for both Mac and PC that plugs in to just about every software application that you are likely to be using.

Native Instruments Studio Collection

The Studio Collection combines the Pro Tools Editions of the B4 organ, the Pro-53 virtual analogue synthesizer and the Battery drum sampler into a powerful collection of creative instruments for use in Pro Tools TDM, LE and Free in HTDM or RTAS formats.

B-4

The B-4 plug-in is, quite simply, the best simulation of the classic Hammond B3 tonewheel organ and Leslie rotating speaker sound that I have ever heard – and I have tried several. Native Instruments have modelled all the tonal characteristics of the original B3 and the Leslie rotating speaker – complete with 'faults' of the original, such as 'key-click', which many musicians regard as a part of the sound. Two views are provided in the B4 window. The Console view shows the organ console with the upper and lower manuals, the bass pedals, the drawbars, the expression pedal and the performance switches. The Control View shows all the controls you need for editing in more detail.

You can play the upper and lower manuals and the bass pedals using three different MIDI keyboards, each set to different MIDI channels – or you can play everything from one MIDI keyboard using the keyboard splits mode. One thing that Hammond players will immediately miss is the hands-on control of the drawbars that many musicians adjust continuously while playing. The good news is that you can hook up any box of MIDI faders, such as the Peavey PC1600x or the Kenton Control Freak 16, to provide real controls for the B4's drawbars. The 'icing on the cake' is the rotating speaker which speeds up and slows down smoothly when you switch between speeds – just like a real Leslie would – and the software even lets you adjust the slow and fast speeds to your liking. The amplifier controls can also produce a wide range of different sounds – from clean to overdrive – and there is even a set of controls provided to emulate the way the 'virtual microphones' are set up around the speaker cabinet. Compared with the original, the B4 has a much wider range of presets and also features velocity-sensitivity when you play the keys (which you cannot get on a real Hammond organ but which can be a useful performance feature). If you are serious about the Hammond sound – then you need the

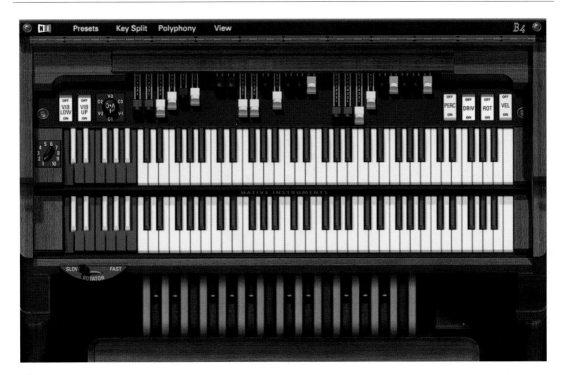

Figure 10.9 Native Instruments B4 Console view.

B4. The bonus is that you can even use the Leslie rotating speaker simulation as a VST insert-effect processor with VST-compatible software. So you can play your guitar through the Leslie as well!

Pro-53

Native Instruments Pro-53 software synthesizer is a full-blown emulation of the classic Prophet 5 synthesizer, right down to the colour of the wooden casing. And, unlike the original, you are not restricted to just five notes of polyphony – you can play as many notes at the same time as your CPU will allow. The Pro-53 comes with eight files each containing eight banks each containing eight presets – providing a total of 512 preset programs or 'patches'. These sound extremely close in character to those of the original Prophet 5. Unlike the original Prophet 5, the Pro-53 will respond to MIDI Velocity messages that you can use to control the filter and output levels according to how fast you strike the keys. What I missed, badly, from the original Prophet 5 was the intimate control of the knobs and buttons on the real thing. What I liked was the convenience of access to the preset library and the excellent quality of the sound. If you liked the sound of the Prophet 5 you will be even more impressed with the Pro-53.

Pro-53, the third generation of the virtual analogue classic, replaces Native Instruments Pro-52 model. The synthesizer emulation has been vastly improved by a new oscillator technology that offers an even warmer and more brilliant sound. New features like a high-pass filter mode and an invertible filter envelope have been added and the control surface has been improved.

Figure 10.10 Native Instruments Pro-53 synthesizer.

Battery

Native Instruments Battery percussion sampler features 32-bit internal resolution and comes with a comprehensive library of 30 sets of percussion sounds. It can be used simply as a sample playback unit, or you can manipulate the samples using velocity layers, individual tunings, volume and pitch envelopes, apply bit reduction and waveshaping and various modulations. Battery can read samples from Akai, SF2, LM4, AIFF, SDII, WAV and MAP sources with any bit resolution from 8 to 32 bits – so you can load in sounds from more or less any common source.

Load a kit by clicking on the 'File' button in the Master section. You will then see the names of the individual drum sounds written in the different cells in the matrix of 'pads'. Click on any 'pad' to hear the sound play back. You can drag and drop samples and parameters between these cells, and underneath the cell matrix you can set various parameters for each cell, such as the MIDI note or note range that will play any particular cell. The Tuning and Shape edit section has four global settings – cell tuning, wave shaping, bit reduction and sample start offset. A wave-form display shows how these controls affect your sounds, and editable envelopes for pitch and volume are also provided. Battery scores over most of its competition by having 54 pads that can be grouped together to play from a particular MIDI note number. The envelopes, bit crusher and so forth make it great for dance music. And you get 16 outputs from one Battery, so you can have your entire drum library set up with separate outputs.

Emagic software for use with TDM hardware

Logic Platinum includes its own set of software instruments, although you have to pay extra to get 'keys' to 'unlock' these. What is not so widely known is that most of Emagic's virtual instruments are available for Pro Tools TDM systems – as are a selection of Logic's signal processing plug-ins.

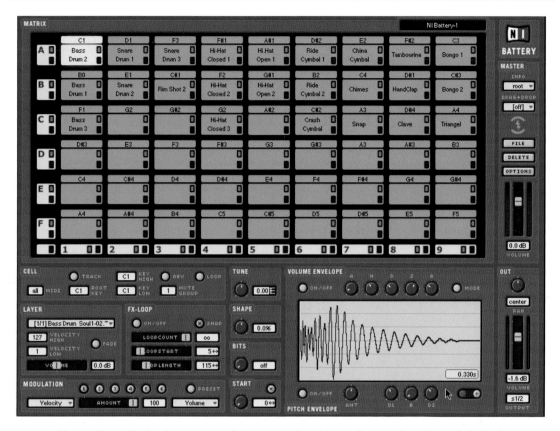

Figure 10.11 Native Instruments Battery plug-in with the Factory Soul Drumkit loaded.

HD Extension

Emagic's Logic Platinum software is often used as an alternative to the Pro Tool software and is compatible with PT|24 and MIX TDM hardware as standard, although to use the Pro Tools|HD hardware you need to buy the Emagic HD Extension.

Host TDM Enabler

If you are using Pro Tools TDM hardware, you will probably also want to buy Emagic's Host TDM Enabler, which lets Emagic's software instruments use HTDM. Using the Host TDM Enabler you can insert Emagic software instruments such as the ES1, ES2, EVP88, EVB3 and EVD6 into aux channels within Logic's TDM mixer so that they play back through your Pro Tools TDM hardware. You can use up to 32 mono or 16 stereo channels (or any combination of either). These plug-ins run on your computer's CPU, not on the Pro Tools TDM cards, but you can process the outputs from any of these directly using any of your TDM plug-ins.

> Tip: The HostTDM Enabler can be used with Logic Platinum, or with MOTU Digital Performer, or with Digidesign Pro Tools software – whether running on Pro Tools|24 MIX, Pro Tools MIX or ProTools|HD systems.

Figure 10.12 EVB3 and EVP88 in the Pro Tools TDM mixer.

Note: The EVOC20 and the EXS24 are not available in HTDM format.

ESB TDM

Another optional extra for Logic users is the Emagic System Bridge (ESB TDM). To put it 'in a nutshell', ESB lets you play your VST virtual instruments and effects plug-ins back through Digidesign TDM cards when you are using Logic Platinum as the host software.

The ESB TDM allows all Logic tracks, native and VST plug-ins (Mac OS9) or Audio Unit plug-ins (Mac OSX), including Audio Instruments, to be routed into the Aux inputs of Logic's TDM mixer via Direct TDM. Additional processing can be applied using the Pro Tools TDM plug-ins that use the TDM system DSPs.

Note: ESB TDM also allows the EXS24 be inserted into the Aux channels of Logic Platinum's TDM mixer.

Typically, Digital Audio Workstations either use dedicated hardware DSPs, as with TDM systems, or use the native processing capabilities of the computer's CPU. The ESB TDM connects the TDM DSP and CPU-based systems within Logic Platinum using 32 sample-accurate audio streams to connect the DSP- and CPU-based worlds via two components: Direct TDM and the EXS24 TDM.

Logic Platinum, in conjunction with the ESB TDM, provides support for multiple hardware devices via an additional audio engine, known as Direct TDM, which runs in parallel with DAE/TDM. As a fully-featured 32-bit audio engine, Direct TDM offers access to all of the facilities that native processing has to offer: up to 96 audio tracks, over 50 built-in Emagic plug-ins, plus support for Emagic's ES1, ES2, EVP88, EVOC20 and third-party VST 1.0 and 2.0-compatible effects and Audio Instruments. The ESB TDM provides up to eight discrete mono, or four stereo, mixer sum outputs from the native Direct TDM audio engine. These can be streamed into Logic's TDM mixer for further treatment with TDM plug-ins.

ESB TDM also allows the direct insertion of up to 32 instances of Emagic's EXS24 Sampler into the Aux channels of Logic's TDM mixer. The stereo output signals of each EXS24 instance can be treated via any of the processing options afforded by the TDM DSP environment, such as TDM plug-ins. Each instance of the EXS24 TDM runs on the host CPU, and places no overhead on the TDM DSPs. The MIDI performances of each EXS24 TDM instance are recorded on TDM Auxiliary tracks, and are controlled directly in Logic Platinum – providing the benefit of sample-accurate playback for all instances of the EXS24 TDM.

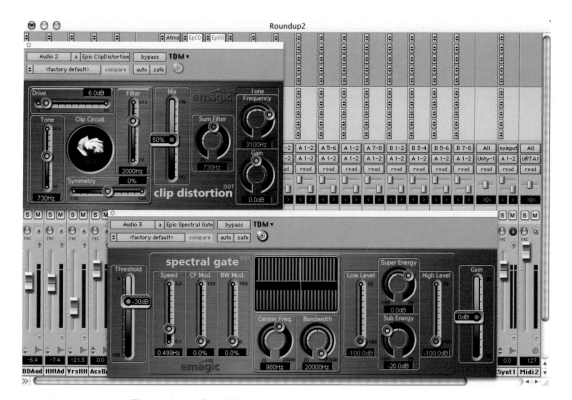

Figure 10.13 Epic TDM plug-ins in the Pro Tools TDM mixer.

Epic TDM

You can also buy Epic TDM – a suite of 15 of Logic's own plug-ins that you can use with TDM hardware. 'In a nutshell', this means that you can use your favourite Logic plug-ins within Pro Tools TDM software – or from within any other TDM host software, such as Digital Performer or... Logic Platinum! That is, from mixer channels in Logic Platinum configured to use the TDM hardware.

The set includes Spectral Gate, SubBass, Autofilter, Modulation Delay, Ensemble, Phaser, Tremolo, BitCrusher, PhaseDistortion, ClipDistortion, Overdrive, Distortion, Enveloper, Tape Delay and the ES1. These versions of the standard Logic signal-processing plug-ins actually run on the DSP chips on Pro Tools III, 24, MIX or HD hardware systems. As Emagic puts it, 'even die-hard Logic Platinum users will benefit from using EPIC TDM'. Particularly if you want to use power-demanding Logic plug-ins such as the Spectral Gate or Ensemble, it makes sense to run these on your TDM system's DSP – thus reducing the load on your host computer.

Logic Pro update

Just as this book was being prepared for printing, news arrived that Apple has renamed Logic and re-jigged some features. The flagship software is now called Logic Pro and it includes Emagic's entire range of plug-ins and virtual instruments, such as the EXS24 sampler and Space Designer reverb, for no extra cost! Support for Pro Tools|HD is now built-in, along with the ESB TDM functionality. Sadly, the Host TDM Enabler and the Epic TDM plug-ins have been discontinued. One giant step forward – one rather smaller step backward!

Summary

Virtual instruments are the next 'revolution' for Pro Tools systems – bringing the music-making equipment under the umbrella of the Pro Tools environment. At the time of writing, this revolution is just 'getting into gear'. It will be interesting to see if sales of the original hardware instruments are affected by the increasing popularity of their software simulations. And they are not all simulations of actual instruments. Many new instruments are being developed – often with features that would be impossible to implement in hardware. You will need plenty of RAM and the fastest (ideally dual-processor) computer you can afford for best results. As processor speeds increase, which they inexorably do, these virtual instruments will become ever more powerful. There can be no doubt that they will play an increasingly significant role in the future.

11 ReWire

Introduction

Digidesign made yet another good move with their version 6.1 software – they added compatibility with Propellerheads Software's ReWire 2.0 technology. This instantly made it possible to use Propellerheads' Reason and Ableton's Live with Pro Tools – both of which can use ReWire and both of which are extremely popular 'tools' for music production.

Using ReWire, you can transfer MIDI and audio data between applications running on the same computer without using any external connections – ReWire makes 'virtual' connections internally. MIDI data sent from Pro Tools, for example, can play back a software synthesizer in Reason – and audio data from Reason can be played back through the Pro Tools mixer. The timing of the linked applications is synchronized with sample accuracy and you can use the transport controls on either application to control the other. Of course, once the audio outputs from Reason's software synthesizers and samplers are routed into Pro Tools you can process these incoming audio signals with plug-ins, automate volume, pan, and plug-in controls, and use the Bounce To Disk command to 'fix' the incoming audio as files on disk.

So how does ReWire work with Pro Tools? Compatible ReWire client applications are automatically detected by Pro Tools and are made available as RTAS plug-ins. When you insert any of these plug-ins into the Pro Tools mixer, Pro Tools automatically launches the corresponding ReWire application if the client application supports this feature – or you launch the ReWire application manually if not. When the ReWire application has been launched, any corresponding MIDI nodes for that application become available as destinations in the MIDI Track Output selector in Pro Tools.

ReWire in action with Reason

To use Reason, for example, you launch Pro Tools first, open a session, then create a MIDI track that you will use to play back one of Reason's rack of synthesizers and an Audio track to receive the audio output from this synthesizer.

In the Mix window, insert the RTAS ReWire plug-in for Reason using the audio track's Inserts section. The Reason ReWire plug-in is available as a Multi-Channel or as a Multi-Mono RTAS plug-in.

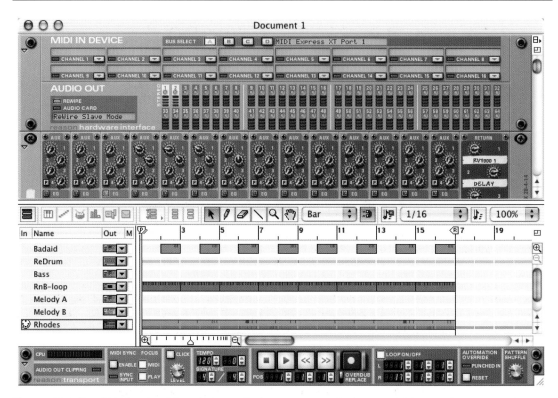

Figure 11.1 The ReWire plug-in's window will also open – ready for you to choose the outputs that you
want to use from Reason.

Reason will launch automatically in the background as soon as you insert the ReWire plug-in.

Select the outputs that you want to use from Reason by clicking on the pop-up list of Reason outputs located at the lower right of the plug-in's window.

Figure 11.2 Multi-Channel (Stereo) Reason ReWire plug-in before choosing Reason outputs.

If you have opted to use the Multi-Channel RTAS plug-in and you are running a stereo Pro Tools session, the only choice available here will be Reason's Mix L–Mix R output pair.

Figure 11.3 Multi-Channel (Stereo) Reason ReWire plug-in after choosing Reason's Mix L and Mix R output pair (the only choice available in this case).

If you have opted to use the Multi-Mono RTAS plug-in and you are running a stereo Pro Tools session, the plug-in looks slightly different, featuring an 'L'/'R' pop-up selector for the left and right channels and a stereo Link button that defaults to being selected.

Figure 11.4 Multi-Mono Reason ReWire plug-in before choosing Reason outputs.

With the two channels linked, this means that when you select an output channel using the Reason outputs selector lower down in the window, whichever channel you select applies to both the left and right outputs – which is almost certainly not what you want! Normally you will need to deselect the Link button by clicking on it. See Figure 11.5.

With the Link button deselected, you can choose a Reason output for the left channel. See Figure 11.6. Then change the pop-up channel selector to 'R' and choose a different Reason output for the right channel. See Figure 11.7. Select the outputs from Reason that you want to use.

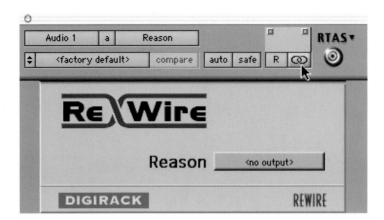

Figure 11.5 Multi-Mono Reason ReWire plug-in with stereo Link deselected.

Figure 11.6 Multi-Mono Reason ReWire plug-in with left channel set to Reason output Channel 3.

Figure 11.7 Multi-Mono Reason ReWire plug-in with right channel set to Reason output Channel 4.

Setting up the Reason synthesizers

Normally, when you launch Reason, the standard default song document opens. This contains a mixer, various effects modules and several Reason synthesizer modules – along with the demo song sequences. If you are using Reason with Pro Tools, it often makes more sense to avoid using Reason's mixer and effects – because you can use the much more full-featured mixer and higher quality effects in Pro Tools itself. In this case you could prepare a custom default song containing just the Reason devices you plan to use with their outputs connected directly to the Reason hardware interface. This hardware interface then feeds Reason's audio outputs directly into Pro Tools via the ReWire RTAS plug-ins.

To set this up, you can set the General Preferences in Reason to open an Empty Rack first. With this Empty Rack open in Reason, you can add whichever Reason devices you anticipate using in Pro Tools.

Each time you create a new device in your Reason rack, a sequencer track with its output set to this device will be automatically added to Reason's main sequencer. See Figure 11.8.

> Tip: If you are not intending to use Reason's main sequencer, you can click and drag on the dividing line between the sequencer and the modules above it until it is completely hidden from view. You can also collapse Reason's Transport controls by clicking the small arrowhead at the far left of this module.

At the back of the Reason rack, you can choose which modules are connected to which outputs. See Figure 11.9.

When you have set up your Reason synthesizer rack the way you want it, save this, say, into your Reason folder, as a new song document. You might call this 'Reason Synths' – or whatever works for you. Now change Reason's General Preferences to open a Custom song as default, and select the 'Reason Synths' song document. See Figure 11.10.

Once your default song is set up in Reason, this is the song that will be opened each time you launch Reason or open a New document from Reason's File menu. Reason expects that you will record new sequences into its own sequencers and that you will save this new document (based on the default song) as a new file – named however you wish. If you do not record any new sequences into Reason or make any changes, then there will be no need to save a new Reason document before ending your session. However, if you have selected different patches for your synthesizers, added or removed any modules to or from your Reason rack, or made any changes to the module settings that you want to keep, then you will need to save a copy of this edited Reason song document – preferably into the Pro Tools folder for your session.

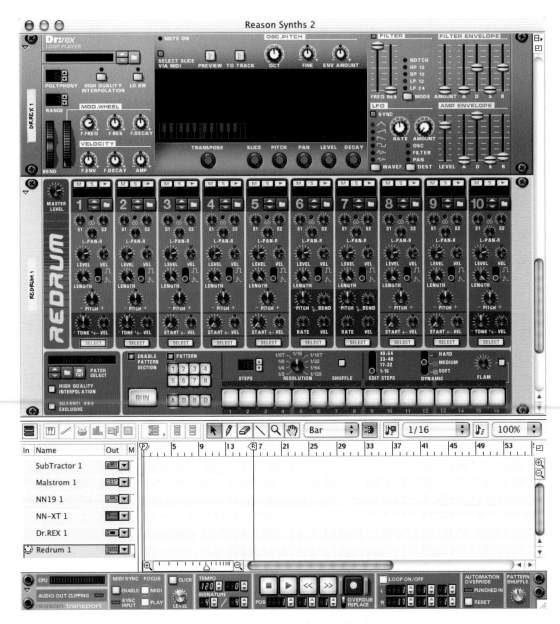

Figure 11.8 Reason Synths Rack.

Figure 11.9 Rear of Reason Rack showing module output connections.

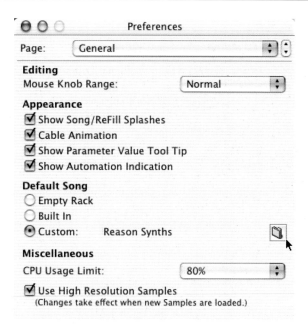

Figure 11.10 Setting Reason's Preferences to open a Custom Default Song.

Routing MIDI from Pro Tools to Reason

When you have inserted a Reason RTAS plug-in and Reason has launched, all the available Reason sound modules in the current Reason document will be listed as Pro Tools MIDI Track Output destinations.

To set up the NN-XT, for example, which I connected to Reason outputs 9 and 10, I had to make sure that I had a stereo ReWire Multi-Mono RTAS plug-in inserted into a stereo audio track with ReWire Channels 9 and 10 selected as the L and R inputs to the plug-in.

Figure 11.11 ReWire Channels being selected.

Then I chose NN-XT from among the Reason devices listed as Track Output destinations for a MIDI track – which I named NN-XT.

Figure 11.12 Setting up Pro Tools to play Reason's NN-XT module.

To play the Reason sound modules from a MIDI keyboard connected to your MIDI interface, all you need to do is to make sure that MIDI Thru is selected in the MIDI menu and that the Record button is enabled in the Pro Tools MIDI track. Now you are all set up and ready to record using the Reason module you have chosen.

Playing ReWire application sequences in sync with Pro Tools

Pro Tools sends both Tempo and Meter data to ReWire client applications, so ReWire compatible sequencers can follow any tempo or meter changes in a Pro Tools session.

Reason has its own built-in sequencer and the Reason document you have opened may contain sequences that you have recorded in Reason. Similarly, you may have recorded sequences using Ableton Live that you want to play back in sync with your Pro Tools session. To play these sequences in sync with Pro Tools, simply hit the space-bar or click the Play button on the Pro Tools Transport.

Note: With the Pro Tools Conductor button selected, Pro Tools always acts as the Tempo master, using the tempo map defined in its Tempo Ruler. With the Pro Tools Conductor button deselected, the ReWire client acts as the Tempo master. In both cases, playback can be started or stopped in either application.

Using Ableton Live with Pro Tools

One way to use Live with Pro Tools is to build sequences and arrangements using Live stand-alone, then open these for playback into the Pro Tools mixer. With this set-up, you can conveniently record the output from Live onto Pro Tools tracks, then quit Live to reduce the load on your CPU. With the Live material in Pro Tools, you can then record additional material alongside this.

Another way to work is to build sequences and arrangements in Live while it is being hosted by Pro Tools, so that you can also record material directly into Pro Tools at any stage during the session. This is a very CPU-intensive way to work, so it will only work well with the very fastest systems.

To use Live with Pro Tools, launch Pro Tools first and open a session, then create at least one Audio track to monitor the audio output from Live.

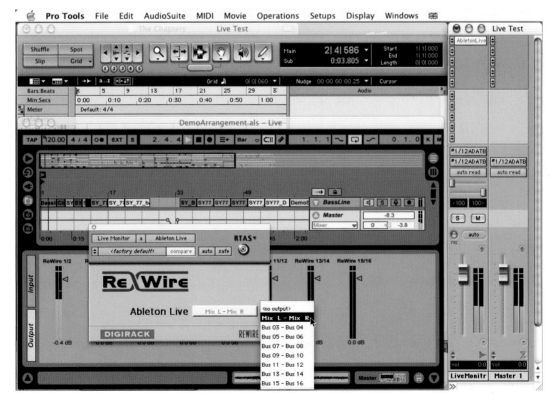

Figure 11.13 Setting up Pro Tools with Ableton Live.

In the Mix window, insert the RTAS ReWire plug-in for Live using this audio track's Inserts section. The Live ReWire plug-in is available as a Multi-Channel or as a Multi-Mono RTAS plug-in.

Unlike Reason, Live does not launch automatically in the background as soon as you insert the ReWire plug-in, so the next step is to launch this manually.

And that's all there is to it! Now you can press Play either in Pro Tools or in Live, and both applications will start to play back.

Recording from Ableton Live into Pro Tools

Recording the audio output from Ableton Live into Pro Tools is similar to recording audio from virtual instruments into Pro Tools. You need to add another audio track (or tracks) to record onto. The audio tracks that you are using to monitor the audio output from Live will not allow you to record this audio to disk. The solution is to route the audio from the audio track (or tracks) you are using to monitor the audio coming from Live to these additional tracks via one or more of the internal busses.

Once this is set up, all you need to do is to record-enable the track or tracks that you intend to record onto, then click Record and Play in the Transport window.

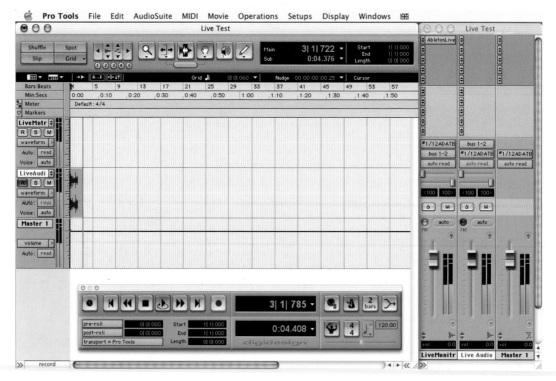

Figure 11.14 Recording into Pro Tools from Ableton Live.

Looping playback

If you want to loop playback in Pro Tools, just select the time range that you want to loop in the Pro Tools Time Line before starting playback. Another way to do the same thing is to set loop or playback markers in the ReWire client sequencer before you start playback.

There are a couple of things to watch out for here. For instance, if you create a playback loop by making a selection in the Pro Tools Time Line, once playback is started, any changes made to loop or playback markers within the ReWire client application will de-select the Pro Tools Time Line selection and remove the loop. Also, some ReWire client applications, such as Reason, may misinterpret Pro Tools meter changes, resulting in mismatched locate points and other unexpected behaviour. So you should avoid using meter changes in Pro Tools when you are using Reason as a ReWire client.

Caveats

One restriction on TDM systems is that ReWire RTAS plug-ins can only be used with audio tracks – not with auxiliary inputs.

Be aware that every ReWire channel you use uses up one of your available Pro Tools voices. For example, if you insert a stereo ReWire plug-in into a stereo audio track, you have used up four Pro Tools voices. And think about this: ReWire supports a maximum of 64 audio streams per application. If you used all of these in a 48 kHz/24-bit session on a 128-voice Pro Tools|HD 2 core system, you would have no more audio channels free – as you would need 64 audio channels to accept the audio from the 64 ReWire channels. The point to note here is that you are going to use up your available Pro Tools voices much more quickly when you are using ReWire plug-ins.

You should also bear in mind that using ReWire at higher sample rates will increase the load on the CPU. Double the sample rate and you double the CPU load. The CPU load at 96 kHz, for example, is double the load at 48 kHz. So, when using ReWire, it is always a good idea to keep an eye on Pro Tools' System Usage window to make sure that you are not overloading your system.

Finally, when it's time to quit your session, don't forget to quit the ReWire client application first, then quit Pro Tools.

Summary

Allowing Pro Tools users to work with two of the most popular music software packages is clearly a 'master stroke'. Previously, some musicians, songwriters and remixers would complain about Pro Tools saying that it was maybe OK for engineers to use for mixing, but that it was not too 'hot' when it came to creative music making. Adding ReWire compatibility now makes it possible for anyone to get musically creative within the Pro Tools environment.

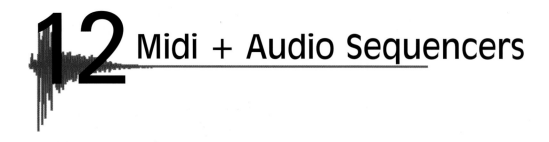

12 Midi + Audio Sequencers

Introduction

Musicians working with MIDI will often choose to use a full-featured MIDI + Audio sequencer as the 'front-end' for their Pro Tools hardware – in preference to the Pro Tools software. Although Pro Tools has reasonable MIDI features, these are no match for the MIDI features in a dedicated MIDI sequencer. For example, Pro Tools does not display MIDI data as conventional music notation – a feature that many composers and musicians find indispensable. There are also various other software packages that creative musicians, composers, producers and remixers often like to use when working with MIDI or with audio loops.

On the Mac, the four main alternatives to Pro Tools that are used for MIDI + Audio recording are Digital Performer, Logic Pro, Cubase SX and Nuendo. Digital Performer and Logic Pro both offer full support for TDM systems – so you can access all the inputs and outputs on your Digidesign TDM hardware from the mixing consoles in Performer or Logic, and you can insert TDM plug-in into these mixers just as you can in the Pro Tools software. Cubase SX and Nuendo do not offer TDM support, but, as with most audio software for OSX, these can use Pro Tools hardware for basic input and output via OSX's Core Audio. BIAS software offers Deck as an alternative to Pro Tools on the Mac – although this only has a very basic MIDI file playback capability. Ableton Live is a popular tool to use with loop-based music that also uses Core Audio with Mac OSX.

On the PC, Cakewalk Sonar is increasing in popularity, although Cubase SX and Nuendo are the leading packages used by professionals. Other packages to watch out for are Sony's ACID Pro and Adobe's Audition (formerly Cool Edit Pro) – which now incoporates both MIDI and multi-track audio recording capabilities.

MOTU Digital Performer

www.motu.com

Mark of the Unicorn's Digital Performer incorporates audio recording, editing, signal processing and mixing facilities alongside its MIDI sequencer to provide an integrated environment for music production. Digital Performer works extremely well as a 'front-end' Pro Tools TDM hardware, although Mark of the Unicorn also makes its own digital audio hardware these days. In fact, MOTU audio systems compete well with the Digidesign hardware by providing sensibly thought-out interface options for Mac and PC that will work with just about all MIDI and Audio software – apart from the Pro Tools software itself. The MotU PCI cards and Firewire interfaces do not offer DSP processing like the Digidesign TDM cards, but they do cost a lot less than the Digidesign systems.

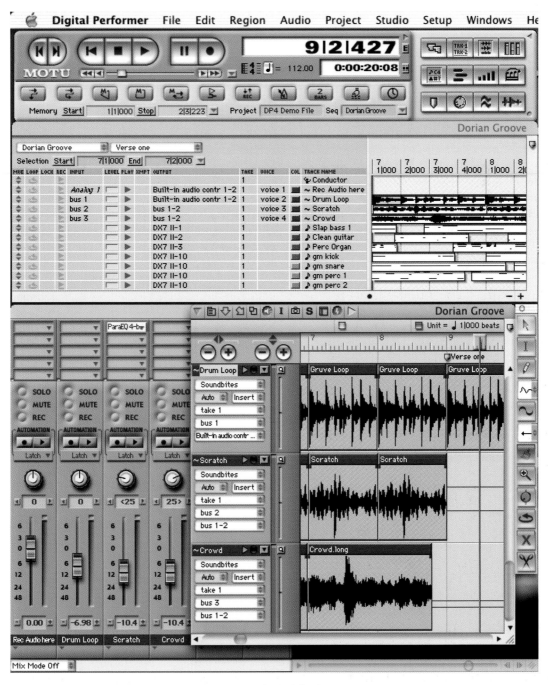

Figure 12.1 Digital Performer.

Digital Performer features a Tracks window that incorporates a graphical representation of the MIDI and Audio data. Originally, this was the main window to work with in Digital Performer. A full-featured mixer window was developed next, for both audio and MIDI tracks. This is similar to the Mix window in Pro Tools, with the usual fader, pan, mute, solo and record-enable controls, and with an area above these that you can use to insert plug-ins. To compete more effectively with software such as Pro Tools, the Sequence Editor window was developed. This is similar to the Pro Tools Edit window, letting you display and edit all the MIDI and audio tracks in one window. As in Pro Tools, both MIDI and audio tracks can be independently resized vertically and you can record multiple 'takes' into any track – choosing between these using a pop-up menu. You can edit both audio and MIDI at a 'macro' level using the Tracks window and you can use the more detailed Sequence Editor for cutting and pasting to create arrangements. And whether you prefer list or graphical editing, writing a conventional score or programming beats, fine-tuning single MIDI events, or moving entire sections around, Digital Performer has an editor to suit. For more detailed, sample-accurate, audio editing, Digital Performer offers its Waveform Editor. For MIDI editing, you can choose from the Graphic Editor, Event List, Drum Editor or QuickScribe Notation Editor. The Event List provides a numerical representation of the MIDI or audio data in each track. The Graphic Editor provides a piano-roll style display for MIDI data with an associated graph of MIDI continuous controller information. Unlike the Event List, which displays a track at a time, the Graphic Editor can display multiple MIDI tracks. You can also edit the Conductor Track using the Graphic Editor, in which case it displays tempo changes. The Drum Editor provides a fixed note grid, which is easier to edit when you are working with MIDI percussion parts. The Notation Editor provides a conventional music notation display for a single MIDI track. The QuickScribe editor displays multiple tracks as a conventional musical score – looking just as it will if you print this score out. Then there is the Waveform Editor, which lets you zoom in to sample level to edit your audio waveforms. Of course, there are still more windows, such as the Chunks window, the Markers window, and Digital Performer's unique 'Polar' window. Polar lets you record audio to RAM, overdubbing on the fly, to quickly build up loops for use in your compositions or arrangements.

Digital Performer supports MOTU's own MAS format plug-ins, along with the new Apple Audio Units plug-ins and virtual instruments, and there are plenty of useful plug-ins supplied with Digital Performer to get you started. Digital Performer also features dedicated virtual instrument tracks that give users access to all AU- and MAS-compatible virtual instrument plug-ins installed on their computer.

Like Ableton Live, Digital Performer automatically converts audio files to the correct sample rate, format and tempo. So, for example, you can drag and drop Recycle 2 REX files or Acid files into your project and they will automatically snap to your project's tempo. And, like Logic Pro, Digital Performer has a 'freeze' tracks feature that allows users to temporarily 'print' to disk any audio tracks that use plug-ins, or any virtual instrument tracks, to free up CPU time.

One of Digital Performer's many strengths is its professionally designed user interface. The attention to detail here is superb compared with any software I have used. And it's not just the way everything looks and the way everything fits so well on screen – it's things like the way that the Markers window works so efficiently, the way that Event List editing is so smooth, and the advanced features that the Undo History has. Then there is the unique POLAR feature that lets you record audio in its special window – drum-machine style. This is great for developing sections that you can loop later on. It is also great for guitarists who want to perfect their solos!

Digital Performer is probably the best choice of all when it comes to working to picture. You can synchronize Digital Performer to any type of video equipment using MOTU's Digital Timepiece (or similar equipment) – or you can import digitized video in QuickTime format and play this back in a Digital Performer window. For post-production, DP has support for Synchro Arts VocAlign dialog replacement technology; you can trim sound bites or edit automation while the QuickTime movie chases to your edits; and you can scrub through the audio waveform to find your exact edit point. If you are composing music, each cue can be written into a separate sequence and these sequences can be played back in order using the 'Chunks' window. The 'Locking Markers to SMPTE' locations command is useful, while the 'Modify Conductor Track' sub-menu commands in the Project menu provide answers to a film composer's prayers – enabling you to adjust meters and tempos to automatically align musical cues to hit points. And support for surround sound is excellent: there are four panner plug-ins and each audio track can be assigned to any surround sound format – from LCRS up to 10.2.

Digital Performer supports the OMF Interchange file format to allow projects to be transferred to and from other OMF-compatible software. To transfer efficiently to and from Pro Tools, you need to use another piece of software called DigiTranslator. DigiTranslator is available free to all registered AVoption and AVoption|XL owners, but otherwise it will set you back $495.00. The Open Media Framework Interchange file format was developed by AVID, Digidesign's parent company, originally to allow video post-production projects to be exchanged between Avid Media Composer digital video editing systems and other work stations in a way that would preserve all the edits. DigiTranslator has plenty of conversion options – such as translation of clip-based volume data and options for sample rate and bit depth conversion – and it can even translate files over a network or batch convert media files in the background while other applications are running. The main thing is that it provides sample-accurate and frame-accurate import and export of audio files between Pro Tools and other DAWs – including Digital Performer.

Digital Performer makes an ideal choice as a software 'front-end' to Pro Tools TDM hardware – especially for MIDI-intensive recording sessions. Nuendo and Cubase SX are not compatible with Pro Tools TDM hardware, so the only other realistic choice is Logic Platinum. Logic suffers from an aging and often-confusing user-interface, while Digital Performer's user-interface just gets better and better. However, Nuendo and Cubase SX are both stiff competition for Digital Performer – especially if you are not using Pro Tools TDM hardware.

Add all this together and Digital Performer gets my endorsement as software of choice for MIDI + audio applications if you are working on the Mac running OSX. It has an even more effective user interface than Cubase SX and has powerful features to rival or better both Logic 6 and Pro Tools 6. It is a close call for me as to whether Nuendo is better than Digital Performer – or not. The choice between these probably comes down to whether you are using Pro Tools TDM hardware or audio hardware from MOTU, Steinberg or other third parties.

Emagic Logic Pro

www.emagic.de

Logic Pro, the top-of-the-range offering from Emagic, combines music composition, notation and audio production facilities into one integrated environment.

Figure 12.2 Logic Pro.

Logic has been a success story since the mid-1990s when it overtook Cubase VST to become the sequencer of choice for professional users. One of the many reasons for its adoption by professional users was undoubtedly its Pro Tools TDM compatibility – which allowed Logic to be used as the 'front-end' for Pro Tools rigs used for music production.

Available for both OS9 and OSX on the Mac, Logic Pro has all the 'bells and whistles' you could ask for. Everything you need to record, edit and mix your music is there. The MIDI features are as advanced as anything you will find elsewhere and the software comes with an impressive range of virtual instruments. These include the widely used EXS24 sampler, the ES1 and ES2 synthesizers, the EVP88 electric pianos, the EVD6 Clavinet and the EVB3 Hammond Organ. The downside is that the user-interface is showing signs of age – despite recent 'facelifts' – and Logic can be very 'fiddly' to use. Recent versions support OMF and OpenTL interchange formats, allowing projects to be interchanged with Pro Tools and with Tascam hard disk systems.

Many people like Pro Tools because its Edit window lets you edit to sample accuracy, you can select whatever you like with its Grabber tool, and you can work entirely within the Edit window (if you prefer) by displaying the mixer channel parameters there. Also, Pro Tools automatically creates a folder for each project and puts all the relevant files into that folder, or into folders within. To compete with these Pro Tools' features, Logic now displays the Channel Strip of the currently selected track in the Arrange page's parameter area. Also, you can edit to sample accuracy when working at higher zoom levels – and make region-independent selections in the Arrange page using the Marquee tool. You can time stretch audio directly in the Arrange page without going into Logic's audio waveform editor – another Pro Tools feature. You can stretch selected regions to match the length defined by the locators, or adjust the region length to the closest whole number of bars. Logic also supports grouping in the Mixer and the Arrange page – and lets you hide tracks that you don't need to see in the Arrange page using the Hide button – yet more features copied from Pro Tools. Although Logic does not put all the files associated with each song into a dedicated folder automatically from the outset, you can use the 'Save as Project…' command to gather everything together later into one folder. This is similar to Pro Tools' 'Save Session Copy In…' command, although Pro Tools is more efficient – combining the functions of Logic's 'Save as Project…' and 'Consolidate Song' dialogs into one dialog window. Pro Tools also allows you to change the audio file type, sample rate, and bit depth using this dialog.

As with the Pro Tools Project Browsers, Logic has a Project Manager that helps you to manage all the files associated with your Logic projects – audio files, samples, sampler instruments and so forth. It establishes a database of media files on your system that you can just drag and drop into your song. You can even attach meta-data comments to files so that you can search for files with these comments attached later. Logic Pro also has useful features such as Offline Bounce and Freeze. The Offline Bounce feature lets you bounce tracks much quicker than using Realtime Bounce. It also enables you to bounce far more complex set-ups that would normally refuse to playback in real time. Logic's Virtual Instrument tracks have a Freeze button that creates a temporary stereo audio file containing a mixdown of the audio from the Instrument track then disables the playback via the virtual instrument – playing the temporary audio instead. The neat thing here is that playing back this stereo audio file takes up a tiny fraction of the CPU resources compared with playing a virtual instrument. The Arrange page doesn't change and the 'frozen' track sounds identical to the unfrozen track. You can still adjust the volume and pan settings – and you can instantly go back to the plug-in version. When you unfreeze, Logic deletes the temporary file and restores everything so that it uses the plug-in again.

Logic's Video thumbnail track displays a series of still images from a QuickTime video horizontally along the track, so you see more frames as you zoom in – helping you to keep track of where you are in the video. Logic's digital video (DV) output capability is also very convenient for those working to picture. Any QuickTime video in Logic will now output DV via the Firewire port so you can show this on a monitor.

Logic Pro incorporates more than 50 plug-in effects, including Adaptive Limiter, SubBass, DeEsser, Phase Distortion, Clip Distortion, Tremolo, Exciter, StereoSpread, Denoiser, Limiter and Multiband Compressor. Logic Audio also makes a very good choice as a partner for Pro Tools TDM hardware. It is compatible with all the TDM hardware and its System Bridge lets you route up to 32 channels of audio from Emagic's EXS24 software sampler into the Aux channels of the TDM mixer, for example.

Logic is probably the leading MIDI + Audio package used professionally around the world today. It has been marketed very effectively and packs in an incredible number of features. Personally, I find the user-interface cluttered and not too friendly – especially compared with Pro Tools. Of course, once you are up to speed with all the keyboard commands and become familiar with the 'logic' of the software, you can work extremely quickly with Logic. It offers good audio features which many people prefer to those available in Digital Performer or Cubase SX and it offers better music notation capabilities than you will find in Digital Performer – although Cubase SX and Nuendo have, arguably, even better notation features than Logic.

Steinberg Nuendo

www.steinberg.net

Steinberg's Nuendo has only been available since the year 2000 – but it has already won the attention of leading engineers and producers working on surround productions. Nuendo 2.0 was released in the summer of 2003 – updated with features that will please the most demanding of users.

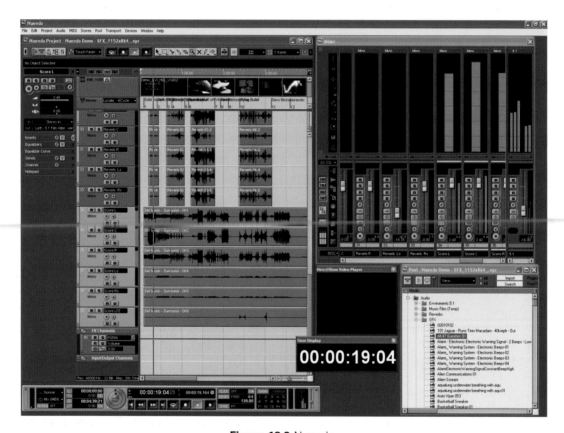

Figure 12.3 Nuendo

Nuendo is serious competition for Pro Tools. It is actually a better choice of work station than Pro Tools for music production and any kind of work that involves working with MIDI or musical scores. And the way you can hide the MIDI features if you are working on an audio-only production is very neat. This will make audio engineers feel even more comfortable with Nuendo – and Nuendo does not lack for technical features compared with Pro Tools. You can use any hardware for which ASIO drivers are available – including the Pro Tools|HD and MIX cards – although there is no DAE support so you cannot use TDM plug-ins. Support for 32-bit floating-point operation, comprehensive support for file formats, bit-depths and sample rates, and the first-rate user-interface design are even more compelling reasons to use Nuendo.

Professional nine-pin machine control facilities let you hook up video recorders and synchroni-zers to work to picture. You can import QuickTime video and this will be displayed as frames in a video track and displayed in a small video window. The audio track is automatically stripped out and placed in a linked audio track that you can use or discard as you please. Nuendo supports all surround formats that you are likely to encounter and comes with a number of multi-channel effects for surround.

Just click on any MIDI part and up comes a graphical MIDI Editor which lets you move notes around individually, with a section provided to graphically edit Velocity and MIDI controller data. And the Browser window in the Project menu provides list editing for each track, with events handily filed away into folders for each track. You can import ReCycle REX files containing drum loops that speed up or slow down as you adjust the tempo, and there is excellent support for ReWire – which lets you stream the audio outputs from compatible applications into the Nuendo Mixer with sample-accurate sync.

The main workspace is the Project window, which corresponds to the Arrange window in Logic or the Edit window in Pro Tools. Ranged across the top of this window there are various 'tools' that will be familiar to Cubase users, such as the 'scissors' and 'glue' tools. And at the left you can open an 'Inspector' area – similar to the one in Cubase SX – with a notepad that you can use to write useful instructions and directions and a display of the Equalizer Curve.

Nuendo's mixer is laid out, like most mixers, using channels with faders and associated pans, mute switches, record-enable buttons and so forth. As with Cubase SX, the mixer can be extended to reveal the EQ, effects sends, or inserts above the faders, and to show or hide the Input/Output settings at the top of each channel strip.

Pro Tools does have some advantages over Nuendo, such as the ability to record alternate takes onto the same tracks. In Nuendo, if you want to preserve an original take on the tracks you have set up to use, you would have to create new tracks for new takes, or use some other work-around, such as recording new takes further along the time line on the same tracks. If you are overdubbing one or two musicians in a project studio, this is probably not too much of a problem. But if you are recording a drummer or a band – or an orchestra – you will quickly realize the problem. Pro Tools also lets you route internal buses to the inputs of audio tracks and offers more flexible ways to group tracks together in its Edit and Mix windows. Pro Tools also makes it easy to re-position the first bar of the session to correspond to a particular time-code location – with the tempo and time signature changes automatically adjusting as necessary. This is something that you may need to do on a film scoring session that can be more awkward in Nuendo.

Nevertheless, excellent features for working to picture with strong support for surround sound, including built-in LCRS coding, make Nuendo a great choice for post-production for film, televi-sion, video or DVD. And now that it incorporates all of Cubase SX's MIDI and Score Edit features, Nuendo also makes a great choice for music production, i.e. composing, arranging, recording, editing and mixing. Of course, Nuendo's music scoring capabilities can be extremely useful on a film scoring session as well – and Pro Tools does not offer this feature.

All things considered, Nuendo is probably the most serious alternative to Pro Tools that is available today. It offers better features for music production and it can be a better choice for some types of surround production.

Using Nuendo with Pro Tools hardware

Nuendo can be used with Pro Tools TDM cards on the Mac as long as you have the correct Digidesign Core Audio software installed. At the time of writing, this was version 6.1.1 and I was using Nuendo version 2.1.0.

You have to make sure that Nuendo is on the list of supported applications in CoreAudio Setup first.

> Tip: If you are not sure where this application is, type Command-F on your Mac keyboard to open the 'Find' dialog. Type 'Digidesign CoreAudio Setup' into this dialog and click on 'Search'. When you find it, double-click its icon to open the CoreAudio Setup application.

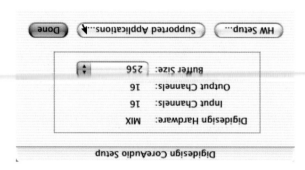

Digidesign CoreAudio Setup

Digidesign Hardware:	MIX
Input Channels:	16
Output Channels:	16
Buffer Size:	256

(HW Setup...)   (Supported Applications...)   (Done)

Figure 12.4 Digidesign CoreAudio Setup utility.

When you click on the Supported Applications button in CoreAudio Setup, a window appears with a list of these applications – see Figure 12.5.

If Nuendo is not in this list, you can click on the Add New Application button to add it to the list. Browse through the disk drives and partitions, folders and files until you locate the application software that you want to add – in this case Nuendo. Click on Nuendo to select it, then click 'Choose' to add this to the Supported Applications list – see Figure 12.6.

> Note: If Cubase SX 2.0 is already in the list of supported applications, there is no need to add Nuendo, as they both link to CoreAudio applications in exactly the same way.

Figure 12.5 Digidesign CoreAudio supported applications window.

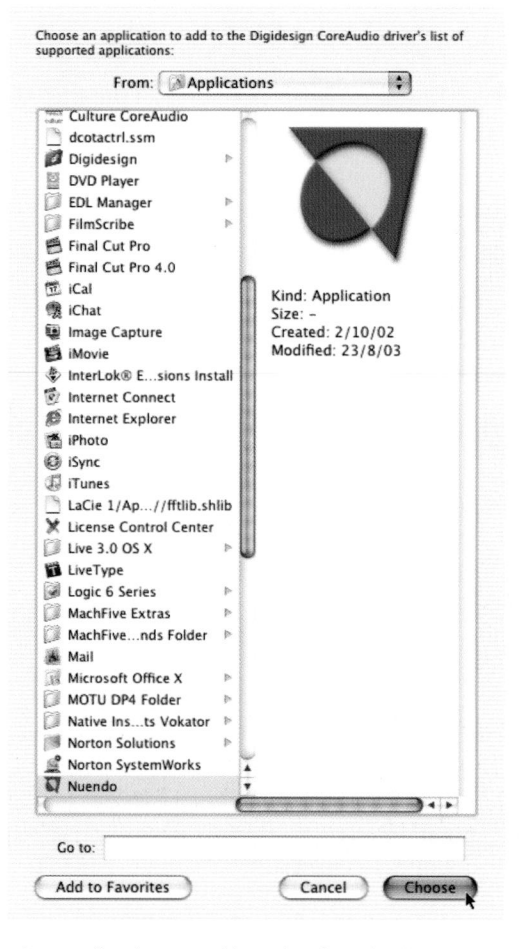

Figure 12.6 Finding the application to add to the CoreAudio supported applications list.

Just one more step is required to get Nuendo to use the Digidesign card – you must select the card using the ASIO Driver pop-up on the VST Multitrack page of the Device Setup window.

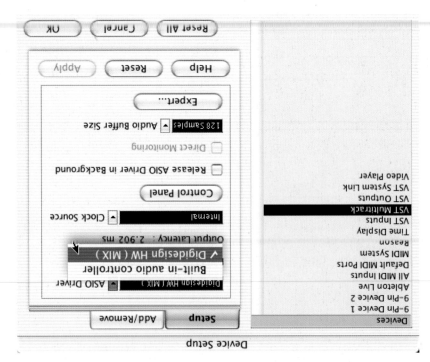

Figure 12.7 Choosing a Digidesign card using the Device Setup window in Nuendo.

Now you can access the inputs and outputs on your Digidesign interface from Nuendo – although you cannot use TDM plug-ins with Nuendo.

Of course, you would not buy a Digidesign TDM card to use with Nuendo unless you wanted to be able to use Digidesign's Pro Tools or other TDM software applications as well. But if you do, or if you already have Pro Tools hardware, it is useful to know that you can run Nuendo successfully with this.

Steinberg Cubase SX

www.steinberg.net

Cubase SX represents a major re-design of Steinberg's popular Cubase VST software. It falls in line with the way that Nuendo works – bringing the user-interface bang up-to-date so that Cubase can compete very favourably with software from Emagic and the rest. Cubase is aimed primarily at users who make heavy use of MIDI and work mostly from a musical per-spective. Nuendo is aimed primarily at professional recording engineers and producers who work mostly with audio, and particularly those working to picture or working with surround sound. Nevertheless, Cubase will do most of what Nuendo will do, and vice versa – and the user-interface designs are so similar that if you learn one you will know 90 per cent of the other.

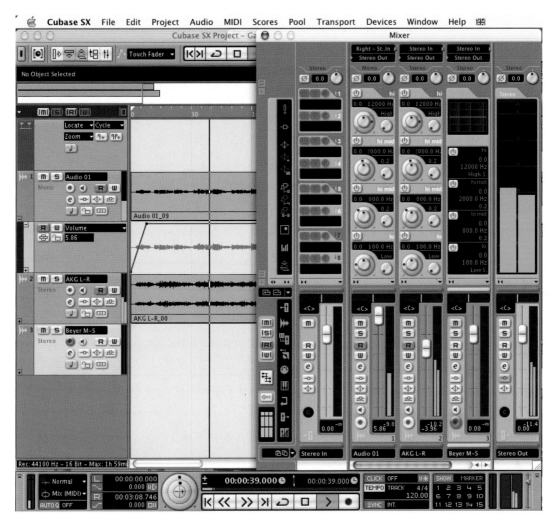

Figure 12.8 Cubase SX 2.0 showing the Project window to the left, the Mixer window to the right and the Transport Panel at the bottom.

Cubase SX has a reasonable selection of audio plug-ins and there are three virtual instruments supplied to get you started – the VB-1 bass guitar simulation, the LM7 drum-machine and the A1 Synthesizer – so you can always record basic drums, bass and synthesizer pads using these. The real-time automation modes in Logic Audio are more comprehensive than those in Cubase SX and the selection of plug-ins and virtual instruments is better. Nevertheless, in contrast to previous versions of Cubase (and to Logic), Cubase SX is much easier to learn and use. And if you need help, it's there in the Help menu as standard help files plus Acrobat versions of the manuals and links to the Steinberg website. Steinberg has clearly decided to make life as easy as possible for Cubase SX users, which is as it should be! Overall, Cubase SX is a serious option for professional users once more.

Compared with Cubase VST, the basic concept of Songs and Arrangements is no longer used. Instead, with Cubase SX you get the more appropriately named Project window. And, as with

Pro Tools, when you create a project in Cubase SX, a project folder is automatically created containing the project file with an associated Audio files folder. Other folders hold audio Edits (edited or processed audio files), Fades, and waveform Images. The project file itself contains references to all the audio and video files, along with playback information, MIDI data, and settings for the project such as sample rate and frame rate. Unlike Pro Tools, you can have several open projects, which is great – although only one can be active at a time.

The Windows dialog helps you to manage all the windows, with buttons to activate, minimize, restore or close the specified window – as does the Devices panel. This is a floating window containing a list of windows that you can leave visible on your desktop and click on to open or close any listed window. You can also create window layouts (similar to Logic screen sets) and subsequently recall these using the Window Layouts sub-menu in the Windows menu.

The Project window is the main window in Cubase SX. This provides you with a graphic overview of the project, allowing you to navigate and perform large scale editing. It is divided vertically into tracks and has a time line going from left to right. You can access Solo, Mute, Record Enable and other controls at the left of each track. To the left of the tracks you will find the Inspector area. This has tabs that you can click on to reveal the various settings you can make in the Mixer window – to save you from having to have the mixer window open. Here you can access the Inserts, Sends and Mixer Channel strip for each track plus the Equalizers for the audio tracks and the Track Parameters (Transpose, etc.) for the MIDI tracks. The area extending to the far right of the Project window is called the event display. This is where you view and edit audio and MIDI events, automation curves, and so forth.

Just double click on any audio in the event display to open the Sample Editor. This features 'Hitpoint' detection, similar to ReCycle, allowing you to create 'slices' for drum loops so that when you change the tempo of your sequence the loops automatically adapt. Hitpoint detection in the Sample Editor automatically detects attack transients in an audio file, and then adds a type of marker, a 'hitpoint', at each transient. These hitpoints allow you to create 'slices', where each slice ideally represents each individual sound or 'beat' in a loop (drum or other rhythmic loops work best with this feature). When you have successfully sliced the audio file, you can change the tempo without affecting pitch; extract a groove map from a drum loop then use this to quantize other events; replace individual sounds in a drum loop; edit the actual playing in the drum loop without affecting the basic feel; or extract sounds from loops.

The Sample Editor is similar to the one in Cubase VST 5.1, but looks tidier. The same goes for the Tempo Track, Key, Score, Drum and List editors. The Project Browser window is new. This provides an editable list of all the events on all the tracks – a feature sorely lacking in Pro Tools, for example.

The Mixer has been completely re-designed to look and work much more like a professional audio mixer and you can mix your MIDI channels alongside your audio channels. From an operational perspective, everything has been rationalized and simplified wherever possible. The Mixer window automatically includes a channel strip for each audio and MIDI track present in the current project. Beside each channel's level fader, there is a level meter that indicates the signal level of audio events during playback. For MIDI tracks, the meters show velocity levels instead. You can also set panning and other parameters.

Each audio and MIDI track in the Track list has an automation track containing all parameters for the track. You can select which parameters to view and edit by opening sub-tracks for the automation track – click the small '+' sign to the left of any track and a linked sub-track will open underneath showing, say, the volume automation. Hit the Pencil tool icon and you can just draw in your volume automation moves. Click the small '+' sign to the left of the sub-track, and another one opens. Set this to, say, Pan and you can draw in your panning moves. It is as simple as that. Of course, you can always enable the automation Write mode and move the fader or pan control in real-time during a mix pass if you prefer. And you needn't worry unduly about making mistakes. Cubase SX offers wide-ranging, multiple Undo, allowing you to undo virtually any action you perform. The Edit History dialog allows you to undo or redo several actions in one go. The Offline Process History allows you to remove and modify applied processing, and is different from the regular Undo in that you don't have to undo processing functions in the order they were performed.

With Cubase SX you don't have to specify the number of audio channels you want to use in a project – you simply create as many tracks as you need, up to the limits of your computer system. And now an audio track and an audio channel are the same thing – all audio tracks have a corresponding audio channel strip in the Mixer. This is a lot less confusing than Logic – which still lets you have several tracks in its Arrange page sharing the same channel in its audio mixer. And, unlike the Play Parameters in previous Cubase versions, track parameters in Cubase SX cannot be applied to individual MIDI parts – they always apply to complete MIDI tracks – removing yet another potential source of confusion.

When you record audio in Cubase SX, an audio file is created on the hard disk and an audio clip that refers to this is created in the Cubase Project file. An audio 'event' is also created. This is the 'object' that you place in a track on the timeline in the project window to play back the audio clip – resizing as you like to play the section of the audio clip you are interested in, and copying as you like to play at different positions in time. You can gather several audio events together into a 'part' to move or copy as one. Similarly, when you record MIDI, MIDI events are created and placed in MIDI 'parts'.

Click on any audio event and a set of blue handles will appear at the top. This is similar to the Volume HyperEdit in Logic Audio, so it lets you adjust the overall volume of the event or create a fade-in or fade-out by dragging the middle, beginning or end handles respectively. And guess what! The size of the waveform pictured in the event increases or decreases to reflect the volume settings! Cool! And if you want to change the shape of your fade, just double-click on the fade line to bring up a fades dialog where you can adjust the settings. Make a mistake and you can undo it – even it was several moves back, and even if you want to keep the intervening edits intact!

Hot on the heels of Nuendo 2, Steinberg released Cubase SX 2 with the look and feel revamped to fall even more closely in line with Nuendo 2, and with most of the technical features from Nuendo added. So Cubase SX 2 has the advanced audio 'engine' with its plug-in delay compensation and comprehensive surround support – although this is limited to six channels. Also, OMFI file compatibility has been added – so you can now exchange audio projects between Cubase SX and Pro Tools, for example. And to compete with Logic, a Freeze function has been added for use with CPU-intensive virtual instruments.

Cubase VST 'fans' will be pleased to hear that the toolbox is back – SX originally made these tools available using the Quick menu that appears when you Control-click within a window, as well as providing fast access to useful menu commands. A preference can now be set to restrict this to the original Toolbox icons. Cubase SX now has three windows that you can use to display different sets of mixer channels – Nuendo has four – and the Cubase mixer now has all the same features as the Nuendo mixer.

As with Nuendo, the user-interface can be customized, although not quite as radically. You can't change the menu options, for instance. But you can change the contents and look of several sections and panels, including the Transport panel, toolbars in the Project window and the editors, and the track controls in the Track list. You can also hide unwanted controls and settings from view and change the order of items on the panels.

When you have recorded audio or MIDI without a metronome click the new Time Warp tool can be used to create a tempo map that fits the recording – or to match positions in a video with positions in the music. Just drag a bar line to line this up with an obvious downbeat in your audio waveform. It's so simple and intuitive I can't imagine why no one thought of this before! And the Tempo track now has a Tempo Record slider. Move this during playback to write Tempo events into the Tempo track – again, simple and intuitive.

And there are even more new features relating to tempo: you can export the current tempo track as a special file for use in other projects and you can import a saved tempo track to replace the tempo track in your current project. The Beat Calculator tool lets you calculate the tempo of any audio or MIDI material and you can insert the calculated tempo into the Tempo track if you wish. Clicking 'At Tempo Track Start' will adjust the first tempo curve point, while 'At Selection Start' will add a new tempo curve point at the selection's start position. You can also tap the tempo you want using the Tap Tempo option in the Beat Calculator, and you can even create a complete tempo track based on your tapping.

Cubase SX 2 also has a new 'Stacked' Cycle Record Mode that works with both MIDI and audio tracks. The way this works is that when you record audio in cycle mode and the 'Stacked' Cycle Record Mode is selected on the Transport panel, each complete recorded cycle lap is turned into a separate audio event, the track is divided into 'lanes', one for each cycle lap, and the events are stacked above each other, each on a different lane. This makes it easy to assemble a perfect take in the Project window by combining the best parts from the different cycle laps. And you can stop the recording between cycle laps if you need to. When you start recording again, you can pick up where you left off, stacking up more takes, as long as you have recorded at least one complete cycle to create an Event or Part on the track. When you play back the recorded section, only the lowest, i.e. the last, take in the stack plays. To hear another take, either mute the lower take (or takes) using the Mute tool or move the takes between lanes by dragging or using the Edit menu's Move to Next Lane/Previous Lane commands. To assemble the best bits, edit the takes so that only the parts you want to keep are heard. You can cut events with the Scissors tool, resize them, mute them or delete them. The sections that will be heard are indicated in green. When you are satisfied with the result, select all the events on all the lanes and choose 'Delete Overlaps' from the Advanced sub-menu on the Audio menu. This puts all the events back on a single lane and resizes the events so that any overlapping sections are removed. This new 'Stacked' mode goes a long way to rivalling, or in some ways bettering, Pro Tools' system of 'takes'.

So what has Nuendo got that Cubase SX hasn't? In a nutshell – all the extra features needed for professional post-production. These include Sony nine-pin machine control to let you run professional VCRs in sync, more advanced surround features with up to 12 channels, Child Busses in the VST Connections window, and more powerful crossfade editing capabilities. If music production is your game, you won't miss these extra features – so Cubase SX will serve you perfectly well.

Using Cubase SX with Pro Tools TDM cards

Like Nuendo, Cubase SX can also be used with Pro Tools TDM cards as long as you have the correct Digidesign Core Audio software installed. At the time of writing, this was version 6.1.1 and I was using Cubase SX version 2.0.1. The procedure is the same as in Nuendo – see the previous section.

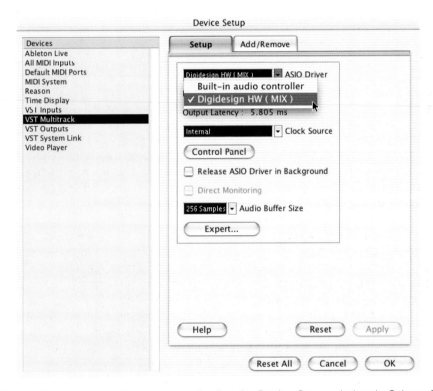

Figure 12.9 Choosing a Digidesign card using the Device Setup window in Cubase SX.

BIAS Deck

www.bias-inc.com

If you are looking for an alternative to Pro Tools and you mostly want to record, edit and mix multiple tracks of audio, then you should check out Deck – one of the original digital audio applications developed for the Mac. In its first incarnation, Deck was used as the 'front-end' software for Digidesign's Pro Tools hardware during the first year of its existence – until Digidesign got around to writing its own. Since then, back in the early 1990s, Deck has been through various changes of ownership until it ended up at BIAS. For many years, Deck simply had maintenance fixes to keep it working with newer OSs. Version 3.5 reversed this trend with a major re-write for OSX – incorporating all the latest developments, such as 5.1 surround sound, and supporting OSX's CoreAudio standard with its super-low latency times and many other benefits.

Figure 12.10 BIAS Deck.

BIAS Deck is a multi-track recording, editing and mixing application like Pro Tools. Deck lets you record tracks, adjust the level and EQ, add effects, and mix down your recording. Deck also offers waveform editing and moving-fader mixer automation. In the Mixer Window you can insert up to four VST effects plug-in per track and Deck ships with over 20 Mac OSX-compatible VST audio effects. You can record and playback up to 64 tracks at the same time – while working with up to 999 'virtual' tracks. Deck is not a MIDI sequencer, although it can replay

MIDI files. Deck makes an excellent choice if you are doing audio-for-picture, offering 5.1 surround sound, OMF import, SMPTE sync to external devices and Chase Positioning.

Using Deck with Digidesign cards

Deck uses Core Audio for audio input and outputs, so, as with Nuendo and Cubase SX, you can use Digidesign cards if you add Deck to the list of supported applications in the Digidesign Core Audio Setup utility then choose the Digidesign card in Deck's CoreAudio Settings window.

Figure 12.11 Core Audio Settings in Deck.

Cakewalk Sonar

www.cakewalk.com

Cakewalk's Sonar software is a good choice for musicians, composers, arrangers and producers who like to work with loops and virtual instruments. You can work with MIDI tracks or audio tracks and a mixing console window is provided with full automation.

Sonar comes with plenty of plug-ins and with more virtual instruments than any of its competitors. It also has good features for working with video. You can import digital video files and replace the audio with new stuff you get together using Sonar. Sonar also has professional SMPTE/MTC sync features and supports Midi Machine Control. And it supports OMF for file interchange – so you can take your work into Pro Tools to finish off.

Ableton Live

www.ableton.com

A relative newcomer that seems to be cropping up more and more is Ableton's Live software. Live is not a MIDI sequencer and it is not a full-blown audio recording, editing, mixing environment – although it shares some of these features. It is a specialized environment for working with audio 'loops' that lets you mix and match samples and short audio recordings that were originally recorded at different tempos, playing these back at whatever new tempo you choose. Import a sample containing a well-defined musical loop of 1, 2, 4 or 8 bars in length and Live will

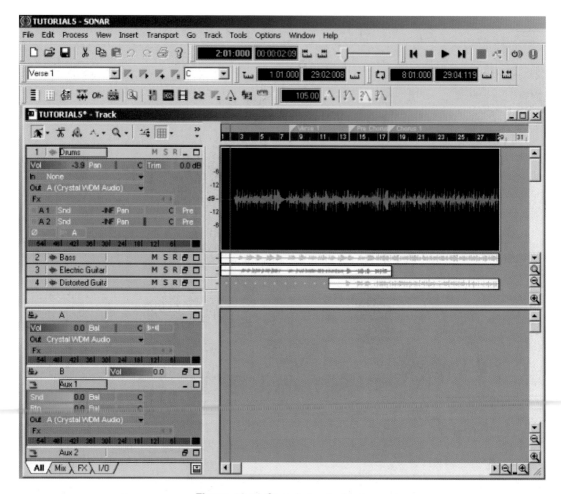

Figure 12.12 Sonar's track view.

automatically time-stretch this to play back at the tempo you have chosen to work with. New audio can be sourced from the inputs to your audio hardware, or from a ReWire application (such as Reason) running at the same time as Live, or from Live itself. Talking about Reason, this makes something of an ideal partner for Live – especially if you want to create music using MIDI as well as audio – and offers great 'bang for your buck'.

Live has a full ReWire implementation so it works either as a ReWire Master or as a ReWire Slave application under Windows, Mac OS9, and Mac OSX. Under Mac OSX, Live versions 2.1 and later can connect via ReWire to Pro Tools, Digital Performer, Cubase, Nuendo and Logic. Live 2.1 includes Direct I/O support – providing integration with the complete range of Digidesign audio hardware.

Ableton's Live software eliminates the boundaries between composition, performance, recording and sound design. You can use Live to combine loops, phrases and songs, improvise song arrangements, instantly drop samples on cue, and interact with other performers, musicians or DJs. And you can instantly change the tempo, melody, groove and sonic signature of your sounds at any time.

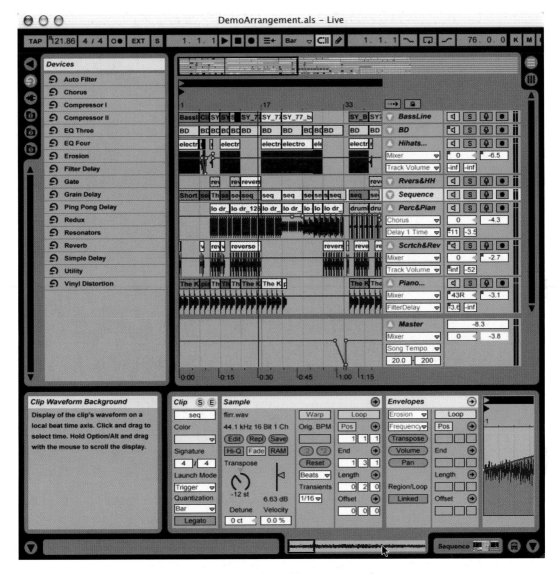

Figure 12.13 Ableton Live: arranger view.

The Control Bar at the top of the screen has controls for tempo, metronome, synchronization, transport and monitoring. At the left of the window, Live can display various browsers: the Effects browser for built-in effects, the Plug-in effects browser and the Content Browser for 'Live Sets'. The Info View at bottom left tells you about whatever is under your mouse. To the right of this, you can either display the Track View to show the selected track's effects, or the Clip view to let you edit the clip properties. Above these, Live has two main views. The 'Arrangement View' displays a visual representation of the audio laid out along a song time line. You can record multiple tracks of audio into this Arrangement, treating it like a tape recorder. The 'Session View' is similar to a mixing console, with volume faders and so forth. It also hosts clips in slots above the mixer controls that you can click on to play back in different ways. You can record the results of your experimentation in the Session view, and the results will appear in the Arrangement – what Ableton call 'Live' playing. This way of working is great for people who like to improvise music, changing patterns on-the-fly as the mood takes them.

Every audio 'clip' in Live contains independent envelopes enabling you to draw pitch, volume, warping, mixer and effect parameters. So what's warping? Well, when the Warp switch is off, Live plays samples at their original, 'normal' tempos, irrespective of Live's tempo. Turn the Warp switch on and your samples, loops or recordings will be 'warped', i.e. time-stretched, to play in sync with the current song tempo.

The built-in effects that can be applied to your samples and loops can change their character drastically. There are 17 of these, ranging from equalizers and dynamic processors to more exotic devices such as 'Grain Delay' or the 'Resonators', and they can be inserted, chained and re-ordered using drag-and-drop – all without interrupting the music so that you maintain your creative flow. You can control just about anything, including effects, via MIDI, and you can apply automation by simply moving any control while in Record.

Sony Acid Pro

http://mediasoftware.sonypictures.com

Acid Pro is an interesting software application that was originally designed to let you work with audio and MIDI loops to build up pattern-based music. It has grown beyond those original aims to the point where it now starts to rival the major MIDI + Audio sequencing packages such as Sonar and Cubase. Not only can you use Acid to create new songs or remix tracks, you can also do music to picture, produce 5.1 surround sound, or create music for websites and Flash animations. The software comes with a library of more than 400 loops in different genres to get you started. Change the tempo and the loops will play correctly at the new tempo – and you can even work with alternative time signatures. You can use Direct-X plug-ins along with VST virtual instruments, and there is virtually no limit to the number of tracks of audio or MIDI you can use. The editing features are optimized for working with loops, but you can easily record vocal or instrumental tracks along with loop sections that you have built up into a complete song arrangement. Acid features both list editing and piano-roll graphical editing. You can open multiple file formats into the same project and Acid supports a wide range of these, including MP3s, Real Networks and Windows Media Audio and Video files. You can also save to a number of audio file formats including WAV, Windows WMA, Real Audio RM and MP3. As with Sound Forge and Vegas Video, you can also burn files to CD directly from within the software.

Summary

Lots of professionals use Logic Pro, while Cubase leads in the educational and hobbyist market. Nevertheless, Cubase SX has many features that all users will value. For example, Steinberg's unique VST System Link. This lets you connect multiple computers (even cross-platform) using VST software and ASIO hardware. So you could run all your VST Instruments on one, all your audio tracks on another, and create your effects on a third. This is a very powerful innovation – especially for those who already have more than one computer.

Cubase SX has no support for Pro Tools TDM hardware – which will be a major disadvantage for many professional users. The real-time automation modes in Logic Pro are more comprehensive than those in Cubase SX and Logic's selection of plug-ins is better and virtual instruments is much better.

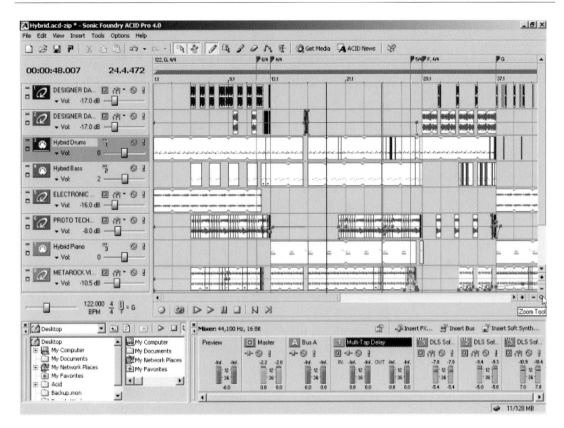

Figure 12.14 Acid Pro.

On the Mac, Digital Performer provides a viable alternative to Logic Pro for Pro Tools hardware owners and is as suitable for film work as for producing the next big dance-floor hit. Pro Tools is the best choice if you are not doing too much with MIDI. You can do basic MIDI stuff OK with Pro Tools, but for anything more demanding you will need to use, say, Digital Performer or Logic Pro instead. Overall, I prefer Digital Performer to Logic Pro, as I value the smoother user-interface that Digital Performer provides. However, Steinberg's Nuendo is the best choice for anyone who wants a cross-platform solution that is a viable alternative to Pro Tools.

On the PC, Sonar XL makes a good choice if you are using loops for music composition and production and it comes with several virtual instruments to get you started. Steinberg's complete range of software is also available for the PC, and Cubase SX is an extremely popular choice. Ableton Live and Sony's Acid Pro are also popular choices on the PC. Pro Tools LE is an increasingly popular choice on the PC, while Pro Tools TDM systems are usually Mac-based.

Appendices

1 Hardware Control

Introduction

One of the dilemmas for Pro Tools users is whether to use the computer keyboard and mouse or whether to go for a hardware control surface that presents itself to the operator more like a conventional mixing console. Experienced recording engineers have developed their craft over many years, getting a 'feel' for moving the faders, reaching out to 'tweak' the EQ while in the middle of a mix, and 'dancing' on the transport controls while 'overdubbing like crazy'. Engineering recording sessions is very much a hands-on affair – and engineers like to keep their hands busy! The four main choices when it comes to adding a hardware control surface to Pro Tools systems are Digidesign's own ProControl, the Mackie HUI, the Digidesign/Focusrite Control|24 and the CM Labs Motor Mix.

ProControl is a relatively expensive piece of equipment that is finding its niche mainly in high-end music studios, video post-production and film studios at present. It is a modular system and connects to the computer via Ethernet. The basic unit has eight faders, but ProControl can be expanded to 48 faders using extra units containing groups of eight faders – and with the Edit Pack to provide additional editing controls. Small project studios are more likely to be interested in the Mackie HUI, which hooks up via MIDI and is very reasonably priced. If you want to record bands in a small project studio, the CM Labs Motor Mix makes a good choice. This is a modular system with eight faders on each unit and three or four of these can easily be linked to provide 24 or 32 faders at a very reasonable cost. The newcomer is the Digidesign/Focusrite Control|24, which probably has the best combination of features for budget-conscious professional studios as well as for more up-market project studios.

Bear in mind that ProControl is designed for and sold to studios that would otherwise go with analogue boards such as SSLs. With ProControl, engineers have the freedom to choose their own premium outboard gear – pre-amps, compressors and other outboard. More top music studios use ProControl than any other control surface for Pro Tools. The patented DigiFaders have the feel of a high-end console and both ProControl and Control|24 have a fader resolution of 1024 steps – much higher than HUI. Also, HUI is controlled via MIDI, whereas ProControl and Control|24 are controlled via Ethernet which provides better performance.

Digidesign ProControl

ProControl was designed to offer the familiarity of a conventional mixing console layout while providing today's engineers with the individual tactile control they need over their Pro Tools

Figure A1.1 ProControl itegrated control surface for Pro Tools.

systems. It not only offers the hands-on control of Pro Tools TDM systems that top recording engineers require for mixing, editing and signal processing, but also includes a comprehensive monitoring section – so it can be used as the main studio mixer. ProControl hooks up to TDM-based Pro Tools systems via a 10BaseT Ethernet connection. This provides the fastest, most reliable performance you can achieve – especially with larger Pro Tools system configurations. The patented DigiFaders feature sealed encoders and servo-controlled motors to provide feel, performance and reliability similar to that of the moving faders found on high-end mixing consoles – but without the high cost. These faders do not pass audio – they simply control the faders in Pro Tools. They have a length of 100 mm and provide 1024 steps of resolution – which can be represented using 10 bits of digital data. The Pro Tools software interpolates these values to provide 24-bit operation – ensuring that the fader moves are handled extremely accurately.

Pro Tools TDM systems are equipped with, arguably, the most powerful mix automation features you will find on any digital console. Using the various modes including Write, Touch, Latch and Read, you can automate all fader levels, pans, sends, mutes, plug-in parameters and so forth – controlling all these from ProControl. You can control many other functions as well, such as the Scrub/Shuttle, transport controls, editing tools and modes; you can edit and control the plug-ins directly from the ProControl; and a trackpad is provided so you can edit graphically on-screen whenever you need to. Control room and studio monitoring facilities are included for both stereo and surround sound.

ProControl modules

The basic unit has two main sections. To the left there is a group of eight mixing channels with the usual faders underneath each strip of channel controls and buttons for routing and so forth to the left of this section. To the right, the control section houses the transport controls, the monitoring controls, and edit buttons. Comprehensive metering is provided in the usual position along the back of the control surface. Channels within the Pro Tools software can be switched in banks of eight onto the actual faders on the ProControl so you can access and control even the largest Pro Tools session from the basic unit. Larger studios can add up to five additional fader sections, each containing eight additional faders with associated controls and eight stereo meters, for a total of 48 channel strips.

Figure A1.2 ProControl fader expansion pack.

Post-production editors will probably want to add the Edit Pack option which features a couple of touch-sensitive motorized joystick panners, a QWERTY keyboard and trackball, dedicated edit switches and encoders, and another eight channels of metering.

Visual feedback is everywhere on ProControl – not only adding to the functionality but also making the console mightily impressive to look at. High-resolution LED displays provide critical system feedback at a glance and the unique 'Channel Matrix' allows fast, intuitive system query and navigation. Large, illuminated Solo and Mute buttons help you find your way around the console even if the studio lights are dimmed. Illuminated channel Select buttons are provided for I/O assignment, automation, grouping and other channel-specific edit functions. There are also dedicated, illuminated switches for EQ and Dynamics editing/bypass control, insert assignment/ bypass and record ready states. Put simply – the thing lights up like a Christmas tree!

ProControl rear panel

The rear panel houses the RJ-45 Ethernet connection to your Pro Tools computer; three DB-25 connectors, two for monitor section inputs and one for outputs; a pair of MIDI 'In' and 'Out' jacks; a power connector and power switch; trim pots, an external mouse connection, a com-

Figure A1.3 ProControl Edit Pack.

port, and a pair of 1/4-inch jacks for 'Footswitch' and 'Remote Talkback'. Finally, the Analog Monitor Section has analogue inputs and outputs via balanced DB-25 female connectors. These run at a nominal operating level of +4 dBu with 22 dB of headroom and can handle input levels up to a maximum of +26 dBu.

ProControl summary

ProControl has to be a serious contender for larger music and post-production studios. You can access virtually everything you need to run a Pro Tools session with minimal use of the mouse and QWERTY keyboard – and if you need to use these a lot, as you will particularly for post-production, then you can add the Edit Pack option. You will also need to have suitable converters and outboard such as microphone pre-amps as none of these are included with the ProControl itself. This means that you will need a fairly large budget to equip a whole studio based around a ProControl. Assuming you have the budget, this can be a very good thing as you can decide exactly how much outboard gear you require for your set-up and choose a selection of state-of-the-art units from Focusrite, Universal Audio and similar companies. Taking this approach allows you to build studio systems with analogue electronics to match those found in the very highest-quality analogue consoles. If you do go down this route, you should be strongly considering the Prism ADA8 or Apogee AD8000 converters to ensure that you maintain the highest quality of sound when converting to the digital domain.

Digidesign/Focusrite Control|24

Digidesign/Focusrite Control|24 – the name says what it is mostly about – you get 24 touch-sensitive motorized faders with associated Solo and Mute buttons to control Pro Tools. Most importantly, there are 16 Focusrite Class A Mic pre-amps – originally developed for Focusrite's Platinum range. These will let you hook up enough microphones to record a moderate-sized band playing 'live' in a studio and you also get a sub-mixer with eight stereo inputs. This lets you connect additional audio sources from keyboards or MDMs, for example. Housed in an ergo-nomically designed console, this control surface is definitely intended to be operated by one

Figure A1.4 Control|24.

person rather than a team. In contrast, the large-format mixing consoles to be found in major studios and even the expanded Digidesign ProControl consoles which are increasingly used for film scoring almost demand that they be controlled by more than one person.

Control|24 control surface

The surface of the Control|24 is raked upwards at a few degrees from the horizontal until you get about half way along the channel strips, at which point it is raked upwards at something like a 30 or 40 degree angle. This design makes it much easier to reach all the controls from the central position without moving your chair to right or left and without standing up or even leaning forward much to reach. Also, the faders are positioned much more closely together than on the ProControl – which some music recording engineers prefer. The arm rest at the front of the console has a rubberized coating which makes it feel comfortable to touch and offers a non-slip, non-shiny surface – a neat touch.

Immediately to the right of the faders you will find a Talkback button, with Transport controls and a Jog/Shuttle wheel to the right of this. Three buttons let you map Pro Tools faders onto Control|24 faders in banks of 24, moving to the left or right, with a Nudge button in the middle which changes the function of the left and right buttons to let you move one fader to the left or one fader to the right when it is engaged. Using these you can access all the Pro Tools tracks you are using, up to the limit of 128 audio tracks, 128 MIDI tracks, 64 auxiliary inputs and 64 master faders. Above this is a numeric keypad with Enter and Clear keys. A dedicated Save button lets you save your work – hit once to bring up the Save dialog and hit the button a second time to OK this. A Cancel button lets you change your mind and a dedicated Undo button is also provided. This right-hand section of the ProControl also contains a matrix of Up/Down and Previous/Next arrow keys which change their function according to which of the associated buttons you have selected: Navigation, Zooming or Selection. You also get buttons for Loop Play, Loop Record, Online, Pre-Roll, Go To Start, Go To End, QuickPunch and so forth. To the left of the faders you will find the modifier controls that you normally have on a Mac or PC keyboard for Shift, Alt (Option), Control (PC) and Command (Macintosh). Above this, there is a section containing the Automation controls.

The upper part of the lower half of the console contains Channel Select and Automation Mode Select buttons for each channel. Immediately above these, each channel strip has four buttons to control inserts. If, say, a dynamics plug-in is inserted, control goes to the dedicated Dynamics button and its green indicator LED lights up if there is a plug-in on this channel. The same system works with the EQ. If another plug-in type is inserted, the Insert button indicator goes green to let you know this is active. Touch one of the active buttons and the 'scribble strip' display on the raked section above changes to indicate the plug-in parameters and the rotary controls beneath the display switch function to let you adjust these parameters. One of the Control|24's most useful features lets you flip the plug-in parameters from the rotary controls onto the main faders. As these are touch sensitive, you can trim the automation for the plug-ins as you would with the mix automation. You can do the same thing with the sends as well – which really helps when you are setting up a cue mix, for instance.

Now let's take a look at the top half of the console in more detail. Running along the top above the channel strips you will see stereo 14-segment bar-graph meters for each channel, switchable pre-/post-fader, with the six output meters to the right of these. These output meters display the first six outputs of the Pro Tools interface you are using – ideal for monitoring 5.1 surround. To the right again there is a display for the Counter which can be switched using dedicated buttons to display SMPTE, Bars and Beats, Feet and Frames or Minutes and Seconds.

At the top of the channel strips you will find switches for high-pass filter and for mic/line with associated peak LEDs. The first two inputs have a third switch position for DI inputs so you can plug a guitar, bass, or whatever, directly in. The other channels have a −10 dB setting instead. The relevant input gain controls are directly below these switches, with the Record Arm buttons for each channel below these. Highlighted with a grey background, the section underneath contains buttons for channel inserts, master bypass, assignments, sends and pans. Over to the right a group of controls is provided for Headphone Level, Talkback Level, Listenback Level, and Monitor to Aux. In case you were wondering, Monitor to Aux takes whatever is in the monitor mix and sends this to an auxiliary output jack which you can use to feed a headphone amplifier for the musicians.

The Control Room Monitor controls are grouped nearby with both Main monitor and Alternate monitor controls so you can feed a pair of main monitors and a pair of nearfields. Mute, Dim and Mono buttons are sensibly provided here, with six buttons below these to let you individually solo or mute the six surround outputs. At the far right there are two groups of three buttons. The first three let you bring in external stereo sources, while the second three let you mute or unmute the three pairs of outputs from your Pro Tools interface. A particularly neat feature lets you choose Single Source or Multi Source modes for these external inputs. Normally, in Single Source mode, if you switch between inputs, the others are muted. Using Multi Source mode, you can use all of these simultaneously and listen to them alongside your Pro Tools outputs.

Underneath this section, at the top right of the lower part of the console, there are various groups of buttons dedicated to editing control. You get one-button access to the Smart Tool and to the Grabber, Selector and other tools. All the Edit modes have dedicated buttons as well – Shuffle, Slip, Grid and Spot. The most often-used commands from the Edit menu are all there – including Cut, Copy, Paste, Separate and Capture. The five zoom presets can all be accessed from individual buttons, as can the Mix Window, Edit Window, Transport Window and so forth. And there's more – you can even control the Groups in Pro Tools using dedicated buttons to Enable, Delete, Suspend or Edit the Groups.

In action, at first I found it a little difficult to make accurate trim adjustments to automated mixes – although I got used to the faders after a little practice. I also found it awkward to interpret the sometimes-cryptic scribble strip displays. These only allow four characters to be displayed which is something of a limitation – although controls are provided to let you scroll the scribble strips to display additional parameters if there are too many to fit in the 24 strips available. However, the scribble strips are also placed too far away on the channel strips for my liking. I believe it would have been better to position these closer to the faders to make identification easier. In action, I found it confusing sometimes when using the plug-ins. The controls for these are not all standardized – so particular controls will come up in different places depending on the plug-in. When this happens, you need to read the legend that says what the control represents in the plug-in – and the displays are abbreviated to the point of confusion at times because there are not enough characters available.

Control|24 rear panel

There are 16 XLR microphone inputs on the rear panel and these can accept maximum input signal levels up to +5 dBu. Channels 1 and 2 are equipped with Instrument Level inputs via 1/4-inch TRS jacks that can accept input signal levels up to +8 dBu. Standard 1/4-inch TRS jacks are used for the line inputs – which can be set for +4 dBm or −10dBu operation. Sixteen line outputs are available via two 25 pin D-Sub connectors. There is also an eight stereo-input into a two-output line sub-mixer intended for use with keyboards, samplers and similar sources. This has balanced inputs via two 25 pin D-Sub connectors and balanced outputs via two 1/4-inch TRS jacks. Another pair of 25 pin D-Sub connectors provide connections for three stereo external input sources, such as DAT, CD and tape, or six outputs from a DVD player, along with connections for three stereo output pairs from your Pro Tools interface. Using these, you can connect the Control|24 to any of the Digidesign interfaces with sufficient analogue inputs and outputs, including the 888|24 I/O, 882|20 I/O or 1622 I/O models.

Speaker outputs for Left, Right, Centre, Sub, Left Surround and Right Surround, with Alternate Left and Alternate Right (for a second set of stereo monitors) are all provided on a further 25 pin D-Sub connector. Inputs and outputs are provided for two Auxiliary channels via 1/4-inch TRS jacks. These have associated gain controls and work with the Mix to Aux function. A pair of additional microphone inputs is provided using XLR connectors with associated phantom power switches and gain controls. These let you connect additional microphones for Talkback from the control room or Listenback from the studio. A line level talk-back or 'slate' output is also provided along with a headphone output with associated level adjust and on–off switch via 1/4-inch TRS jacks. Finally, another pair of 1/4-inch TRS jacks allows connection of GPIs for assignable switch functions.

Control|24 summary

Control|24 was engineered and built by Focusrite specifically for Pro Tools in collaboration with Digidesign. As a control surface for Pro Tools it offers hands-on access to nearly every recording, routing, mixing and editing function. Sitting at the Control|24 is like sitting at a conventional mixing console – yet it puts control of Pro Tools right at your fingertips. Just what the designers intended – and what they have achieved remarkably well. You can easily reach all the controls from a central position, although you will need to arrange to have a computer keyboard and mouse on a sliding tray underneath the front of the console to use for some of the things best-controlled using the computer's normal hardware interface. The microphone pre-amps are a distinct improvement on those in my Yamaha 02R, for example, although they are no competition for Focusrite's Red 1 or ISA430, for example. Also, the 'feel' of the faders is not as good as those on Digidesign's ProControl and nowhere near as good as those on a high-end mixing console. Nevertheless, they work fine and are similar to those on comparably-priced digital mixers. In many ways, the Control|24 is similar to the Mackie HUI, although it has more faders and some extra buttons and controls specific to Pro Tools. The main differences are that it uses Ethernet rather than MIDI to communicate with Pro Tools – so it can handle busy mixes more responsively – and it has those all-important Focusrite microphone pre-amps. Priced around £5500, the Control|24 costs just over twice as much as the HUI, but when you take the 16 microphone pre-amps into account, you are getting these for less than £200 each – something of a bargain for equipment with the Focusrite name and consequent 'seal-of-quality' attached! All-in-all, the Control|24 has just the right mix of features and quality to serve the needs of its target user-group in professional project studios working on stereo music mixes or on 5.1 surround.

Mackie HUI

Figure A1.5 Mackie HUI.

Mackie's Human User Interface (HUI) was jointly designed by Digidesign and Mackie to provide tactile control over the recording, editing and mixing features of Pro Tools. Physically, it resembles a small digital mixing console with mix bus and hardware I/O assignment switches at the far left, eight assignable channel strips more or less in the centre, and various other controls to the right. At top right there are controls for plug-ins, with a switching matrix for the automation functions below. In the middle of this section you will find the control room monitoring controls, a built-in talkback mic, and a keypad to let you enter Mac data. A set of transport controls and a jog wheel with associated navigational switches are located below the monitoring section. In the upper right-hand corner there is an alphanumeric display for controlling the parameters of DAE-compatible plug-ins while to the left of this there are eight stereo meters. On the back you get a pair of MIDI ports to connect the HUI to Pro Tools via a MIDI interface along with two pairs of general purpose input/output jacks for triggering Stop/Play/Punch, On-Air, external console solo, or such-like. There are three pairs of jacks for control room audio input and output, along with a headphone output that shares the third output jack. There is a talkback mic pre-amp that you can use to hook up a remote talkback mic rather than using the built-in mic, and two balanced high-quality Mic pre-amp inputs with switchable phantom power are also provided. These have stereo jack inserts to let you strap a compressor or EQ across the inputs, along with balanced/unbalanced line level jack outputs. In short, just about everything you need to have at the centre of your studio set-up.

So why use a HUI instead of a computer keyboard and mouse? Well, Pro Tools has an on-screen simulation of a mixing console as well as a sophisticated waveform editor which you normally control using the computer's mouse. The problem here is that the mouse is not the ideal device to use to control a mixing console. It is difficult to drag the faders smoothly using the mouse, and 'turning' knobs on-screen can be awkward, to say the least. Several hardware controllers are available which you can use with Pro Tools from CM Labs, Penny & Giles, JL Cooper and others, but these only provide a limited number of controls compared with the HUI. The HUI, on the other hand, gives you hands-on control of virtually all of Pro Tools' parameters.

At first glance, the control surface looks rather like a compact digital mixer, with eight assignable faders, associated Select switches, and an electronic 'scribble strip' which picks up the channel names from your Pro Tools project and displays these handily above the faders. The faders are motorized so they move to automatically reflect the positions of Pro Tools' on-screen faders. The HUI also has a built-in meter bridge with eight pairs of dual LED 'ladders'. With mono Pro Tools tracks, just the left LED 'ladder' lights up – while both are used with stereo channels. Having these meters just where you need them is great – especially if you are using an interface that doesn't have any meters.

The big deal with the HUI, though, is that it gives you all the extra facilities you need to run your Pro Tools system without using an external mixer – at least for smaller studio facilities. Most Pro Tools interfaces do not have microphone inputs, so you would normally use an external mixer (or, for state-of-the art recording, high-quality microphone pre-amps such as the Focusrite Red 1) to provide these. You would also normally feed your monitors from an external mixer rather than directly from your Digidesign interface – and many studios use two or even three sets of monitors, such as a pair of nearfields, a large pair of main monitors, and maybe a set of playback monitors in a studio area. You will often want to connect headphones for the musicians when they are overdubbing, or for the engineer to check things without disturbing the playback. Additionally, an external mixer will usually have a talkback system with a small microphone built-in to the mixer's control surface – so that the engineer can communicate with the musicians in the studio area. To cater for these requirements, the HUI has a built-in control room section. This provides a convenient way to monitor your mixes without having to use a separate mixer. It has three stereo inputs and three stereo outputs, plus a headphone output, and there are two high-quality microphone pre-amplifiers whose outputs you can feed at line level into the analogue inputs on your Pro Tools interface. These mic pre-amps are similar to those found on Mackie's professional analogue mixers and have plenty of gain, insert patching and phantom power for condenser mics. A third mic pre-amplifier is available which is intended for use with a remote producer's talkback mic or for 'slating' the recording. [Slating is the term used to describe vocal announcements recorded by the engineer or producer just before the recorded tracks to identify the 'take' or to add other audible session information.] You can use the three stereo inputs in a variety of ways. For example, you might just have one stereo pair of outputs from Pro Tools and use the other two pairs to connect a CD player and a cassette or tape recorder. Hooking up a CD player is probably the quickest way to test your audio connections through to your monitoring system and you can make sure that the talkback is working by simply plugging in a set of headphones, speaking into the mic and raising the master volume level till you can hear yourself OK.

Once you have the HUI wired up correctly, getting hooked up to Pro Tools is pretty straightfor-ward stuff. You need a couple of MIDI cables to connect the HUI to your MIDI interface and then you set up your OMS configuration to include the device settings for the HUI. You also need to hook up a couple of audio cables from your Pro Tools interface to feed a pair of analogue outputs to HUI's Monitor Input 1 and then load up a suitable Pro Tools project to use to check out the system – I used the demo session from the Pro Tools CD-ROM for this. The Pro Tools software has a HUI 'personality' file in the DAE controller folder in your System folder and you can select this in the Peripherals dialog in your Pro Tools software. As soon as HUI is selected as a peripheral, the Pro Tools software will start communicating with HUI, which will immediately display timecode if everything is hooked up correctly. Now all you need to do is hit 'Play' on the HUI and you should be in action.

Getting into the details

Each channel strip has a 100 mm touch-sensitive motorized fader that you can assign to control channel level, aux return level, MIDI track level or master fader level. Above the fader there is a channel select switch to use when assigning groups and so forth, with a four-character LED display above this where you can display the channel's name, group status, input and output source, send and insert status, or pre/post status for sends. Three buttons above this are provided for muting, soloing and enabling automation, with a Pan/Send 'V-Pot' and an associated selection switch above these which can also be used to choose items from scrollable I/O assignment lists or to choose send destinations. So what's a 'V-Pot'? Well, this is just Mackie's jargon for 'virtual potentiometer' – chosen to reflect the fact that these pots can perform various functions depending on the software being used with the HUI.

Looking to the left of the channel strips there are several groupings of buttons. Again, going from the bottom up, the first group of eight buttons lets you control various keyboard com-mands using HUI – such as Command, Shift, Undo and Save. Above this, six buttons are provided to let you select the various windows in Pro Tools – Edit, Mix, Transport and so forth. Next up there are a pair of channel switches and a pair of bank select switches which you can use to move single faders or banks of eight faders onto the HUI channel strips. The top group of control switches includes the input/output assign switches, a Suspend switch which lets you temporarily disable the automation globally, a Default switch which lets you set selected channels back to their original settings and the Select/Assign switches which you use to choose what a track's V-Pot will control. The default setting is Pan, but you can also assign the V-Pots to control any of up to five auxiliary sends, Mute or Shift. The last couple of switches at the top are for Record/Ready (i.e. to enable or disable all the tracks for recording) and for Bypass (which lets you bypass any channel inserts, whether hardware or DSP plug-ins, on any selected channels).

Looking to the right of the channel strips there are five main groupings of controls – the DSP Edit/Assign section, the Switch Matrix section, the Control Room section, the Talkback section and the Transport section. The DSP Edit/Assign section is at the top right, and above this there is a 40-character wide two-line vacuum fluorescent display. This shows up to four plug-ins or up to eight plug-in parameters and is also used to display general HUI text information. The DSP Edit/Assign controls include an Assign switch which lets you assign a plug-in to a channel strip; a Compare button which lets you compare the current DSP parameter setting with the previous one; and a Bypass switch to let you bypass the plug-in parameters or any plug-in assigned to the channel – as applicable. The Insert/Parameter switch lets you toggle the display between the plug-in assigned to a particular insert or the plug-in parameters for editing. A rotary Scroll control is used in conjunction with this to either toggle the display between Inserts 1–4 and Insert 5, or to scroll through control parameter pages for the currently active plug-in. There are also four assignable rotary controls with associated Select switches. You use these to assign plug-ins and edit plug-in parameters corresponding to the HUI display and Pro Tools software screen displays.

The Switch Matrix section contains switches for global information enabling, mode selection, group creation and so forth. There are eight Function (F) keys that let you access special HUI functions. F1 clears clip and peak holds from the meter; F2 activates Relay Outputs 1 and 2 which let you remotely control Play and Record functions on other equipment using HUI's Play and Record buttons; F3 lets you disable the audible click function for the V-pots below the fluorescent display; F4 displays the version number of the HUI 'personality' file currently installed in the host computer; F5, 6 and 7 are reserved for future expansion; and F8 acts as an escape switch to cancel any assignment mode or on-screen dialog. Underneath the F-keys there are four groups of switches. The first group, labelled Auto Enable, includes fader, mute, pan, send, plug-in and send mute switches to let you globally enable these functions. The next group of switches, labelled Auto Mode lets you enable or disable automation on individual channels or channel groups. Options include Read, Touch, Latch, Write, Trim and Off. Next are the Status/Group switches. These are used to query automation, monitor and group status, and to create or change groups – with switches labelled Auto, Group, Monitor, Create, Phase and Suspend. The last group of switches lets you perform standard editing functions including Capture, Separate, Cut, Copy, Paste and Delete. Below these there is a timecode display which shows the current time location in timecode, feet and frames, bars and beats, or simply in minutes and seconds. To the right of this you will find a numeric keypad which can be used to control the locate feature.

Underneath the timecode display the Control Room section lets you control your input and output sources and the Master Volume. There are three input source switches to let you choose between monitor inputs 1, 2 or 3. You can choose whether HUI's inputs act as three stereo pairs assigned to any stereo output or as six discrete mono inputs assigned to their corresponding outputs for surround mixing using the Discrete switch. A Mono switch is also provided to sum all the signals via the Master Volume control. There are individual output level controls for the three stereo output pairs, along with switches to mute these. A separate Mute switch lets you mute all three outputs simultaneously, while a Dim switch is provided which lowers the monitor output level by a set amount – the default being 20 dB. Below these controls there is a talkback enable switch and an associated talkback level control, along with the built-in talkback mic.

Finally, the Transport controls are situated at the lower right of the control surface. Switches are provided to let you set the In and Out points for the punch-in, Audition the section you've selected, and set the Pre- and Post-roll amounts before and after the punch-in. A second row of switches is provided to let you Return to zero, go to the End, put Pro Tools Online, engage Loop playback, or enable the QuickPunch feature. A third row of larger buttons let you control Pro Tools' Rewind, Fast Forward, Stop, Play and Record functions. A set of four switches is also available to let you navigate, zoom, and make selections in the waveform display. An associated mode switch lets you choose whether these switches will act as horizontal/vertical view expanders/contractors or whether they can be used to locate the cursor – as an alternative to using the tab and arrow keys on your computer keyboard. There is also a large jog wheel with a pair of associated buttons to switch this between Scrub and Shuttle modes.

Rear panel

On the back panel there is a standard IEC connector next to a power on/off switch. To the right of these you will find a pair of MIDI sockets and two pairs of Apple Desktop Bus connectors to let you feed your computer's mouse and keyboard connectors via the HUI – provided so you can site your mouse and keyboard near to the HUI rather than for direct control of the HUI. A couple of nine-pin connectors are provided to allow connection to other external devices and one of these is switchable between RS232 and RS422 operation. To the right of these a pair of 1/4-inch jack sockets are provided for foot-switches to control functions such as play or record, along with a pair of relay output jacks which you can hook up to external equipment such as 'On-air' or 'Recording in progress' indicator lights. To the right of these are six 1/4-inch jacks which you can use to feed balanced or unbalanced line level signals into the monitor section and to the right again another six 1/4-inch jacks provide balanced or unbalanced line level outputs which you can feed to your monitor amplifiers or wherever. Beneath these there is a single stereo 1/4-inch headphone output jack whose signal is derived from Monitor Output 3 – so a pair of headphones can be plugged in simultaneously.

At the far right of the rear panel you will find three microphone input channels, all with XLR sockets, +48 V phantom powering and trim controls to cater for a wide (60 dB) range of input signals. The first of these is intended for use with an external talkback microphone and a switch is provided to disable the internal microphone. A 'trigger' jack input/output is also provided. This can be used with a foot-switch so that a producer can remotely trigger the HUI's talkback function, or can be connected to a mixing console to enable its talkback function when the talkback switch is pressed on the HUI. The other two input channels are intended for microphones or instruments that you want to record into Pro Tools. Both of these have 1/4-inch TRS insert points so you can connect signal processors such as compressors across the channels – and they can also be used to provide direct outputs for additional flexibility. Each of these two mic channels also has a 1/4-inch jack socket, which will normally be used to let you feed unbalanced or balanced signals to a Pro Tools interface. These microphone pre-amplifiers are definitely up to Mackie's usual high standard and provide an excellent way of connecting low-level signals to a Pro Tools interface.

Monitoring

The HUI's monitoring facilities deserve a special mention as these can be in mono, stereo or up to six discrete channels of surround sound with overall volume controlled by the Master Volume knob. In stereo monitor mode, with the Discrete switch turned off, any of the three stereo input sources can feed any of the three stereo output pairs and all the outputs can be active at once, so you could send outputs to various combinations of monitor speakers or to external recorders such as cassette machines. Keep in mind, however, that these outputs are designed for monitoring purposes only – not for mixing. When the Discrete switch is engaged, the control room section becomes a discrete surround matrix capable of either standard 4.1 (L/C/R/S plus sub-woofer) surround operation or 5.1 (L/LS/C/R/RS plus sub-woofer) operation. With this set-up you can connect up to three pairs of stereo 'stem' outputs from the Pro Tools audio interface to the HUI's monitor inputs and feed these to the three output pairs, in which case the level controls for the three HUI outputs can be used to trim the output levels of the respective mix stems. These surround-monitoring features make the HUI particularly suited to post-production work for video or DVD which increasingly use surround formats.

Summary

Compared with using the mouse to control Pro Tools, using the HUI everything simply works much more smoothly and you can get to everything you need so much more quickly. The jog wheel is much smoother in action, the faders feel much smoother, and if you want to hit Solo or Mute you just reach out and touch! Also, the way the plug-ins' parameters are displayed numerically and the way you can control these using the V-pots lets you run your session much more efficiently than when you have to mouse around the graphical plug-in displays on your computer screen. The HUI does provide a viable alternative to using an external mixer as long as you have enough DSP for your mixing requirements and can live with just a pair of microphone pre-amps. And for smaller post-production studios working with surround sound the HUI makes a superb choice as a partner to Pro Tools systems.

Hardware control summary

The marriage between these hardware control surfaces and Pro Tools software running on a personal computer is not a perfect one as yet. There are still some operations best controlled using keyboard and mouse, and I find myself wanting to have these positioned centrally for best access – yet this is where the faders are on these hardware controllers. So an element of compromise is necessary. Traditional music recording engineers will find the Control|24 to be the best choice, while smaller MIDI-based project studios will value the Mackie HUI or CM Labs Motor Mix. Larger music or post-production studios will find the Digidesign ProControl offers a comprehensive range of features for music production or to suit working to picture.

2 File Management

Introduction

Pro Tools 6.1 (and later versions) feature a specialized database referred to as DigiBase. This database is accessed using various types of DigiBase 'browsers' – the Workspace, Volume, Project and Catalog. A toolbar at the top of each browser window contains the Browser menu, a Search icon and five view presets.

> Tip: Arrange any browser window the way you want it, then Command-click (Mac) or Control-click (Windows) on a view preset to store this arrangement for future recall.

The fixed pane at the left of the browser window contains the Items List that displays the contents of a volume, folder, session or Catalog database. Each column in the items list displays meta-data such as the file name and format for items in the list. Columns can be resized by dragging the column border or rearranged by dragging the column headers. To the right of the fixed pane, there is a scrolling window pane with lots of columns. Columns that you want to keep in view can be dragged to the fixed pane, while columns you view less frequently can be placed in the scrolling pane.

Pro Tools needs to index the volumes, folders and items so that they can be searched for quickly. The contents of each folder are indexed automatically when you open its browser, although this can take some time. If you have a large folder of sound effects or clips, for example, you should make sure that this has been indexed before you begin to work with the browser (whenever this is possible).

> Tip: The browser menu command 'Update Database for Selected' lets you manually index any selected folder or volume including all sub-folders.

You can sort items in Browsers by clicking on any column header to sort under that category. A number '1' appears at the far right of the column header to indicate that this is the primary sort category. You can Option-click (Mac) or Alt-click (Windows) a second column to carry out a secondary sort within another category. A '2' appears at the far right of this column header to indicate that this is the secondary sort category.

You can search the DigiBase Browsers or Catalogs for files by name, modification date, kind or using multiple criteria. When you click on the Search icon, various 'Find' fields appear above the columns in the Browser or Catalog windows. You can enter a Name, Date or Kind in the appropriate find Field, and the DigiBase Pro for TDM systems lets you search using many other criteria such as bit-depth or sample rate.

You can drag and drop audio files and sessions from browsers and Catalogs into the current session and if you have DigiTranslator 2.0 installed, you can drag and drop OMF or AAF sequences and files from DigiBase browsers.

> Tip: When you Shift-drag a session to the Timeline, the Import Session dialog appears to let you select various import options.

For each file in the database, you can enter comments into two types of comments fields – the File Comments and the Database Comments. File comments, as the name implies, are stored with the metadata of the file itself, although not all file types support these. Indexing stores File Comments in the database, so these can be searched and viewed even if the file is offline. File Comments can also be edited in browsers (unless they are marked as read-only).

Database Comments in a Catalog are stored in the Catalog database, while Database Comments in the Project Browser are stored in the Session. Database Comments (up to 256 characters long) are searchable, cross-platform and editable.

The Waveform column shows the waveforms for any audio files and audio files can be auditioned from within DigiBase browsers by clicking and holding the Speaker icon to the left of the waveform display or by simply pressing the space-bar on your computer's keyboard if the file's browser is the current window.

The Workspace Browser

You can think of the Workspace Browser as an alternative to the Macintosh Finder or the Windows Explorer. Like these, it lets you view and access all your available data storage volumes so you can use it to find project folders, session files, media files – or any other files or folders for that matter. So why not just use the Finder or Explorer? The reason is that the Workspace Browser has various features specific to Pro Tools. For example, all sorts of data about the files is displayed in the browser – not just the name, date modified, size and kind, but also the Unique ID, any time-stamp information, the sample rate and so forth. You can also type in comments for each file, and search for or sort files.

Data storage volumes (such as hard disk partitions, CD-Rs or optical storage disks) may be classified as 'Record' (for recording and playback), 'Playback' (playback only) or 'Transfer' (for storage or copying) of audio and other media files, such as video.

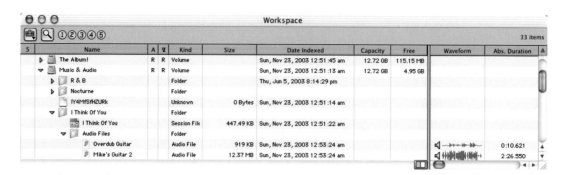

Figure A2.1 The Workspace Browser.

If you need to look for particular files, the Workspace Browser is the best place to do this from. Click on the magnifying glass icon and the 'Find' field appears to let you type the name of a file, and a 'Results' area opens in the lower part of the window to display the files found.

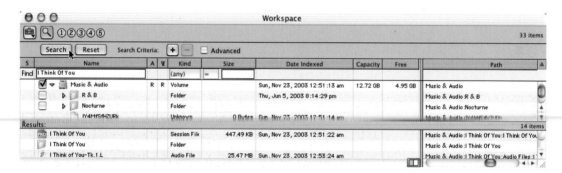

Figure A2.2 Finding a file in the Workspace Browser.

Volume Browsers

A 'volume' is any formatted partition on a physical drive such as a hard drive or optical drive, CD-R, DVD-R, DVD-RAM drive or similar data storage device.

If you double-click on any volume listed in the Workspace Browser, a Volume Browser opens in a separate window. You can use this to view, manage, audition and import individual items on the volume, and to update a database for the contents of the volume.

Catalogs

You can also create, view and access Catalogs in the Workspace Browser. These are collections of frequently used items or favourite files that can be grouped together regardless of where they are actually stored – in contrast to the way Volume Browsers group files according to which volume they are on. A Catalog can be thought of as similar to a folder full of aliases to particular files of interest. You can open several Catalogs in the Workspace Browser if you like, using the New Catalog command from the browser menu. Once created, you can fill the Catalog by dragging and dropping folders or files onto its icon from any other Catalog or browser.

The Clip Names that appear in the timeline and Audio Regions List when a file is imported into a session can be edited in a catalog (without affecting the file names).

The Project Browser

The Project Browser displays and manages all the files used in your current session – no matter which volume they are actually on. This browser contains all the files associated with a session and allows you to add metadata in the form of comments to each file. As with other Pro Tools browsers, you can use this browser to search and sort and to identify, select, copy or relink files.

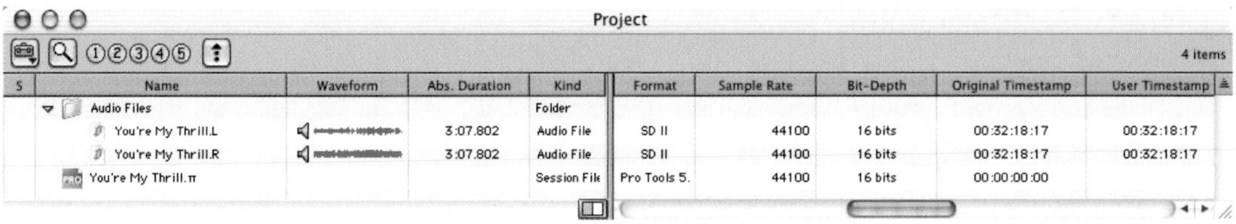

Figure A2.3 The Project Browser.

> Tip: When the 'Show Parent in Project Browser' command in the Audio Regions List menu is enabled, the Project browser automatically highlights the parent file of any file or region selected in the Audio Regions List.

> Note: Remember that the Project Browser only contains files that are in the current session. So, to import a file from another browser into the current session, you must drag files to the Timeline or Regions List in the Pro Tools Edit window. You cannot import a file into a session by dragging it to the Project Browser. (It does work the other way, though: you can drag files from the Project Browser into Timeline of your current session.)

Unique file IDs

A problem that comes up occasionally when moving sessions around is that Pro Tools can't find some of the session files or doesn't know which disk these are stored on.

Pro Tools tags each audio file in a session with a unique file identifier to allow it to distinguish a particular file even if its name or location has changed. Older versions of Pro Tools do not create these unique file IDs though, so if the unique identifier is not present, Pro Tools can identify an audio file using other file attributes including sample rate, bit depth, file length, and creation or modification date.

In Pro Tools 5.3.x and lower, if Pro Tools cannot find the audio files needed for the session, it will open a Find File dialog to let you search for files based either on the file names or on their unique file IDs. If files with similar attributes are found, these are then presented in a list of Candidate Files from which you can select the most likely choices.

Missing files

With Pro Tools 6.x, you manage 'links' to audio and other media files using the Relink window. If any files are missing when you open a project, or if they are not on a Performance volume, Pro Tools can find these for you. This process is referred to as 'relinking' as it allows you to relink the files – i.e. make them available again to your session.

Files located on CD-ROMs or on drives attached to other computers on a network are referred to as Transfer files and are not suitable for playback. If you open a session that links to any such Transfer files, Pro Tools posts a warning and you can either click 'Yes' to use the 'Copy and Relink' dialog to copy these to a Performance volume, or you can click 'No' to go ahead and open the session with the Transfer files off-line – in which case you cannot play these back.

If a file is missing when you open an existing session, Pro Tools opens the Missing Files dialog. Here, you can skip these files, leaving them off-line when the session opens, or manually find and relink these, if you know where they are. Otherwise, you can select the 'Automatically Find and Relink' option to have Pro Tools search all the Performance volumes for your missing files. In this case it relinks based on criteria including the file name, the unique file ID, the file format and length.

Figure A2.4 Missing Files dialog.

Relinking missing files in an open session

If you are working with a session that has missing files and you decide at some point that you need these files, you can open up the Relink window by selecting the files you want to relink in the Project Browser and choosing 'Relink Selected' from the browser's local menu.

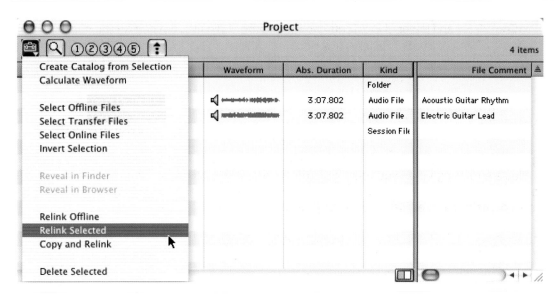

Figure A2.5 Choosing the Relink Selected command from the Project Browser's local menu.

When the Relink window opens, make sure the files you want to relink are selected and that the drives you want to search are ticked.

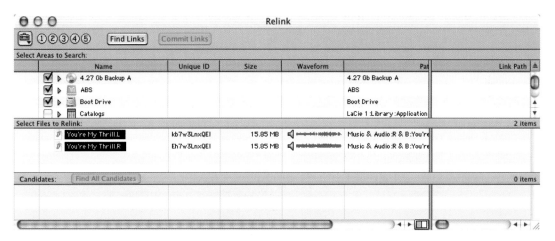

Figure A2.6 Relink window.

Then click the 'Find Links' button. Up comes a Linking Options dialog that lets you choose whether to find by name or by file ID or other criteria.

Figure A2.7 Linking Options dialog.

Pro Tools searches the selected drives to find 'candidate' files that match the selected files. For example, there may be multiple copies of the same files on different drives, or there may be files with the same ID but different names. The candidate files are shown in the lowest area of the window and a link icon appears next to each file as Pro Tools finds and links each candidate to the missing file. If you are not sure whether Pro Tools has found the correct versions of the files, you can click on the 'Find All Candidates' button to find all the possible candidates that could match a selected file. When you are satisfied that you have found the right candidates, click the 'Commit Links' button at the top of the window.

3 Transferring Projects

Pro Tools transfers

If you need to transfer your Pro Tools session to a different Pro Tools system, perhaps at a larger studio, or onto a portable system, or across platform Mac-to-PC or vice versa, then you should prepare the session carefully first to make sure that everything will transfer just the way you want it.

You should also make a copy of the project and use this for the transfer – deleting any tracks you don't intend to use. This way, your original files are left untouched, so you can always go back to these source files in case anything goes wrong later on.

Getting ready to make the transfers

Typically, after a busy Pro Tools session you may have several tracks containing alternate takes that you subsequently decide will never be used. You might have other tracks that you used for one reason or another along the way, such as an auxiliary track to monitor a metronome click, or a stereo version of the demo that you have been using as a guide to work to. Before making your transfer, you should save an edited version of your session with any such extra tracks removed.

It is also quite likely that you have a number of unused regions clogging up your regions list as a result of trial edits, false takes and so forth. The quickest way to get rid of most of these is to invoke the 'Select Unused Regions Except Whole Files' command from the pop-up menu at the top of the Regions List.

> Note: You should choose the 'Unused Regions Except Whole Files' option because otherwise any whole files that you had edited to create shorter regions from would become selected, ready for deletion – which, of course, you would not normally want.

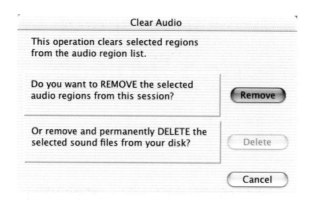

Figure A3.1 Pro Tools Clear Audio Regions dialog.

The next step is to choose the Clear command from this menu, which brings up the Clear Audio dialog. Here you can choose whether to simply remove the selected regions from the list or whether to completely delete these from your hard drive. In this case, you probably should choose the Remove option – unless you know for sure that all these unused regions belong to files for which you have no further use.

You should also carefully go through the 'whole file' regions, listed in bold type in the list, to get rid of any false takes or other stuff that you will never need again – auditioning each of these carefully first if you are at all unsure. When you identify any such files, you can use the Clear Audio command to get rid of these, choosing the 'delete' option this time.

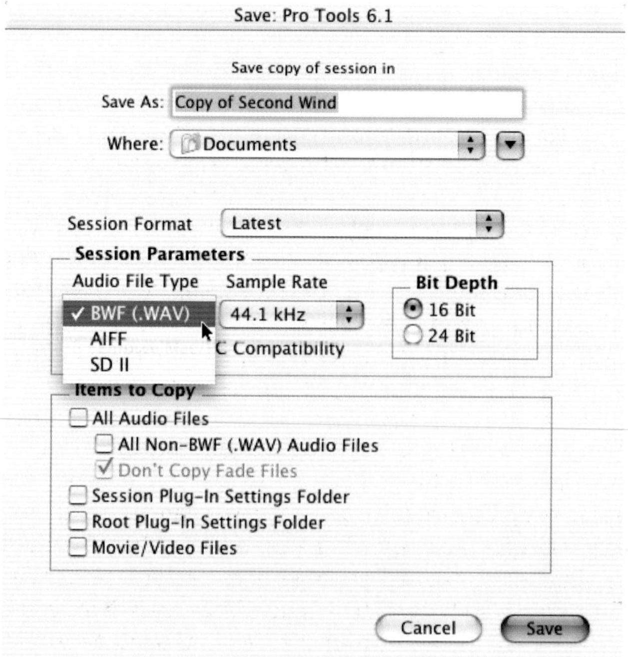

Figure A3.2 Pro Tools 'Save Session Copy In...' dialog.

When you have tidied up your session to this point, your next move should be to make a copy of the session – complete with any associated files. Choose the 'Save Session Copy In...' command from the File menu. In the dialog that appears, you can choose the sample rate, bit-depth, and audio file type for the target system to which you are making your transfer. For example, to transfer to Pro Tools running on a PC, you would choose the BWF audio file type.

You can use the Session Format pop-up to choose an older session format if you are transferring to an older Pro Tools system.

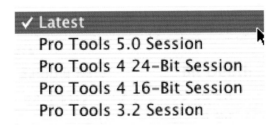

Figure A3.3 Session Format pop-up.

And you should tick the 'Items to Copy' box for 'All Audio Files' to make sure that all the audio files associated with your project will be copied – even any files that may exist on a different hard drive or in a different folder on the same hard drive. You can also include various other files and settings if necessary.

Transferring MIDI

If you need to transfer MIDI sequences into or out of Pro Tools, just consider for a moment what is involved. If you have recorded MIDI tracks using a particular synthesizer or sampler, will this be available, with the sounds you used, after you make your transfer? If you are simply moving between Pro Tools and, say, Logic on your own system, then this will probably be the case. But if you are taking the project to another studio or bringing in a project from another studio, it almost certainly will not be the case. You may get lucky – some of the sound modules and patches might be available – but it is unlikely that everything will be available. And the MIDI equipment is unlikely to be connected to the same interface ports as on the system it came from or is going to – so a fair amount of re-configuration of MIDI track channel/port outputs may be necessary. And if none of the modules and sounds are available, then a lot of time will be spent choosing alternative sounds from whatever is available – which are unlikely to be too close to the sounds originally used. And this can upset the whole balance of the recording – maybe even requiring extensive editing of the MIDI parts, or even re-recording of the MIDI parts to make these work with the new set-up. In short – a nightmare situation! A partial solution is to record the actual patch data from your synthesizer as SysEx data at the beginning of each MIDI track that uses this synthesizer. Then you can at least be sure that if you have access to that synthesizer you can put the correct sound or sounds back into it. However, this is not a practical option with samplers and can be awkward or impossible with some models of synthesizer.

> Tip: To avoid this nightmare altogether, it makes much more sense to record your MIDI tracks as audio tracks as soon as this is a practical option during the production process. You can keep your MIDI tracks hidden in your Pro Tools session in case you need to re-visit these to make edits.

Nevertheless, there can be many circumstances where it makes sense to import or export MIDI files – and the good news is that this does work as advertised. All the MIDI tracks and track names that are not muted can be saved, along with tempo and meter information – and with any markers that you have set up.

> Tip: One thing that is necessary with sequencers that allow sophisticated, possibly nested, looping, 'ghost' tracks and so forth is that you must create versions of your MIDI tracks to export without any of this trickery going on. The MIDI file format will only transfer regular MIDI tracks.

Importing MIDI files

If you have recorded MIDI sequences using another MIDI sequencer, you can transfer these into Pro Tools by saving them as Standard MIDI Files. You can use the 'Import MIDI to Track' command to place the imported MIDI directly onto new tracks. Alternatively, you can use the 'Import MIDI' command in the MIDI Regions List pop-up menu to place the data into the MIDI Regions List, from where it can be dragged onto existing tracks. If you want to import the MIDI file's tempo and meter tracks, select the 'Import Tempo From MIDI File' option in the Import MIDI dialog. This option overwrites existing meter and tempo events in the current session. If you don't want to overwrite the existing meter and tempo events, select 'Use Existing Tempo From Session'. Another thing to watch out for is that if the Standard MIDI File contains markers, these will only be imported if the current Pro Tools session does not contain any markers.

Exporting MIDI files

You may wish to export MIDI tracks from your Pro Tools session so that you can use these with another MIDI application, or even to transfer these to an external hardware sequencer. Pro Tools lets you export MIDI tracks as a Standard MIDI File using the 'Export MIDI' command from the File menu. Exported MIDI information includes notes, controller events, program changes and System Exclusive data, as well as events for tempo, meter and markers. Just make sure that you mute any MIDI tracks that you don't want to export. Then, when you export, all the unmuted MIDI tracks will be merged to a single, multi-channel track (Type 0), or they will be saved as multiple tracks (Type 1) – depending on which option you select in the 'Export MIDI' dialog's 'MIDI Version' field.

Importing tracks and track attributes

You can import entire tracks from other Pro Tools sessions into your current Pro Tools session using the Import Session Data command.

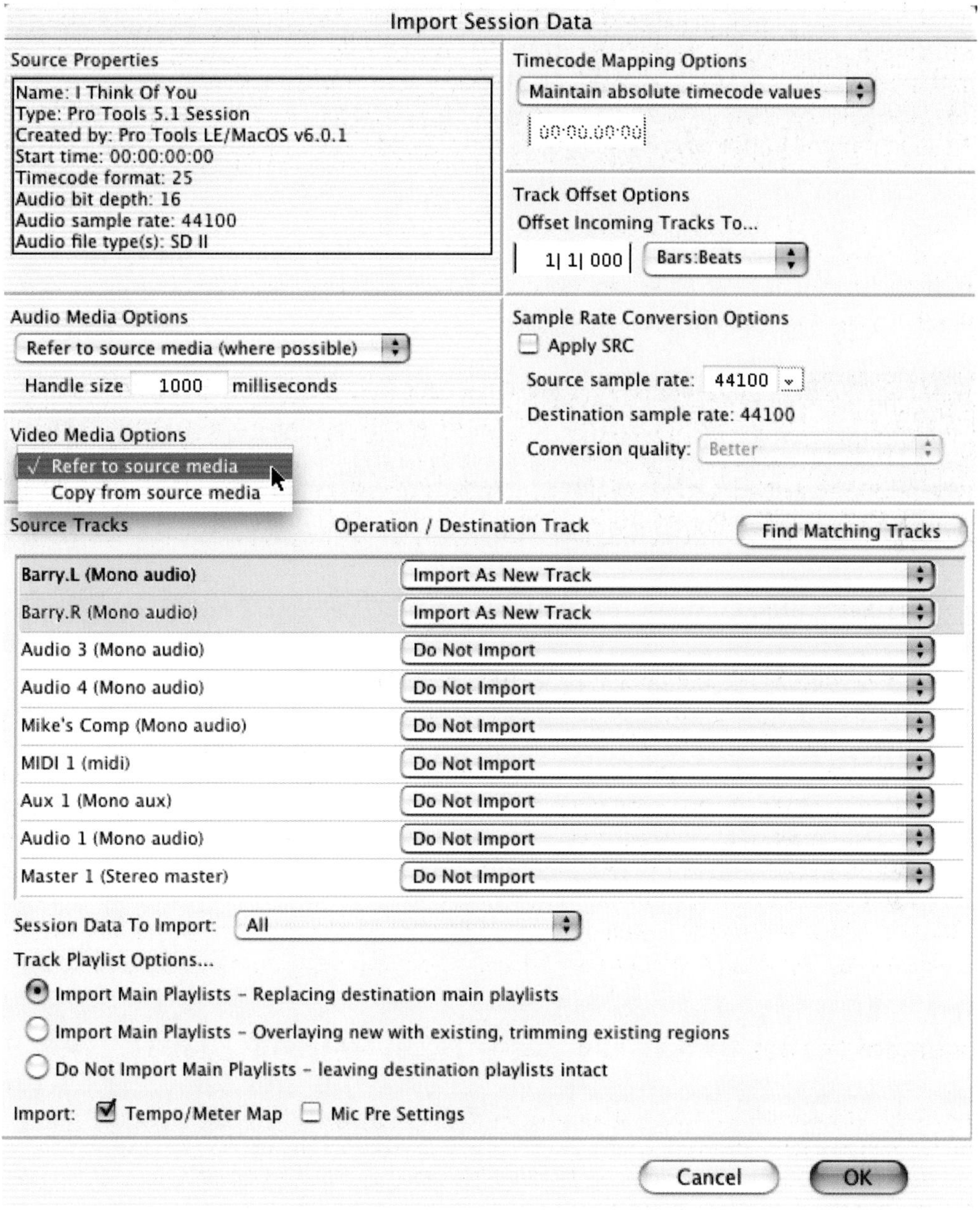

Figure A3.4 Import Session Data dialog.

On TDM systems, you can choose which attributes of those tracks you want to import. For example, you can choose to import only the track's audio into your current Pro Tools mixer. This is analogous to 'changing the tape reel' in a traditional studio set-up with a tape machine and mixing console.

Or, you can choose to import all of a track's mixer settings without its audio, effectively importing a channel strip and using it on a track in your current session. By importing mixer settings for all of the tracks in a session or session template, you can re-use an entire Pro Tools mixer on all the sessions in a project.

To import tracks or their attributes choose 'Import Session Data' from the File menu, select the session to import data from, and click Open. Or you can just drag the session file whose tracks or attributes you want to import from a DigiBase browser into the track playlist area in the current session's Edit window. Select the tracks to import by clicking the track names in the Source Tracks list.

Note: If the current Pro Tools system does not support surround mixing, surround tracks are not displayed in the Source Tracks list.

You can choose to import each track as a new track, or choose a destination track from the corresponding pop-up menu.

Tip: If you already have tracks in your session with the same names as the tracks you are importing, you can simply click 'Find Matching Tracks' to automatically match the source tracks to destination tracks with the same names. Tracks must have the same name, track type and channel format to be automatically matched.

The options provided in the Import Session dialog give you lots of flexibility to decide how you want to import the source tracks. You can copy files from the source media into your session, or you can simply refer to the original source files if you prefer. You can convert the sample rate, bit

Note: When using the Import Session Data command, the Tempo/Meter Map is not imported if the destination session is in Manual Tempo mode. The fix for this is simple enough – just disable Manual Tempo mode when you are importing a Tempo/Meter map from another session.

Tip: You can use the 'Consolidate from Source Media' option to just copy the regions of the audio files actually used in the source tracks – without copying any unused audio. When you choose this option, you can also choose the size in milliseconds of a 'handle' that will be applied to the consolidated audio. This 'handle' is an amount of the original audio file that is preserved before and after each region in case you need to make any edits to the new regions.

depth and file type of the source audio to match your session, as necessary. You can also import mic pre-settings or meter and tempo maps from the source session.

Time code mapping options

At the top right of the Import Session Data dialog, you can specify where the imported tracks are placed in the current session using the Time Code Mapping and Track Offset options.

'Maintain Absolute Time Code Values' places tracks at the locations where they were located in the source session. For example, if the current session starts at 00:01:00:00, and the session from which you are importing starts at 10:00:00:00, the earliest imported tracks can appear in your session is 9 hours and 59 minutes after the start of the session.

'Maintain Relative Time Code Values' places tracks at the same offset from session start as they had in the source session. For example, if the source session starts at 01:00:00:00 and contains a track that starts at 01:01:00:00, and the current session start is 02:00:00:00, the track will be placed at 02:01:00:00 in the current session.

'Map Start Time Code to hh:mm:ss:ff' places tracks relative to their original session start time. For example, on a TDM system, or Pro Tools LE system with DV Toolkit, if the current session starts at 00:01:00:00, and the session from which you are importing starts at 10:00:00:00, you can reset the start time code to 00:01:00:00, to avoid placing files 9 hours and 59 minutes from the start of your session.

Track playlist options

You can select from three options to control how the main playlist from each source track is imported to the destination track in the current session using the 'radio' buttons in the lower part of the Import Session dialog – see Figure A3.4.

If you choose 'Import Main Playlists – Replacing destination main playlists', the main playlist in the destination track is deleted and replaced with the imported playlist. You also get to choose which, if any, of the source track's attributes (fader levels, pan settings, plug-in assignments and so forth) to import using the 'Session Data to Import' pop-up.

If you choose 'Import Main Playlists – Overlaying new with existing, trimming existing regions', any existing playlist data that overlaps data imported from the source track is trimmed and replaced with the imported data. Any playlist data in the destination track that does not overlap remains in the destination track. This is useful if you have recorded, say, a guitar part for one version of the song and you decide that the guitar part in, say, the chorus section was better in a previous or different version, where the guitar only played in the choruses. Now you still want to use the new guitar in the verses, but you want to replace the new guitar in the choruses by the old guitar.

If you choose 'Do Not Import Main Playlists – Leaving destination playlists intact', only the attributes selected in the 'Session Data to Import' list are imported to the selected tracks – no audio is imported. You might choose this if you wanted to use all of the source track's input, output, send, insert and plug-in attributes to save you the trouble of setting these up in your current session.

Note: The Import Session dialog imports the Tempo/Meter map by default, which may not be what you want – so watch out!

Selecting session data to import

You can select All, None, or any combination of the listed attributes to import. For example, you might only want to import regions and media into your existing Pro Tools session. This way, you can set the volumes, pans and other track attributes the way you want them to be for this session rather than the way they were set in the session they came from.

Figure A3.5 Pop-up Menu: Session Data to Import.

'All' imports all of the source track's playlists and all of the attributes in the Session Data to Import list, while 'None' imports only the source track's main playlist. Be aware that importing any of these settings will replace the corresponding settings in the destination tracks – which may not be what you want to do.

Note: When you import a track's input, output, send output or hardware insert assignments, any custom path names and I/O configurations from the source session are not imported. However, you can import path names and I/O configurations by importing I/O Setup settings.

Transferring audio

Transferring audio into or out of Pro Tools used to be a tedious business. The only practical way was to make sure that every audio track in Pro Tools started at the beginning of the session – Bar 1, Beat 1 or whatever. If you had just recorded a few seconds of audio right at the end of a track, for example, you would have to bounce this to disk to create a new track that started from the beginning. This would be filled with silence all the way up until the few seconds of actual audio recorded at the end of the track. A waste of disk space and a waste of time! The advantage being that when opening a new session in Cubase, Digital Performer, or whichever, you could simply import these audio tracks and place them all into your arrangement starting at Bar 1, Beat 1 (or at your chosen session start point) and they would all line up correctly in sync with no more ado.

You could not transfer your actual Pro Tools session files to any other software, nor could any other software transfer its project files, arrangements or whatever they were called, into Pro Tools. So you always had to set up a new session file from scratch in Pro Tools or Logic – and all the edits you had made, that you might not have finalized, would be lost. And if you wanted the tempo and meter information, markers and MIDI tracks, you would have to use MIDI files for this.

And then came OMF. AVID had developed the OMF Interchange (OMFI) file format to allow projects from their video editing systems to be transferred to and from rival systems. And OMF allowed projects to transfer complete with the edits. And video has associated multi-track audio. Naturally, it wasn't long before AVID/Digidesign added OMF capabilities to Pro Tools to allow the audio from these video projects to be edited using Pro Tools systems. It still took some time before it occurred to other audio software manufacturers that this would be a great thing to use to allow audio projects to be transferred between rival systems in the same way. The 'penny dropped' first for Mark of the Unicorn, who added OMF capabilities to Digital Performer a couple of years before anyone else got the message. The good news is that, today, every major MIDI + Audio software manufacturer has added OMF support to their professional packages – including Cubase SX, Nuendo, Sonar, Logic Platinum and Digital Performer.

> Note: Pro Tools is now the only MIDI + Audio software that does not support import and export of OMF files 'out of the box'. Digidesign require you to purchase the DigiTranslator 2.0 software option for Pro Tools TDM or Pro Tools LE systems to add this functionality – at a price of around $495. Specifically, you cannot open OMF files into Pro Tools unless you have purchased and installed the DigiTranslator 2.0 software option for Pro Tools. Also, you cannot export tracks from Pro Tools as OMF/AAF files unless you have the DigiTranslator option installed.

More recently, rival interchange formats have also been developed. For example, AES31 was developed as an interchange format by the Audio Engineering Society that will take information about the positions of events, fades and so forth. AES31 uses Broadcast Wave files as the default audio file format. Another relatively new interchange format is the Advanced Authoring Format (AAF), available since 2001, sponsored by the Advanced Authoring Format Association (a large group of professional media manufacturers) – see

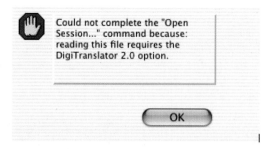

Figure A3.6 Error message that appears in Pro Tools 6.1 when attempting to open OMF files without DigiTranslator.

www.aafassociation.org. The metadata incorporated into OMF and AAF files includes information about each media file, such as sample rate, bit-depth, region names, the name of the video tape from which the media file was captured, and even time code values that specify where a file was used in the project. It also includes information about what files are used, where they appear in a time line and automation.

Currently, of the popular MIDI + Audio applications, only Pro Tools supports AAF, while Nuendo and others support AES31, for example. No doubt, in due course, they will all support every interchange file format – as this is the only strategy that makes any sense. After all, the whole point is to facilitate interchanges – not prevent people from interchanging projects.

Preparing for OMFI transfers

Just as when you are transferring projects between Pro Tools systems, if you are transferring files via OMF to or from Logic Pro, Digital Performer, Nuendo, Cubase SX or SONAR, you should prepare your project files carefully first. Then make a copy of the project and use this for the transfer – deleting any tracks you don't intend to use. If the project uses any MIDI tracks, you can either record these as audio tracks or export them as a MIDI File. If you have tracks or regions that use automation, you may need to bounce these to disk first. Similarly, if you have used any plug-in effects, you should process your audio using these before exporting – in case these plug-ins are not available on the target system. And if your project uses more tracks than the target system can handle you may need to reduce the number of tracks – making sub-mixes if necessary. If you are preparing a project for transfer to another studio, then you should ask whoever will be using the files to specify the format and specifications for the audio files you will supply, such as Wave or AIFF. And if you know this in advance, it makes sense to record your project using the sample rate and bit depth that matches that of the target system – otherwise you will have to convert the file formats at some stage, which takes time and can lose quality. And it would be wise to make sure that the transfer works properly by checking it out at the target studio at least a day before it is actually needed – to allow time for any problems to be sorted out. It is also a good idea to create a text file that you can keep with the transfer files containing essential information such as the tempo and meter and any special instructions for your project.

Exporting OMFI files from Pro Tools

With the DigiTranslator 2.0 (or higher) Option installed, Pro Tools lets you export individual tracks or an entire Pro Tools session in OMFI format (Pro Tools 5.1.3 and higher) or AAF format (Pro Tools 6.1 and higher).

> Note: The DigiTranslator option is contained within the Pro Tools 6.1 application software as a demo version that can be authorized for permanent use if you buy an iLok key. When it is authorized, the File Menu command, 'Export Selected Tracks as OMF/AAF...' becomes active to let you export projects, and the 'Open Session' command allows you to import projects in OMF format.

With DigiTranslator installed, you can export individual tracks or an entire Pro Tools session in OMFI format or AAF format using the 'Bounce to Disk', 'Export Selected as Files', and 'Export Selected Tracks as OMFI' (Pro Tools 6.0 and 5.x) commands. In Pro Tools 6.1 you can also export AAF and OMFI sequences using the 'Export Selected Tracks as OMF/AAF...' command.

> Note: AAF and OMFI sequences and files exported from Pro Tools only retain time code addresses, region names and definitions. AAF and OMFI sequences exported from Pro Tools *do not* retain information about plug-ins' assignments or parameters, routing or grouping – so you should render any effects before exporting. Pro Tools *does* export volume automation to AAF and OMF, but only if the 'Quantize edits to frames boundaries' option (Avid Compatibility Mode) is disabled.

Using Pro Tools 6.1 for example, just select the tracks you want to export and choose 'Export to OMF/AAF' from the File menu. A dialog appears that lets you choose the file formats and set various other options.

The idea of transferring projects to and from Pro Tools is extremely appealing, and when it works, it's great. But there are lots of things to watch out for. For instance, DigiTranslator 2.0 only supports OMFI version 2 files – OMFI version 1 files are not supported. Also, AAF import and export is only supported in Pro Tools 6.1 – earlier versions of Pro Tools do not support AAF. Embedded video is not supported with OMF. If you do encounter an OMF file that has embedded video, you will be prompted to find the video file. In this case, you should select 'Skip all Video' to open just the audio portion of the OMF. And Pro Tools with DigiTranslator cannot import or export AAF with embedded media – Pro Tools with DigiTranslator only supports AAF import and export with external file references. This means that you need to transfer the audio files that you have used in your session along with the AAF file. Finally, Pro Tools cannot import or export AAF sequences that reference Sound Designer II (SDII) files – so the most sensible choice for compatibility is now the .wav file format, although AIFF is an option.

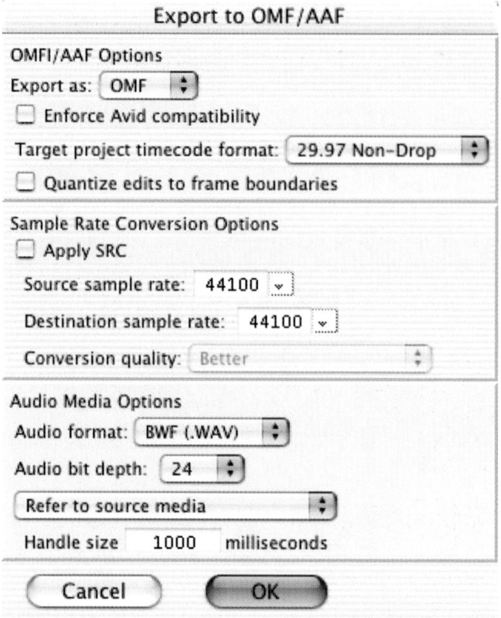

Figure A3.7 Pro Tools Export to OMF/AAF dialog.

Importing OMFI files into Pro Tools

You can import OMFI or AAF sequences directly into Pro Tools with the DigiTranslator 2.0 Option. Specifically, you can import AAF sequences with their referenced audio files, or OMFI media files and sequences including those with embedded audio, into Pro Tools using the Open Session, Import Track, Import Audio to Track and Import Audio (to Regions List) commands.

You can also drag and drop media or sequence files from MediaManager into a Pro Tools session (Windows XP or Mac OS9 only). In Pro Tools 6.x, you can drag and drop OMF sequences or audio files from any DigiBase browser. In Pro Tools 6.1 you can also drag and drop AAF sequences from any DigiBase browser.

> Note: Pro Tools does not support AAF files with embedded audio (or video), so AAF sequences with embedded media will display 'Kind Unknown' in the Type column in DigiBase browsers.

In Pro Tools 6.1, when opening an AAF or OMFI sequence the 'New Session' dialog opens, prompting you to name and save the sequence as a new session. Once you name and save the new session, the 'Import Session Data' dialog opens. In the 'Import Session' dialog you can configure DigiTranslator translation settings and other import options.

When you use the 'Import Audio to Track' command from the File menu, Pro Tools imports the audio, creates a new track, and places the imported audio into the newly created track as well as into the Regions List. The imported region takes its name and time stamp from the OMF data.

Let's take a look at how this works in practice. When you use the 'Open Session' command, for example, a dialog appears to let you look for the file you want to open.

Figure A3.8 Pro Tools Open Session dialog.

Select an OMF file to open and the New Session dialog will appear. Here you can name the converted session and choose where to save it. You can also choose a new audio file format, sample rate and bit-depth for your new session, if necessary.

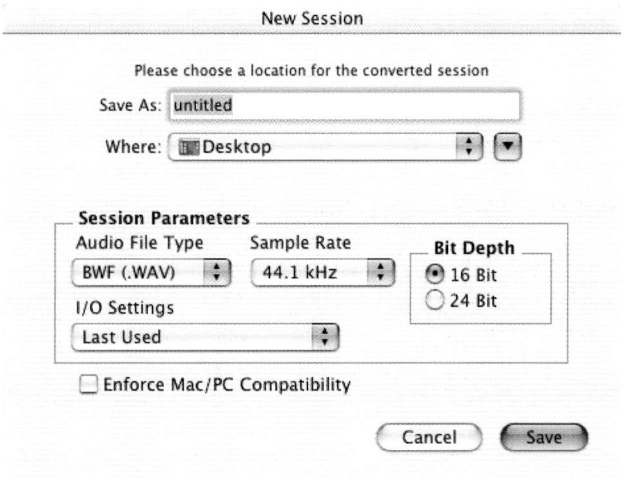

Figure A3.9 Pro Tools New Session dialog.

For the sake of optimal compatibility, always enable Enforce Mac/PC Compatibility in the New Session dialog. This ensures that all Pro Tools files are saved in the AIFF or BWF (.WAV) format, and that no illegal characters are used in file names. You can also configure existing sessions to default to AIFF or BWF (.WAV) format, or convert SDII files upon export.

Click 'Save' in the 'New Session' dialog and the 'Import Session Data' dialog will appear. Here you can specify exactly which tracks you want to import, what additional data to import, how Playlists will be handled, and so forth. One thing to watch out for is that the default setting for the source tracks is 'Do Not Import', so you need to specifically change this to 'Import as New Track' for every track you want to import.

> Tip: To change all the Source track settings to 'Import As New Track' at once, hold the Option/Alt key on your computer keyboard while changing any one of the tracks.

Moving sessions between platforms

If you know that you are going to move a Pro Tools session from Mac to PC, or vice versa, at some point, then you should check the option to 'Enforce Mac/PC Compatibility' in the New Session dialog, and choose either AIFF or BWF (.WAV) file formats when you first open a new session.

Of course, you may only decide this after the session is under way. In this case, you can select the above options if you use the 'Save Session Copy In...' command. With any type of back-up or transfer, the safest way is to save a copy of the session you want to transfer first, using the 'Save Session Copy In...' File menu option. Using this dialog you can make sure that all the media files associated with your session are copied from wherever they may be residing into a destination folder of your choice. You can also take this opportunity to change sample rates, bit-depths and file formats using the options provided.

If you have all your Pro Tools sessions on a Mac HFS or HFS+ formatted disk drive, you may wish to simply connect this to a PC. Normally, this is not possible, as PCs cannot read Mac-formatted drives. However, you can use MacOpener software to mount Mac HFS or HFS+ formatted drives on a PC to use as Transfer drives. And a demo version of this software is included on the Pro Tools Installer CD. There are a few limitations, however. For example, the session needs to be Pro Tools 5.1.1 or later and it cannot contain mixed audio file formats. Also, when you are working with a session on your PC that references audio files on a Mac-formatted drive, you cannot bounce new audio files to this disk – the bounce destination must be a Windows FAT, FAT32 or NTFS drive. And opening sessions from a Mac-formatted drive in Windows is slower than from a Windows-formatted drive.

Moving between PT LE and PT TDM

One of the most useful features in Pro Tools is the session interchange facility provided between LE and TDM systems. The way this handles sessions which use TDM plug-ins is particularly neat. As you will appreciate, you can't use TDM plug-ins with LE systems – so you could end up losing lots of useful signal processing that you have used to build your session.

Fortunately, there are Real Time Audio Suite equivalents for many TDM plug-ins, so if you have used any of these in your TDM session, all the plug-in settings will transfer to the LE system – where it will use the RTAS equivalent. If there is no RTAS equivalent, the settings will be

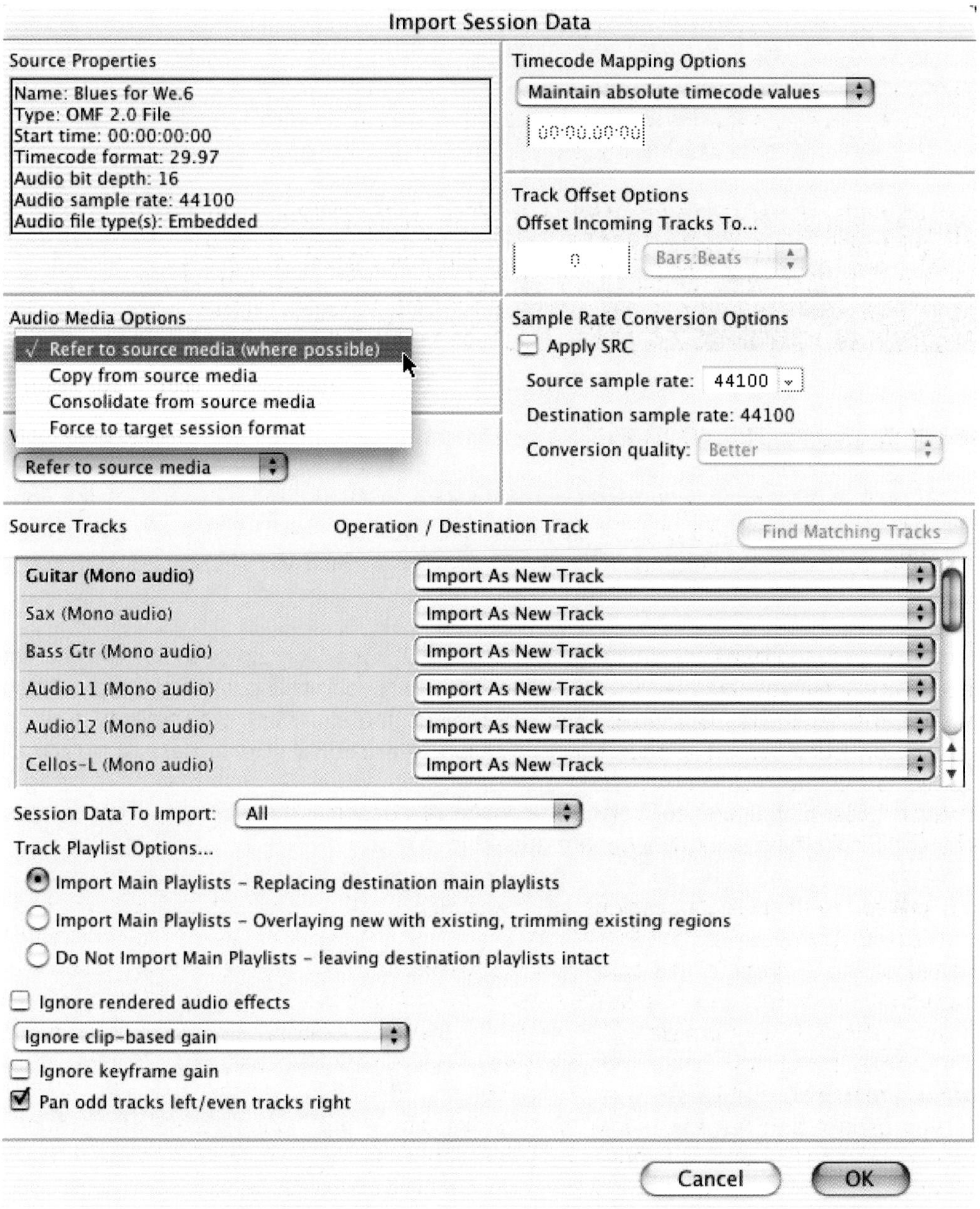

Import Session Data

Source Properties

Name: Blues for We.6
Type: OMF 2.0 File
Start time: 00:00:00:00
Timecode format: 29.97
Audio bit depth: 16
Audio sample rate: 44100
Audio file type(s): Embedded

Timecode Mapping Options

Maintain absolute timecode values

00:00.00:00

Track Offset Options

Offset Incoming Tracks To...

0   Bars:Beats

Audio Media Options

√ Refer to source media (where possible)
   Copy from source media
   Consolidate from source media
   Force to target session format

Refer to source media

Sample Rate Conversion Options

☐ Apply SRC

Source sample rate: 44100

Destination sample rate: 44100

Conversion quality: Better

Source Tracks   **Operation / Destination Track**   Find Matching Tracks

Guitar (Mono audio)	Import As New Track
Sax (Mono audio)	Import As New Track
Bass Gtr (Mono audio)	Import As New Track
Audio11 (Mono audio)	Import As New Track
Audio12 (Mono audio)	Import As New Track
Cellos-L (Mono audio)	Import As New Track

Session Data To Import: All

Track Playlist Options...

⦿ Import Main Playlists – Replacing destination main playlists

◯ Import Main Playlists – Overlaying new with existing, trimming existing regions

◯ Do Not Import Main Playlists – leaving destination playlists intact

☐ Ignore rendered audio effects

Ignore clip-based gain

☐ Ignore keyframe gain

☑ Pan odd tracks left/even tracks right

Cancel   OK

Figure A3.10 Pro Tools Import Session Data dialog.

transferred to the LE version, but the plug-in will be made inactive. When you transfer this back to the TDM system the TDM plug-ins will become active again and the settings data will still all be there.

Of course, Pro Tools LE only supports 32 tracks (24 tracks on older systems), so if you have used more than this in your TDM session you will have to make some decisions about how to handle these. One way is to save a version of your TDM session that only includes the 32 most

Note: This even works when transferring between two differently-configured TDM systems – Pro Tools will automatically deactivate the unavailable routing assignments and plug-ins while keeping all settings and automation. So when you move any Pro Tools session to any different Pro Tools system, Pro Tools will automatically deactivate the necessary items while retaining all related data (including references to plug-ins not installed on the current system).

important tracks. Another way is to bounce combinations of tracks together, such as a stereo mix of all the drums, to get the track count down to 32 before you make the transfer. You could even bounce everything as a stereo mix and add a further 30 new tracks on your LE system – then transfer these back to the TDM system and combine the two sessions using the Import Track feature. The Import Track feature will seamlessly transfer audio and MIDI tracks between sessions with all mixer settings, tempo maps, plug-ins and automation intact, so you could open either the original TDM session and import the new tracks from the transferred LE session – or vice versa.

So what happens if you just open your TDM session with more than 32 mono tracks into Pro Tools LE? In this case, only the audio tracks assigned to the first 32 voices will open. If you subsequently save the TDM session using Pro Tools LE, any audio tracks beyond the first 32 will be lost. This is because Pro Tools LE does not support virtual tracks or 'inactive tracks'. Therefore any work that is carried back and forth needs to be limited to 32 tracks – a relatively small price to pay for the otherwise-excellent session transportability.

Now what if you don't have enough hard disk space to hold the audio and video files that you are working on in the studio – but you want to do some work at home using MIDI, for example? No problem here, as Pro Tools will allow you to open and edit sessions even if none of the audio or video files used in the original sessions are available.

Summary

Transferring projects between different Pro Tools systems and computer platforms or to and from other popular MIDI + Audio software is now relatively easy – thanks to the support for interchange file formats such as OMF.

4 Pro Tools 6.4

Automatic Delay Compensation is now available with HD systems – Pro Tools automatically compensates for inherent delay issues caused by plug-in latency, bussing and routing within the Pro Tools mixer. Another long-overdue feature – hierarchical plug-in menus, organized by plug-in category – has also been added. And Pro Tools now gives each track a fixed sequential Track Position number, enabling operators to better organize their sessions, as well as to quickly locate their GUI and control surfaces to a selected track. The new +12 dB (over 0 dBfs) fader gain gives the user more latitude while mixing than with the previous +6 dB (over 0 dBfs). This is particularly useful when balancing recordings made at lower levels. With a new taper above 0 dB, the faders also have a more familiar 'console' feel.

Lots of features have been added or improved to make Pro Tools more suitable for professional studio, post-production and film work.

For example, with HD systems plug-in clipping is now displayed on the D-Control work surface and on the GUI send labels for easy diagnosis of gain structure level problems while mixing. Plus, a new 'clear clip' key command and a new 'clip hold' preference provide additional display enhancements.

TrackInput for HD systems allows per-track switching between input source and playback from disk. Foley Record Mute for HD systems mutes the record input upon 'stop' and prevents unwanted loud noises during Foley record sessions.

TrackPunch for HD systems allows Pro Tools to be used as a recorder on film stages, rather than only as an editing/playback device. For music, TrackPunch enhances current QuickPunch capabilities for recording by allowing the user to arm tracks on the fly. Operators can punch in and out of tracks by using individual track record buttons instead of one global command. Combined with TrackInput, TrackPunch allows workflows similar to traditional tape-based multi-track recorders.

The 23.976 fps support for HD systems (and for LE with the DV Toolkit option) enables post customers who work with high-definition video to use this new frame rate specific to the medium, providing proper synchronization to the source. Pro Tools now offers complete recall of all editing, processing, mixing console and machine control parameters, while working entirely within the new 24 and 23.976 frame rates required for high-definition video production.

Pro Tools HD (and LE with DV Toolkit) now lets you set a 'zero feet+frames' point anywhere in the session and, with multiple feet+frame rates, allows the feet+frames timeline to remain in sync with all workflow variations.

RecordLock for HD systems allows audio with discontiguous ('broken') time code to be loaded, so an assistant no longer has to continually re-arm Pro Tools when loading in different takes from film shoots. This also prevents missed record takes on shooting stages, where the distributed time code stops and starts between takes.

Pro Tools 6.4 software also provides exclusive support for three new Digidesign products: ICON with D-Control, Command|8 and AVoption|V10. Command|8 is a small, affordable control surface for project studios, while ICON is Digidesign's new mixing console for large professional studios. Icon includes Digidesign's new D-Control tactile work surface, Pro Tools|HD Accel as its core DSP engine, and modular HD audio I/O interfaces to form an integrated mixing and recording system. The AVoption|V10 hardware supports Pro Tools playback of all Avid-created video media. It can even mix different video resolutions in the timeline, so you can accept projects cut on the full line of Avid DNA products, including Xpress Pro and Media Composer Adrenaline. The legacy Meridien and ABVB-based editing systems and video captured with Digidesign's AVoption and AVoption|XL are also supported, along with Component, Composite, S-Video, and Serial Digital Interface Input/Output.

Glossary

Access Time   This is a measure of the speed of a hard disk drive. The access time is equal to the seek time – how long it takes the read/write arm to find a file on the hard disk platter – plus the latency – the amount of time it takes for the disk to spin the data around to the read/write head.

A/D or ADC   These are abbreviations for Analogue-to-Digital Converter. An Analogue-to-Digital Converter converts analogue audio into a stream of 1s and 0s referred to as binary digits or bits.

ADAT Optical Format   As used on the popular Alesis ADAT recorders, this format is increasingly to be found in DAW interfaces and compact digital mixers such as the Mark of the Unicorn 2408 and the Yamaha 02R, as well as on popular digital I/O cards for personal computers such as the Korg 1212. Sometimes called 'lightpipe', this carries eight channels of digital audio on a single fibre-optic cable. Each channel works at 44.1 or 48 kHz with up to 24-bit resolution with 64 bits of sub-code information available. However, particular devices may be restricted to 20- or even 16-bit resolution in practice. Cable lengths of up to 15 feet are possible. The ADAT format uses an embedded clock signal that provides the same functionality as a standard word clock and the timing signal is also transmitted on a separate nine-pin sync cable. This sync cable also carries transport control commands.

ADAT Sync   ADAT recorders send and receive digital audio sync via their optical connectors but can also send and receive transport control messages using a special multi-pin connector. These can be used to connect to a BRC unit or to several of the popular synchronizers from Mark of the Unicorn and others.

ADB   This is an acronym for Apple Desktop Bus – the original system used to connect the Mac's keyboard and mouse to the computer using a four-pin connector. It is also used to connect hardware 'dongles' from Emagic, Steinberg, Waves and others.

ADSR   This is an acronym for Attack–Decay–Sustain–Release. It refers to the parameters used to describe the amplitude envelope of a sound. In the Attack portion of the envelope, the amplitude of the signal rises to its maximum. After this it will Decay fairly rapidly, then Sustain at a fairly constant level for a certain length of time. Finally, the sound will die away during the Release portion of the envelope.

AES/EBU Clock   The timing signals used in a AES digital audio stream can alternatively be supplied via a separate XLR connector which just carries timing signals – no audio. This is provided on some, although not all, high-end professional digital audio equipment.

AES/EBU Format Digital Audio   Developed by the Audio Engineering Society and the European Broadcast Union in 1985, the AES/EBU format transmits two channels of digital audio serially (one bit at a time) over a single cable at resolutions up to 24-bit with sampling rates up to 48 kHz. This format normally uses balanced 110-ohm cables fitted with XLR connectors which will work over distances up to 100 metres. Some AES/EBU devices use balanced 1/4-inch connectors or even unbalanced 75-ohm video cables with BNC connectors. AES/EBU signals are self-clocking, as they use an embedded word clock. A single cable carries two channels of audio plus the word clock. Alternatively, a separate master clock signal can be used and these are normally carried on 75-ohm coaxial cables with BNC connectors.

AIFF   The Audio Interchange File Format, for which the acronym is AIFF, is a standard 16-bit multi-channel file format containing linear PCM audio. It is often used to transfer files between computers as it can be used by a wide range of software running on different computer platforms.

Algorithm   This is a method or way of carrying out a calculation in a computer. For example, the mathematical operations carried out on a digital audio signal to add simulated reverberation are often referred to as a 'reverb algorithm'.

Amplitude   The amplitude of a waveform is the amount of displacement above and below the zero level – in other words, the distance between the highest and lowest levels. In the case of sound pressure levels, the amplitude is the amount of pressure displacement above and below the equilibrium level.

Array of Hard Disks   Two or more hard disks can be linked via software to form a drive array. Such an array can be used either to increase the throughput and access time or to provide redundancy. A RAID system is a Redundant Array of Inexpensive Drives that holds the same data on two or more drives so that if one drive fails the data is immediately available from another drive.

Bandwidth   Bandwidth is the width of the 'band' or range of frequencies that a system or device can handle, i.e. the range between the lowest and highest frequencies contained within the system or device's frequency response. A device capable of producing frequencies ranging from 20 kHz to 100 kHz, for example, would be said to have an 80 kHz bandwidth. Note that when transferring digital data between devices, the bandwidth of the communications path determines the speed at which data can be transferred – i.e. the throughput. So available bandwidth dictates the speed at which modems, networks, storage devices, and computer busses can communicate data.

Black Burst   Also known as colour black, crystal sync, and edit black, a 'black burst' signal is a video signal that contains the colour black at the standard level of 7.5 units. This provides reference points for the colour-burst pulse and a black reference, along with timing sync pulses. It is used often as a basic signal with which to format tape, and is commonly used as a sync reference signal when locking two or more video and/or audio tape machines together.

Burned-in Time Code A time code address can be 'burned in', i.e. mixed with the original video signal to form a new image containing the time code address superimposed on top of the original video, and this can often be positioned wherever you like on the screen. Devices that can do this, such as the Digital Time Piece, contain a character generator that produces the numbers you see on the screen. It is important to make sure that the unit generating the video signal containing the visible time code is 'genlocked' to the video recorder onto which it is being recorded. With burned-in code you can always see which frame the video is at – slowing or stopping to read it as necessary.

CD This is the acronym for Compact Disc – the major commercial distribution medium for audio recordings – and typically holds up to an hour or so of audio sampled at 44.1 kHz with 16-bit resolution.

CD-ROM CD-ROM is the acronym for Compact Disc Read-Only-Memory – a Compact Disc that contains 600 Mb or so of digital audio, digital video, software, word-processor files or other digital data.

Clipping When a signal exceeds the amplitude levels that can be handled by any device or system, the parts of the waveform that exceed these maximum levels will be cut off or 'clipped'. These waveforms can be identified in a waveform editor at maximum zoom-in by the flat portions that will be seen at the top and bottom of any clipped waveform cycles. A waveform that is clipped in this way will inevitably be distorted. You may not hear this so clearly if only occasional waveform cycles are clipped, but the sound quality will be subliminally degraded to produce a less-satisfying listening experience long before such clipping becomes clearly audible. When it is audible, the sound is not at all pleasant and you will almost certainly regard this as unusable. However, you should make sure that little or no clipping is ever present in your audio if you want people to listen to this. Be especially careful not to introduce clipping into your final masters before distribution to the public as it is all too easy to clip waveforms in the attempt to make the sound louder compared with the competition.

Communications Communication of, i.e. transfer of, information.

Complex Wave A complex waveform contains a mixture of many different frequencies at different amplitudes which, added together, typically have an irregular shape. Certain complex waveforms do have recognizable shapes, such as Square Waves, Sawtooth Waves and Triangle Waves.

CPU This is the acronym for Central Processing Unit – the 'heart' of any computer system.

D/A or DAC These are abbreviations for Digital-to-Analogue Converter. A Digital-to-Analogue Converter is a device that converts digital audio into analogue audio.

DAT This is the acronym for Digital Audio Tape.

Data Throughput The speed at which data can be transferred across a communication path – for example from hard disk or CD-ROM to computer, or from computer to computer via local area network or the Internet.

Decibel The bel was originally developed for use by telephone engineers and 1 bel represents a power ratio of 10:1. This measurement unit is named after Alexander Graham Bell – the 'father' of modern telephony. A more practical unit for use in audio measurements is the decibel – one tenth of a bel. It makes sense to use a logarithmic scale of this type because sound levels cover a very wide range which would be more awkward to represent using a linear scale. The abbreviation used for decibel is 'dB' which is the scale marking you will see in use on Sound Pressure Level (SPL) meters and on mixing console faders and EQ boost and cut controls.

A few rules of thumb about decibels are worth remembering – especially when balancing sound levels during a mixing session, for instance. For example, a change of 1 dB represents the smallest difference in intensity that a trained ear can normally perceive at a frequency of 1 kHz. Also, a doubling in sound power, by combining the sound of two instruments, for instance, results in a 3 dB increase in sound pressure level. Zero dB SPL has been established as the threshold of human hearing – the quietest sound the average human being can hear. It is also worth bearing in mind that human perception of loudness and frequency is not the same at different volume levels. A set of graphs called the Fletcher–Munson curves (after the researchers who developed these) shows perceived loudness throughout the audible bandwidth at a range of different volume levels – revealing, for example, that lower and higher frequencies sound quieter in relation to mid-range frequencies at lower listening levels. The 'loudness' button that you will find on many hi-fi amplifiers compensates for this effect when you are listening at lower volume levels by boosting the bass and treble frequencies.

It should also be kept clearly in mind that volume and loudness are not the same thing – although in everyday usage these are often used interchangeably. Volume level is an objective measurement expressed in decibels of Sound Pressure Level whereas Loudness is a subjective measurement of how loud a human listener perceives a sound containing particular frequencies at particular Sound Pressure Levels. A measurement unit called the Phon may be used to represent loudness levels.

Distortion   Distortion is said to occur when a signal is changed in any way from the original. Electronic signals can be affected by various kinds of distortion including non-linear, frequency and phase distortion. If the output signal does not rise and fall directly in proportion with the input signal, non-linear distortion is said to have occurred. Non-linear distortion can be further categorized as Amplitude, Harmonic or Intermodulation distortion. Amplitude distortion occurs when the changes in amplitude at the output are not in scale with those at the input. Harmonic distortion occurs when the levels of certain harmonic frequencies are changed in proportion to the amplitude of the input signal. Intermodulation distortion is said to occur when new frequencies are produced and added to the signal due to interaction between frequency components and these will not normally be at harmonic frequencies. Frequency distortion is said to occur when the output contains frequencies that were not present at the input. Similarly, Phase distortion is said to occur when the phase relationships between the frequency components are not the same at the output as at the input.

DSP   This is the acronym for Digital Signal Processing.

Dynamic Range   This is the range in decibels between the smallest signal which lies just above the noise floor and the largest signal which can pass through the system undistorted – taking into account the 'headroom' which normally exists in audio equipment above the nominal operating level to allow for transient peaks in the program material. Put another way, the dynamic range of any system or part of a system is the range of levels from weakest to strongest that the system can handle without distortion. The dynamic range of a standard audio cassette is about 48 dB, for example, while the dynamic range of the human ear is around

90 dB. Compact Disc can theoretically deliver 96 dB of dynamic range. Note that the term 'dynamic range' is often used interchangeably with 'signal-to-noise ratio'.

EBU Time Code   The European Broadcast Union (EBU) have adapted SMPTE Time Code as a standard running at 25 frames per second to match European video formats – hence EBU Time Code.

Equalization (EQ)   Originally Equalizers were used in telephone systems to compensate for (or equalize) losses occurring in transmission lines between one part of the telephone network and another. The term 'equalizer' has since been extended to cover devices that provide creative or corrective control of frequency response – such as the tone controls on a typical hi-fi. In professional recording, more sophisticated Parametric and Graphic equalizers are used which offer more detailed control.

Fidelity   The word 'fidelity' means faithfulness – whatever the context. Fidelity in the context of audio means how faithful the output is to the input. For example, studio monitors are High-Fidelity (Hi-Fi) while portable transistor radio speakers are Low-Fidelity (Low-Fi). The frequency response, signal-to-noise ratio and distortion characteristics all affect fidelity.

Flam   When two musical notes are played together almost, but not quite simultaneously, they are sometimes said to be 'flamming' together. This terminology is most often used in relation to drums and percussion, where the use of deliberate 'flams' is a useful performance technique.

Floppy Disk   A disk of magnetic recording material, similar to magnetic tape, usually held in a floppy or flexible protective casing, with a sliding cover over a slot in this casing. When the disk is inserted into a disk drive, the cover is slid back, and a read/write head travels over the radius of the disk, touching the disk surface, similar to the way a gramophone arm travels over a record. The disk is rotated by the drive mechanism, and the read/write head moves to locate the sector of the disk containing the information you want to read, or to find a blank sector to write to. The advantage of this system over tape storage systems is that it is much quicker to find the sector you want on the disk than it would be to wind or rewind a tape to find the spot you wanted. Most popular floppy disks are now 3.5-inch in diameter, and are encased in a fairly rigid hard plastic container. They are still called floppy disks because the magnetic disk inside is still floppy. Typically these hold up to 1.4 megabytes of information – although a 100 Mb floppy is available for the Imation SuperDrive.

Frequency   The number of cycles of an audio waveform which take place each second is called the frequency and is measured in Hertz (Hz). This frequency determines the 'pitch' of the sound. High-frequency sounds are said to be high-pitched, and, conversely, low-frequency sounds are said to be low-pitched. For example, a high frequency (say, 10 000 Hz) has a high pitch, while a low frequency (say, 100 Hz) has a low pitch.

Frequency Response   The lowest and highest frequencies that can be transmitted or received by an audio component, communications channel or recording medium are referred to as the frequency response of that component, channel or medium. The typical frequency response of an audio amplifier would ideally be not less than 20 Hz to 20 000 Hz – the theoretical frequency response of the human ear. If the frequency response matched this exactly throughout the frequency range, the response would be said to be 'flat' within the range – as the graph would be a perfectly flat line. In practical devices the response will never be perfectly flat so most technical specifications quote the number of decibels that the amplitude may rise above or fall below the 'flat' frequency response. A specification of ±3 dB would be considered a good response for a loudspeaker, for example.

Generation Loss   Each time you copy an analogue tape there is an increase in noise and distortion as a result of the copying process which is referred to as 'generation loss'. This even occurs when bouncing tracks on a multi-track analogue recorder. Digital recorders offer virtually zero generation loss.

Genlock   Genlock is short for generator lock and refers to the situation whereby video timing signals are synchronized to an external device. For example, it is important to make sure that a time code generator is locked to the timing of the video signal when striping time code onto a videotape to allow other devices to be synchronized with this tape. If the time code generator is not locked to the video signal, there will be no direct correspondence between the time code on the tape and the video on the tape – so you won't be able to successfully sync other devices to the tape. The best procedure is to lock both the video signal and the timecode generator to a stable external video sync source.

Graphic Equalizers   Graphic equalizers divide the frequency range into a number of bands and provide a vertical slider control for the amplitude of each filter band. When you boost or cut any filter band, you immediately see by looking at the sliders what the effect will be on the overall frequency response – hence the name 'graphic' equalizers. Different versions of these equalizers divide the frequency range into different numbers of bands – typically not less than four and not more than 36. You will commonly encounter bandwidths of 1/3-octave or 1-octave.

Hard Disk   A hard disk is a sealed disk storage medium which can hold many megabytes or gigabytes of information – unlike a floppy disk, which usually holds less than 1 MB. The disk rotates much faster than a floppy disk and the head is separated from the magnetic recording media by a layer of air just a few micro-inches thick, thus avoiding contact with the disk. Data access times are much faster than with floppy disks – in the order of tens of milliseconds.

Headroom   The difference between the maximum level that can be handled without distortion and the average or nominal operating level of the system is called 'headroom'. For example, the amplifiers used in professional recording studios can handle signals as high as 26 dB above the nominal operating level – so they are said to have 26 dB of headroom. With most types of electronic equipment, the average signal levels you will work at are set by the manufacturer to allow a reasonable amount of headroom. This is necessary to take account of occasional peak signal levels that would otherwise distort the signal.

Host TDM Plug-ins   Host TDM or HTDM plug-ins works in real-time and can be used with the TDM mixer, just like TDM plug-ins, but the processing is carried out by the host CPU.

Interface   An interface is something that goes in-between or connects two things together. For instance, an audio interface, or a video interface allows you to connect one piece of audio or video equipment to another. This is done using physical cables and connectors, such as 'phono' connectors for semi-professional audio or composite video, BNC connectors for professional video signals, and XLR connectors for professional audio signals. The format of the signal is also included in the interface specifications. For instance, phono connectors use unbalanced circuits for audio, whereas XLR connectors use balanced circuits that reject electromagnetic interference better – allowing longer cables to be used.

I/O   This is a commonly used abbreviation for Input and Output.

LCD   This is the acronym for Liquid Crystal Display.

LED   This is the acronym for Light-Emitting Diode.

Looping   Looping is used to repeat selected material. The material plays through once, then 'loops' back to the beginning and plays again. For example, you can loop a MIDI sequence containing a figure such as a drum pattern that repeats identically for any duration in most MIDI sequencers. You can also loop a sampled sound, or part of the sound, in a sampling keyboard or module so that it will sustain until you release the key that plays it. Pro Tools can loop round a selected number of bars, but does not have the more sophisticated looping features that you will find in MIDI sequencers such as Digital Performer, for example. If you need to loop sections within a Pro Tools session you can achieve this simply by repeating the same section as necessary using copies of the same region or regions.

Master Recording   The term 'master' may be applied to a recording at various stages in the production process. The original multi-track recording which contains the source material may be referred to as the 'master' multi-track to distinguish it from any copies which are made. When you make one or more mixes of a particular recording, just one of these will be chosen as the 'master' mix that will be used for replication and distribution. This mix then goes forward to the next stage in the production process – typically carried out in a so-called 'Mastering' studio. This is a specialized studio that typically carries out the process more correctly known as 'pre-mastering' – as it is the stage immediately prior to making the 'glass master' used in the replication process at CD pressing plants. This final production of the glass master is normally carried out at the plant using the pre-masters prepared at the specialist Mastering studio – or sometimes from master recordings supplied directly from the recording studio.

MIDI   MIDI is the acronym for the Musical Instrument Digital Interface. The MIDI standard was agreed between a group of rival manufacturers in 1982 to ensure that electronic musical equipment from different manufacturers could be successfully linked together. The first application was to allow a 'master' keyboard to play another keyboard, or a rack-mounted keyboard-less sound module. It was quickly realized that by recording the MIDI messages into a MIDI sequencer, musical performances and arrangements could be perfected by editing the data recorded into the sequencer. At first, stand-alone sequencers became popular, then computer-based sequencers were developed – such as Performer on the Mac and Cubase on the Atari. Timing messages can also be sent via MIDI to synchronize sequencers and drum-machines.

MIDI File   This is a standard interchange file format, first developed by Opcode Systems, which has now been adopted by virtually all MIDI software publishers. It contains MIDI Sequence data that can be used by any MIDI sequencer software, along with other information such as Markers or System Exclusive data.

MIDI Interface or Adapter   This is a device that passes MIDI information from connected devices to a computer. Simple interfaces just feature a couple of inputs and outputs, while more advanced interfaces offer multiple MIDI data streams from multiple outputs. The standard for these was originally established by Mark of the Unicorn with their MIDI Time Piece (MTP) interfaces which have eight inputs and eight outputs, each of which can carry separate streams of MIDI data – each with 16 MIDI channels available. The MTP interfaces, for example, also provide SMPTE/MTC conversion and can both read and generate SMPTE Time Code.

MIDI Merger   A MIDI merger takes two incoming MIDI signals and merges these together to form one output signal containing both sets of data. This feature is often available in the more advanced MIDI Interfaces or Patchbays.

MIDI Patchbay   This is a type of electronic patchbay that lets you connect any MIDI input to any MIDI output. Typically, this will be an 8 × 8 matrix with switches on the front panel and MIDI sockets on the back. Some units allow you to store and recall routing 'patches'. Some larger units are available and some units can be linked together if greater numbers of inputs and outputs are needed.

MIDI Sequence   A MIDI sequence is a sequential set of MIDI data that can be recorded in real time or entered in step-time into a 'sequencer' for subsequent editing and replay.

MIDI Sequencer   A MIDI sequencer can be a stand-alone hardware device, or can be incorporated into a MIDI keyboard or drum-machine – or can be implemented in software on a personal computer. Pro Tools incorporates MIDI sequencing capabilities into its software environment. Other popular software sequencers include MotU Performer and Emagic Logic. The data recorded in the sequencer can be edited and played back, and many sequencers also offer a way of entering data step-by-step as an alternative to playing in from a MIDI controller. A MIDI sequencer does not record audio. It simply records a sequence of key-presses on a MIDI keyboard (or data from any other MIDI controller) and then replays these back into the MIDI keyboard or some other MIDI sound module which actually creates the sound.

MIDI Sync or MIDI Clock   MIDI clock signals can be used to keep drum-machines, sequencers and other MIDI devices in sync with each other.

MMC   This is the acronym for Midi Machine Control. MIDI Machine Control uses MIDI System Exclusive messages to remotely control MMC-equipped tape machines and VCRs. All the basic functions such as Stop, Start and Rewind are available in all MMC systems – while some implementations offer much more detailed control.

Modem   Modem is an acronym for modulation/demodulation. A modem is a device that encodes computer data as audio signals so the data can be transmitted over telephone networks. On reception, another modem is used to decode the audio signals back into computer data.

Monophonic   This means 'single sounding' and refers to a piece of music (or an instrument) with only one note sounding at a time. See Polyphonic.

MTC   This is the acronym for MIDI Time Code. MIDI Time Code encapsulates SMPTE Time Code addresses with information about hours, minutes, seconds, frames and user bits and can also contain MIDI 'cueing' or 'transport' messages to tell the system to stop, start or locate. There are two types of MTC messages – full-frame and quarter-frame. Full-frame messages are used when the system is rewinding or fast forwarding to provide appropriate location information as needed – rather than continuously per quarter-frame. Quarter-frame messages are used during normal playback – to reduce demands on MIDI bandwidth. A single frame of time code contains too much information to be represented by a standard three-byte MIDI message so it is split into eight separate messages. Four of these are sent for each time code frame so it takes eight of these quarter-frame messages to represent one complete SMPTE frame address. Consequently, MTC itself can only resolve down to quarter frames while SMPTE can resolve to 1/80 of a frame (1 bit). You may be worried that MTC is not accurate enough, but it turns out that this is not really a problem – because all MTC receivers will interpolate incoming timing data to whatever accuracy the designers of the equipment have chosen. That's why some MIDI sequencers let you specify the SMPTE times of events down to 1/80 or 1/100 of a frame.

Multi-timbral   Used to describe synthesizers and samplers, the term multi-timbral means that the device can act as though it contains a number of separate sound modules. These can be played back individually (and simultaneously) by allocating different MIDI channels to each, and different samples or synthesizer patches can be allocated to each. Multi-timbral devices may contain between two and 16 separate multi-timbral 'instruments'.

Noise   Noise is any unwanted sound that finds is way into program material. Audio cassettes and tape introduce background noise typically called 'tape hiss'. This noise is aperiodic in character and contains a mixture of all audible frequencies of sound. Noise with an equal mix of sounds at all frequencies is called 'white noise'. Electronic circuits and other audio components also introduce background noise. Other 'noises' can include unwanted sounds such as 'spill' from headphones between vocal or instrumental phrases on a recording. Distortions may also take the form of noise – such as the harsh cracking sounds heard when signals are badly 'clipped' on a recording.

Nominal Operating Levels   These are chosen to place signals sufficiently below the maximum undistorted output level to avoid distortion while keeping them well above the noise floor – the idea being to optimize the signal-to-noise ratio.

Parallel Data Transfer   Information is transferred several bits at a time using one wire for each parallel bit. Typically 8 bits, or 1 byte, may be transferred at a time, and in this case eight parallel wires would be needed to carry the information. The advantage here is speed, but the cabling costs are much greater than for serial transfer.

Parametric Equalizers   Parametric equalizers are so-named because the parameters which define these are available to the user to set as required. Typically, these parameters include: the centre frequency around which the filter operates, the filter gain (i.e. the amount of peak or dip at that frequency), and the bandwidth or Q of the filter. The parameters can be continuously varied and there can be any number of filter sections – typically between three and eight.

Phono/RCA Connector   Known as an RCA connector in the USA and as a phono connector in Europe, this type of connector is widely used for audio inputs and outputs on consumer and video equipment – and for digital I/O on 'prosumer' audio equipment.

Pitch   Pitch is the subjective term which corresponds to the objective measurement of the fundamental frequency of a tone or note. If the frequency of the fundamental is higher, the pitch sounds higher and if it is lower the pitch sounds lower.

Polyphonic   This means 'many sounding' and refers to music with several parts or voices playing at the same time. When this term is used with reference to instruments such as synthesizers it refers to the number of notes or voices that can be played at the same time. See Monophonic.

Port The term 'port' when used in relation to computer equipment refers to a physical socket that can be used to input or output data. Ports may be serial or parallel and can be used to connect MIDI interfaces, printers and other peripherals.

Quantization Quantization is one of the stages used in analogue-to-digital conversion In the converter, the analogue waveform is represented by a series of discrete signal levels known as quantization levels. Each level is represented as a binary number which is then passed through the system and may be stored on a suitable medium such as hard disk. A digital-to-analogue converter can subsequently be used to convert this data back into analogue audio signals. The resolution of this quantization depends on the number of bits available to represent the different levels. Eight bits can represent up to 256 levels while 16 bits can represent up to 65 536 levels.

MIDI data can also be 'quantized' in a MIDI sequencer. In this case, each selected note is moved to the nearest timing value or step (quantum) within the sequence that has been specified by the user. This feature makes it easy to correct mistakes in performance.

RAM This is the acronym for Random Access Memory. RAM 'chips' are used on the motherboard of a personal computer to hold data that is being used by the CPU and they provide extremely fast access to this data. Data cannot be accessed from a computer's hard drive anywhere near as quickly, as the access times of hard drives are relatively slow compared with access times from RAM.

Real Time If you perform a piece of music on a MIDI keyboard and record this into a MIDI sequencer, you are said to be recording in 'real time' – as opposed to the step-time method which lets you enter one note or chord then another – pausing as long as you like between entries. If you are using a TDM audio processing plug-in in Pro Tools, the processing calculations are carried out fast enough that they can be applied while the music is playing back at its normal tempo and this is said to be 'real-time' processing. If you use Audio Suite plug-ins, these normally require that playback is stopped while the computer carries out the processing (writing a new, processed, file during the process). This type of processing is sometimes referred to as 'non-real-time' processing. Similarly, other processes that can be carried out while during playback, such as editing of MIDI data, may be referred to as 'real-time' processes.

RTAS Plug-ins RTAS stands for Real Time Audio Suite. RTAS plug-ins can be used with the Pro Tools TDM mixer, but the processing is carried out by the host CPU – not by the TDM processing cards. RTAS plug-ins can also be used with Pro Tools LE systems.

Samplers Samplers are devices that record short portions (called 'samples') of audio such as drum sounds or individual notes of other instruments. These are then 'mapped' onto the notes of a MIDI keyboard which can be used to replay these 'samples' of audio. In the case of drum samples, each note or group of notes on the keyboard plays a different drum sound. In the case of instruments, each sample covers a range of pitches with the same sample replayed faster or slower to make the pitch higher or lower. It is only possible to replay a few semitones higher or lower in pitch before the sound becomes too unnatural, so multiple samples are taken from the original instruments, spaced apart by several semitones each. If there is enough RAM memory available for recording, a sample could be created for each note of the instrument. In practice this is rarely the case, which is why a smaller selection of notes is usually sampled with each replayed at a range of different pitches. Samplers can also be used to record and replay sound effects or even short sections of vocal or dialogue. Typical samplers come in the form of a MIDI keyboard or a MIDI-controllable rackmount unit with popular models available from Emu, Akai,

Roland and others. Software samplers are currently in vogue – such as the Emagic EXS24 and Nemesys GigaSampler for the PC.

Sampling   Sampling is a process used in analogue-to-digital converters to represent analogue audio waveforms as digital bits. This process is used in all digital recorders, including MIDI samplers. With CD-quality audio, for example, 44 100 samples or 'snapshots' of the audio waveform are recorded each second – and played back each second during digital-to-analogue conversion.

Sampling Rate   This is the number of samples used each second by a digital recording system. CDs use 44.1 kHz, DAT uses 48 or 44.1 kHz, while more recent systems can use 96 kHz or higher rates. The sampling rate needs to be at least twice the highest frequency present in the analogue audio for the system to successfully represent the audio.

Sampling Resolution   The sampling resolution is determined by the number of bits available in the system to represent the amplitude of each sample. With 8 bits you can represent 256 different amplitude levels, while with 16 bits you can represent 65 536 different levels. The 'steps' between the amplitude levels can be compared to the markings on a ruler that are used to measure length. If the ruler only measures in inches, for example, you cannot make accurate measurements in-between. If it measures in thousandths of an inch you will be able to make much more accurate measurements. Similarly, using the many thousands of levels available in 16-bit systems allows for much greater accuracy of representation of the digitized analogue signals than the 256 levels available in 8-bit systems.

SCSI   This is the acronym for Small Computer Systems Interface – pronounced 'scuzzy'. SCSI is a standard protocol for computer equipment to allow transfer of data at high speeds between attached devices.

SDII   This is the acronym for Sound Designer II – the file format used by Digidesign audio systems and other professional audio software and hardware.

Serial Data Transfer   During Serial Data Transfer, each bit is sent sequentially (or serially, in a series), i.e. one after another until all the bits have been transferred. MIDI data uses Serial Data Transfer, for example, as do the serial ports on personal computers used to connect printers and other peripherals. A simple pair of wires is all that is needed for this type of data transfer, so the cabling costs are low (cf. Parallel Data Transfer).

Signal   A signal is information – such as audio – which is transferred through a system as a varying electrical voltage that represents the audio information.

Signal-to-noise Ratio   The abbreviation S/N is sometimes used for signal-to-noise ratio, which is, quite simply, the ratio of any (normally undistorted) signal to the noise present at the output – expressed in decibels. S/N is often taken as the largest undistorted signal that can be handled by a system and is often used interchangeably with the term dynamic range. All circuitry and media add a certain amount of background noise to the audio signal and this is often referred to as the 'noise floor'. You can easily hear the background noise if you play a blank tape or listen to a monitoring system with the volume up high with no audio present. Obviously, the greater the difference between the signal and the background noise, the better.

Sine Wave   A waveform that conforms to the equation $y = \sin x$ has a so-called sinusoidal shape when viewed on a graph of the waveform. This shape looks like the letter 'S' turned on its side and placed symmetrically about the x-axis. The x represents degrees and the y-axis can represent voltage or sound pressure level, for example. A Sine Wave represents the waveform of a single frequency (cf. Complex Waveform).

SMPTE   This acronym stands for the Society of Motion Picture and Television Engineers – a committee that sets standards for TV, video and film production.

SMPTE Synchronizer   This is a hardware device that can be used to read or write SMPTE time code at various frame rates. Some of these devices, known as machine synchronizers, can also be used to control the transport and speed of audio or video tape machines. The speed of the 'slave' devices is controlled by a 'master' device – usually the machine synchronizer – although this may be 'locked' in turn to a 'master' clock source. The machine synchronizer typically has transport controls to let you remotely control the play, stop, rewind and locate functions of the slave machines.

SMPTE Time Code   SMPTE Time Code is an electronic signal that can be used to synchronize audio with video or film frames. A unique time code 'address' identifies each hour, minute, second and frame within a 24-hour period. This code can be recorded onto one of the audio tracks of a video player as Longitudinal Time Code or LTC – or embedded within the video signal as Vertical Interval Time Code or VITC, pronounced 'vitsy' – and used to synchronize other equipment to the video frames. It can also be used to synchronize two audio tape players together via a synchronizer unit that controls the speed of the 'slave' tape player. SMPTE to MIDI converters may be used to convert SMPTE code into MIDI timing clocks to synchronize MIDI equipment to either video or audio tape players. The whole idea of using time code came about because TV stations needed to lock their video tape recorders together in perfect synchronization throughout each 24 hours of operation. This is why there is a separate time code 'address' for each second of the 24 hours. Any particular SMPTE number can be regarded as corresponding to a physical location on a video or audio tape, so it can be thought of as the 'address' of that particular location. Time code was developed to comply with standards set by the Society of Motion Picture and Television Engineers (SMPTE) in the USA and the European Broadcast Union (EBU) in Europe. There are 25 video frames used each second in European video – and 30 in the USA. Each timecode frame may be further subdivided into 80 sub-frames or 'bits'. Just to muddy the waters a little here, some equipment actually displays sub-frames as divisions of 100 – which is a little easier for we humans to deal with than 80ths of a frame – but the underlying sub-divisions are still 80ths. Another complexity is the variations of time code used in the USA and some other countries for different TV standards – with colour video actually running at a rate of 29.97 seconds. It is also worth noting that some people refer to time code running at 25 fps as EBU time code on the grounds that this is the European standard while they refer to time code running at 30 fps as SMPTE time code, the American standard. Many people just call it 'SMPTE' or 'time code' – whatever the frame rate.

Sound   Sound is the sensation in the ear caused by vibrations of the air. Sound can also be transmitted through other materials such as metal, paper, wood and water.

S/PDIF Clock   Timing signals are carried within the S/PDIF data stream, but can be supplied via an S/PDIF connector with no audio data present – from the Digital Timepiece, for example.

S/PDIF Format   The Sony/Philips Digital Interface Format (S/PDIF) is the 'prosumer' interface which can be found on the more expensive consumer equipment, as well as on a fair number of more professional devices. This handles various sampling rates up to 48 kHz and can work at up to 24-bit resolution. However, many devices only support 20-bit or 16-bit resolution via S/PDIF. The S/PDIF format also includes (buried in sub-code bits) a Serial Copy Management System (SCMS) which prevents multi-generation copying. S/PDIF uses unbalanced 75-ohm coaxial cables fitted with RCA/phono type connectors of the type commonly found on consumer hi-fi equipment. S/PDIF is also provided on some equipment using an optical format called Toslink which uses small fibre-optical cables. Word clock signals are embedded in the S/PDIF data and devices with S/PDIF connections don't normally have separate word clock connections so they can't normally be configured to use a master system clock.

Super Clock   This is a word clock sync signal used by Digidesign and some other manufacturers to sync their digital audio workstations. It runs at 256 times the rate of standard word clock so it is sometimes referred to as 'Word Clock ×256'.

TDIF Format   Tascam's DA88 series of eight-track digital tape recorders use a proprietary format known as the Tascam Digital Interface Format – or TDIF for short. This is a bi-directional interface which uses a single cable to carry eight channels of data in both directions. TDIF uses multi-wire unbalanced cables with 25-pin D-connectors with a recommended maximum length of 5 metres. As with the ADAT format, TDIF is becoming a popular format on digital mixers, DAWs and so forth. TDIF supports all the standard sampling rates up to 48 kHz with up to 24-bit resolution.

TDIF Sync   TDIF is intended to work with a separate clock signal in a master-clocked system using a separate 75-ohm cable with BNC connectors to carry a standard word clock signal. However, TDIF signals also contain a clock signal known as Left-Right Clock (LRCK). This runs at the same rate as standard word clock but also defines the odd and even channels within a pair. So an LRCK clock signal can be used as an alternative to separate word clock – but only if the particular devices used support this.

TDM   TDM stands for Time Division Multiplexing and refers to the system used to carry audio signals to and from the Pro Tools audio cards. 'TDM' is also used as a label to distinguish Digidesign's high-end audio cards from their lower-priced products that do not have on-board DSP.

The Time Code Word   The time code information recorded within each audio or video frame is called the 'time code word' and each of these 'words' is divided into 80 sections called 'bits'. Each word contains a single, unique time code address corresponding to a particular video frame or a particular digital audio 'frame' or physical location on audio tape. In the case of digital audio, the samples are further grouped into 'frames' of data having a certain length so you sometimes encounter the term 'audio frame' in this context.

Timbre   Timbre (pronounced 'taamber', as it is a French word) refers to the tonal quality of a sound and varies according to the spectral content and envelope characteristics of the sound. Timbre is the characteristic of a sound which lets you distinguish it from another sound with similar pitch and loudness. Anyone can tell the difference between a middle 'c' note played on a trumpet and on a guitar, for example.

Transients Something which is transient in nature occurs for just a short amount of time. Applied to audio, transients are said to occur during the initial 'attack' of the sound where the waveform may rise 20 or 30 dB above the average signal level for a fraction of a second, after which time the signal level drops back to the average level and sustains at this level for some variable length of time before releasing from this sustained portion and decaying back to silence. These initial transients typically contain vital audible information that distinguishes similar-sounding instruments from each other. The sustained portions of many sounds can sound quite similar, but the attack portions will sound very recognizably different. When passing audio signals through any audio system, sufficient headroom needs to be available to avoid clipping these transients, or the sound will be distorted and unnatural.

USB This is the acronym for Universal Serial Bus – the system used on current G4 Macs and PCs to connect various peripheral devices. The mouse and keyboard connect to the G4 using USB, for example, as do the latest MIDI interfaces.

Virtual Memory Computer operating systems typically have a feature called Virtual Memory. This enables free space on the hard drive to be used to store some of the data that would normally go into RAM. This is a way of extending the available memory space that is cheaper than adding extra RAM. It works fine for word-processor files and the like, as these do not make such great demands on the speed of access to the data in RAM as do audio and video applications. If you are working with audio or video you will normally need to turn this feature off.

Volume The term volume in relation to audio refers to the loudness, or intensity, of the sound. It is sometimes, more loosely, used to refer to what is more correctly described as the magnitude of the amplitude of the audio waveform. Loudness is actually a subjective term and each individual human listener may interpret this slightly differently. A system of Volume Units and a VU meter have been developed to measure volume. One Volume Unit corresponds to a change of 1 dB in the case of a sine wave, although most sounds actually consist of complex waveforms which will produce a greater or lesser change than 1 dB.

Waveform This is a visual representation of a sound that shows a graph of signal amplitude against time. The signal amplitude is typically measured as sound pressure level or voltage, as appropriate. A simple waveform containing a single frequency takes the shape of a sine wave – like an 'S' on its side on the graph. A complex waveform containing a number of different frequencies at different amplitudes will have an irregular shape. Some complex waveforms do have recognizable shapes, such as the square wave, sawtooth wave, triangle wave and pulse wave.

Wavelength The distance that a sound travels as the waveform completes one cycle is known as the wavelength of the sound, i.e. the distance between the start points of successive wave cycles. Low frequencies, of say, 50 cycles per second, have relatively long wavelengths compared with higher frequencies, such as 5000 cycles per second.

Word Clock All digital audio signals contain an embedded clock signal called a Word Clock to provide a common timing reference for the sending and receiving devices. It is called a Word Clock because it also defines the number of individual samples (words) sent per second. This embedded clock signal allows any receiving device to lock its timing to the internal clock of any sending device so that audio can be transferred successfully between these. Word Clock signals can also be carried on a separate cable. Professional devices normally have separate Word Clock inputs and/or outputs so that all the devices can be synchronized to a high-quality stand-alone clock source – or to the device containing the highest-quality clock (such as a high-quality set of A/D or D/A converters). Separate Word Clock signals typically use BNC-type connectors. This BNC (British Naval Connector) is the same as the one often used for composite video connections.

Zero Crossing   The position on a waveform graph or display where the waveform crosses the zero-amplitude line is known as the zero-crossing point. This is the best position at which to edit the waveform, for example, when you want to join two sections together or make a cut. If you make a cut at any other position, i.e. when the amplitude of the waveform is at some positive or negative value, the amplitude of the waveform will have to jump to some new value. This may produce an audible pop or click.

Bibliography

Books

Other books by the author

Mike Collins (2002). *Pro Tools 5.1 for Music Production*. Focal Press.
(A comprehensive guide to recording, editing and mixing with Pro Tools TDM software.)

Mike Collins (2003). *A Professional Guide to Audio Plug-ins and Virtual Instruments*. Focal Press.
(This book explains the different plug-in formats; talks about how these were developed; then presents a wide-ranging roundup of audio plug-ins and virtual instruments with useful tips and hints along the way.)

Mike Collins (2004). *Choosing and Using Audio and Music Software*. Focal Press.
(This book serves as a guide to selecting and using the major software applications for music and audio on the Mac and the PC. It also contains lots of useful tutorial material.)

Other books of potential interest

Music technology

Jim Aikin (2003). *Software Synthesizers: The Definitive Guide to Virtual Musical Instruments*. Backbeat Books.
(A useful guide.)

Stephen Bennett (1998). *Fast Guide to Emagic Logic Audio*. PC Publishing.
(Basic tutor book for Logic Audio.)

Stephen Bennett (2003). *Virtual Instruments: A User's Guide*. Music Sales.
(How to set up and use Logic Audio's Virtual Instruments.)

Marc Cooper (2002). *Steinberg Nuendo*. PC Publishing.
(Useful tutor book focusing on the original Nuendo version for the PC.)

Roger Derry (2003). *PC Audio Editing* (2nd Edition). Focal Press.
(Based on Cool Edit Pro version 2.)

David Franz (2001). *Producing in the Home Studio with Pro Tools*. Berklee Press.
(Based around Pro Tools LE, this book is aimed at home studio users just getting started.)

John Keane (2002). *A Musician's Guide to Pro Tools Book 1*. Supercat Press.
(A useful, although fairly short, step-by-step tutorial for Pro Tools 5.1 and Pro Tools Free.)

Colin MacQueen and Steve Albanese (2002). *Pro Tools Power*. Muska and Lipman.
(An excellent book with a wealth of useful tips and tricks for all Pro Tools software up to version 5.3 and hardware up to Pro Tools|HD.)

Simon Millward (2001). *Fast Guide to Cubase VST* (3rd edition). PC Publishing.
(Comprehensive tutor book focusing Cubase VST 5 on the PC.)

Roman Petelin and Yury Petelin (2002). *Cakewalk Sonar: Plug-ins and PC Music Recording, Arrangement, and Mixing*. Music Sales.

Richard Riley (2002). *Audio Editing with Cool Edit*. PC Publishing.
(For Cool Edit 2000 and Cool Edit Pro version 2.)

Steven Roback (2002). *Pro Tools for Macintosh and Windows – Visual Quickstart Guide*. Peachpit Press.
(Covers Pro Tools LE in a very thorough manner, taking a step-by-step instruction list approach.)

Ashley Shepherd (2003). *Pro Tools for Video, Film and Multimedia*. Muska and Lipman.
(The best book available about using Pro Tools for video, film and multimedia.)

José 'Chilitos' Velenzuela (2003). *The Complete Pro Tools Handbook*. Backbeat.
(A 'big' book that covers all Pro Tools systems and includes lots of useful step-by-step examples.)

Studio technology

Tomlinson Holman (2000). *5.1 Surround*. Focal Press.
(How to produce, master, and engineer in the 5.1 surround format.)

Philip Newell (1995). *Studio Monitoring Design*. Focal Press.
(A 'bible' of professional studio monitoring practice.)

Philip Newell (1998). *Recording Spaces*. Focal Press.
(Discusses the acoustics of rooms intended for musical performances.)

Philip Newell (2000). *Project Studios*. Focal Press.
(Essential information about setting up project studios.)

Michael Talbot-Smith (2002). *Sound Engineering Explained* (2nd edition). Focal Press.
(Useful introduction to audio technology for students.)

Digital audio

Baert, Theunissen and Vergult (1988). *Digital Audio and Compact Disc Technology*. Heinemann.
(Technical reference.)

Ken Pohlmann (1985). *Principles of Digital Audio.* Howard Sams.
(In-depth coverage of the subject.)

Francis Rumsey (1996). *The Audio Workstation Handbook.* Focal Press Music Technology Series.
(Covers all the basics very well.)

John Watkinson (1988). *The Art of Digital Audio.* Focal Press.
(Technical reference.)

Microphone techniques and theory

Martin Clifford (1986). *Microphones.* TAB.
(Good reference.)

David Miles Huber (1988). *Microphone Manual.* Howard Sams.
(An in-depth treatment.)

Alec Nisbett (1983). *The Use of Microphones.* Focal Press.
(A simple guide to choice and usage.)

Mixing techniques

David Gibson (1997). *The Art of Mixing.* MIX Books Pro Audio Series.
(A unique visual approach.)

Bobby Owsinski (1999). *The Mixing Engineer's Handbook.* MIX Books Pro Audio Series.
(The best text available on this subject.)

Music production techniques

Craig Anderton (1978). *Home Recording for Musicians.* Guitar Player Books.
(Useful reference.)

Craig Anderton (1985). *Digital Delay Handbook.* AMSCO.
(Useful reference.)

Bruce Bartlett (1986). *Introduction to Professional Recording Techniques.* Howard Sams.
(An excellent general text.)

Bruce Bartlett (1989). *Recording Demo Tapes at Home.* Howard Sams.
(Useful guide.)

Bruce and Jenny Bartlett (1998). *Practical Recording Techniques* (2nd edition). Focal Press.
(A hands-on practical guide for beginning and intermediate recording engineers, producers and musicians.)

Bill Gibson (1999). *The Audio Pro Home Recording Course – Volumes I, II and III.* MIX Books Pro Audio Series.
(An excellent series which includes audio examples on CD.)

David Miles Huber and Robert E. Runstein (1986). *Modern Recording Techniques*. Sams.
(A good reference manual.)

Fred Miller (1981). *Studio Recording for Musicians*. Consolidated Music Publishers.
(An overview of multi-track recording.)

William Moylan (1992). *The Art of Recording*. Van Nostrand Reinhold.
(A look at the creative and musical aspects of recording.)

Bruce Nazarian (1988). *Recording Production Techniques for Musicians*. Amsco.
(An overview of multi-track recording.)

Audio recording techniques and theory

Glyn Alkin (1981). *Sound Recording and Reproduction*. Focal Press.
(A quick reference guide.)

John Borwick (1994). *Sound Recording Practice Fourth Edition*. Oxford University Press.
(Comprehensive text.)

John Eargle (1980). *Sound Recording*. Van Nostrand Reinhold.
(Comprehensive text.)

John Eargle (1986). *Handbook of Recording Engineering*. Van Nostrand Reinhold.
(Comprehensive text.)

David M. Howard and James Angus (1996). *Acoustics and Psychoacoustics*. Focal Press Music
Technology Series.
(Covers all the basics very well.)

Alec Nisbett (1979). *The Technique of the Sound Studio*. Focal Press.
(Sound recording techniques for Radio and TV.)

John Woram (1982). *The Recording Studio Handbook*. Elar.
(Comprehensive text.)

John Woram (1989). *Sound Recording Handbook*. Howard W Sams.
(Advanced and comprehensive text.)

MIDI recording techniques and theory

Craig Anderton (1986). *Midi for Musicians*. Music Sales.
(Highly recommended introductory text.)

Craig Anderton (1988). *The Electronic Musician's Dictionary*. Amsco Publications.
(Includes invaluable definitions of over 1000 terms.)

Michael Boom (1987). *Music Through Midi*. Microsoft Press.
(A musical approach to MIDI.)

Steve DeFuria and Joe Scacciaferro (1986). *The Midi Implementation Book.* Third Earth Publishing.
(Lists of MIDI Implementation charts for popular gear.)

Steve DeFuria and Joe Scacciaferro (1987). *The Midi Book.* Third Earth Publishing.
(An introductory text.)

Steve DeFuria and Joe Scacciaferro (1987). *The Midi Resource Book.* Third Earth Publishing.
(Midi 1.0 Specification + advanced topics.)

Steve DeFuria and Joe Scacciaferro (1987). *The Midi System Exclusive Book.* Third Earth Publishing.
(Lists of SysEx data and formats for popular gear.)

Steve DeFuria and Joe Scacciaferro (1989). *MIDI Programmers Handbook.* M & T Publishing.
(MIDI programming reference book, for Macintosh, IBM, Atari, Amiga.)

David Miles Huber (1991). *The MIDI Manual.* Sams.
(An in-depth treatment.)

Jacobs and Georghiades (1991). *Music and New Technology.* Sigma Press.
(This is a good buyer's guide covering a broad range of topics.)

Paul Lehrman and Tim Tully (1993). *Midi for The Professional.* Amsco.
(Highly recommended reference and technical guide.)

Lloyd and Terry (1991). *Music In Sequence.* Musonix Publishing.
(A complete guide to MIDI sequencing for beginners.)

Jeff Rona (1990). *Synchronization from Reel To Reel.* Hal Leonard, USA.
(An excellent treatment of this often-tricky subject.)

Robert Rowe (1993). *Interactive Music Systems.* The MIT Press.
(A survey of computer programs that can analyse and compose music.)

David M. Rubin (1992). *The Audible Macintosh.* Sybex, USA.
(An excellent overview of Macintosh audio and MIDI software.)

David M. Rubin (1995). *The Desktop Musician.* Osborne.
(Covers everything from sequencing to digital audio recording using desktop computers.)

Francis Rumsey (1994). *Midi Systems and Control* (2nd edition). Focal Press.
(A technical approach to MIDI.)

Christopher Yavelow (1992). *MACWORLD Music and Sound Bible.* IDG Books, California.
(An unbelievably comprehensive 'bible' of just about all available Macintosh audio and MIDI software.)

Synthesis and sampling

Steve DeFuria (1986). *The Secrets of Analog and Digital Synthesis*. Third Earth Productions. (Another highly-recommended 'how to' book.)

Steve DeFuria and Joe Scacciaferro (1987). *The Sampling Book*. Third Earth Publishing. (Highly-recommended 'how to' book.)

Howard Massey, Alex Noyes and Daniel Shklair (1987). *A Synthesist's Guide to Acoustic Instruments*. Amsco.
(How to synthesize a range of acoustic instruments using Sampling, Analogue synthesis or Digital synthesis techniques – comparing the different methods.)

Jeff Pressing (1992). *Synthesizer Performance and Real-time Techniques*. Oxford University Press.
(The best text available on synthesizer techniques.)

Martin Russ (1996). *Sound Synthesis and Sampling*. Focal Press Music Technology Series.
(Covers all the basics very well.)

Film sound

Rick Altman (ed.) (1992). *Sound Theory – Sound Practice*. Routledge, New York and London.

Dan Carlin (1991). *Music in Film and Video Productions*. Focal Press.
(An excellent insight.)

David Miles Huber (1987). *Audio Production Techniques for Video*. Howard Sams.
(A comprehensive text.)

Fred Karlin and Rayburn Wright (1990). *On the Track*. Schirmer Books.
(A guide to contemporary Film Scoring.)

Marvin M. Kerner (1989). *The Art of the Sound Effects Editor*. Focal Press.
(Sound Editing for TV and Film.)

Roy M. Prendergast (1992). *Film Music 'A Neglected Art'*. WW Norton, New York and London.

Elisabeth Weis and John Belton (eds) (1985). *Film Sound 'Theory and Practice'*. Columbia University Press, New York.

Magazines

AudioMedia   Regular equipment and software reviews and feature articles about MIDI and audio recording.
www.audiomedia.com

Electronic Musician (USA)   Regular equipment and software reviews and feature articles about MIDI and computer programming.
www.emusician.com

EQ Magazine (USA)   Regular equipment and software reviews and feature articles about MIDI and audio recording.
www.eqmag.com

Journal of the Audio Engineering Society   Occasional articles about MIDI and digital audio.
www.aes.org

Keyboard Magazine (USA)   Regular equipment and software reviews, and articles about MIDI.
www.keyboardmag.com

Macworld (UK)   Regular product reviews and occasional feature articles about audio and MIDI software and hardware.
www.macworld.co.uk

Macworld (US)   Regular product reviews and occasional feature articles about audio and MIDI software and hardware.
www.macworld.com

MacUser (UK)   Regular product reviews and occasional feature articles about audio and MIDI software and hardware.
www.macuser.co.uk

Mix Magazine (USA)   Articles about audio and MIDI recording.
www.mixonline.com

Pro Sound News Europe   Industry news and views.
www.prosoundeurope.com

Sound on Sound   Regular equipment and software reviews and feature articles about MIDI and audio recording.
www.sound-on-sound.com

Index

A/D (analogue to digital) conversion/converters, 40–1, 343
AAF (Advanced Authoring Format):
  about AAF, 28, 335–6
  exporting files, 337–8
  importing files, 338–40
Ableton Live:
  about Ableton Live, 299–302
  Digidesign Edition plug-in, 254
  with Pro Tools, 279–80
Access Virus Indigo TDM Synthesizer plug-in, 255–6
Acid Pro, Sony, 302–3
ADAT Bridge optical format/I/O, 13, 343
Adobe Video Collection software, 36–7
AES/EBU I/O, 23, 344
Akai samplers, 8
Amplitude LE plug-in, 254
Analogue tape, mixing for, 208
Analogue technologies, 1–2
  A/D conversion/converters, 40–1
Anti-aliasing filters, 104, 105
Apogee AD8000 interface, 27
Apple PowerMacs see Power Macs
ATA/IDE hard drives, 39–40, 45
Atmosphere virtual instruments, 259–60
Audio compression, 209
Audio plug-ins see Plug-ins
Audiophile-quality recording, 104
AudioSuite plug-ins, 232–3, 237–8
Auto fade-in and fade-out, 169–70
Automation:
  about automation, 199–203
  Auto Latch mode, 203, 205
  Auto Touch mode, 203, 205
  Automation Enable window, 201, 217, 219
  automation writing modes, 202–3, 204–6
  breakpoints, manually set, 219–20
  and Loop Playback, 219
  mixer automation, 214

Mute Automation, 221
playlists, 203–4
with plug-ins, 238
snapshot automation, 206, 217–19
Trim mode, 205–6
Autosave, 134–5
Auxiliary Inputs, 195
Auxiliary routings, 222–6
Avid Mojo, 29
Avid Unity MediaNetwork shared storage system, 29
Avid Xpress Pro software, 36
AVoption systems, 3
AVoption|V10 (Avid picture) option, 29
Avoption|XL (Avid picture) option, 29

Backup systems, 46–50
  author's strategy, 50
  CD-ROM/CD-R, 48
  CD-RW, 48
  DAT (Digital Audio Tape), 50
  DLT (Digital Linear Tape), 49
  DVD-RAM, 47–8
  DVD-RW/+RW, 48
  Exabyte, 49–50
  Mezzo, 46
  Retrospect, 46
  Sony AIT, 48–9
  Super DLT (Digital Linear Tape), 49
Bandwidth, and hard drives, 40–2
Bar|Beat Markers, 147–51
Batch mode, 169
Beat Detective, 113, 151, 178–80
Bias Deck, 297–9
Bitheadz Unity Session, 263–4
Bounce to Disk, 207, 209–11
Breakpoints, and automation, 219–20
Browsers:
  about Browsers, 320–1
  Catalogs, 322–3

Browsers (*continued*)
Project Browser, 323
Volume Browsers, 322
Workspace Browser, 321–2
Bypass button, 200

Cakewalk Sonar, 299
Catalogs, 322–3
CDs:
CD-ROM/CD-R as backup, 48
CD-RW as backup, 48
making, 209
unreadable in Macs, 64–5
Cedar Noise reduction system, 11
Channels *see* Voices, channels and tracks
'Click' facilities, 113–15
Click plug-in, 248–9
Collection mode, 179
Compression:
about audio compression, 209
compression/expansion facilities, 175–7
Waves Renaissance compressor, 227
Computer monitors, 35
Computer processing power *see* processing
power
Computer systems:
choice of, 33
leaving on?, 38
see also PCs (personal computers); Power
Macs
Control|24, Digidesign/Focusrite:
about Control|24, 310–11, 313–14
Automation Mode Select buttons, 312
Channel Select buttons, 312
Control Room Monitor controls, 312
control surface, 311–13
editing control, 312
operational limitations, 313
rear panel inputs/outputs, 313
Talkback button, 311
CoreAudio utility, 290–1
Crossfades, 160
Cubase SX, Steinberg:
about Cubase SX, 292–4
features, 294–6
Mixer, 294
Nuendo comparison, 296–7
with Pro Tools TDM cards, 297
Sample Editor, 294
Windows dialog, 294
Cubase VST (Virtual Studio Technology), 12
Custom keyboards, 28

D-Verb RTAS plug-in, 239

DAT (Digital Audio Tape):
applications, 8
as backup, 50
mixing for, 208
Decibel, 345–6
Deck, Bias:
about Deck, 297–9
with Digidesign cards, 299
DeEsser plug-in, 249, 250
Delay:
about delay, 242
Long and Extra Long Mod Delay II, 242
Low-Pass Filter (LPF) control, 242–3
Medium Mod Delay II, 242
Mono In and Stereo Out Mod Delays, 244–5
with plug-ins, 236, 238
problems with plug-ins, 236
Short Mod Delay II, 242
Slap Delay II, 242
tempo, meter, duration and grove facilities,
243–4
Destructive record, 115
Developments:
year 2000 items, 16
year 2001 items, 16
year 2002 items, 16–17
year 2003 items, 17
Digi 001 small project studio system, 16
Digi 002 system, 16–17
Digidesign:
developments 2000 to 2003, 16–17
early drum machines, 8
early history, 8–9
first digital audio systems, 9
Q-sheet software, 8–9
see also ProControl
Digidesign virtual instruments:
Access Virus Indigo TDM Synthesizer plug-in,
255–6
Prosoniq Orange Vocoder RTAS plug-in, 257–8
Waldorf Q TDM, 256–7
Digidesign/Focusrite Control|24 *see* Control|24
Digital media, mixing for, 208
Digital Performer:
MOTU, 282–5
OMF interchange (OMFI) file format, 285
user interface, 284
Digitizing audio, 103–4
DigiTranslator 2.0 option, 28, 337
Disk drives *see* Hard drives
Distortion, 346
Dither plug-ins:
about dither, 239–40
Dithered Mixer plug-ins, 214

and mixing, 211–12
  multi-channel RTAS, 239–41
  POW-r Dither, 241
DLT (Digital Linear Tape) as backup, 49
Downbeat, editing before, 186–9
Drop-in, 108
Drum patterns, with MIDI, 91–3
Drum punch-ins, 109
DVD-RAM as backup, 47–8
DVD-RW/+RW as backup, 48

Edit tools:
  Grabber tool, 85–6, 93, 182–3, 200
  Pencil tool, 85–6, 168, 201
  Scrub tool, 167–8
  Selector tool, 167
  Smart tool, 167
  Trim tool, 86, 137–8, 166–7, 186–9
  Trimmer TCE tool, 177
  Zoom tool, 166
Edit window, 31
  counters, 159
  features, 158–9
  set-up, 159–60
  view options, 110–11
Editing:
  about editing, 158, 194
  accuracy, 170
  auto fade-in and fade-out, 169–70
  Beat Detective, 178–80
  crossfades, 169
  Grid mode features, 174–5, 177
  keyboard commands and short cuts, 165–6
  linked sections, 177–8
  Markers window set-up, 160–1
  note timing correction, 180–3
  Playback Cursor locator, 163
  playlists, 190–2
  region editing features, 173, 180
  scrolling options, 162–3
  section arranging, 189–90
  Slip mode button, 171
  Spot mode, 192–4
  Tab to Transients button, 171–2
  Time Compression/Expansion (TCE) plug-in,
    175–7
  time stamping, 192–4
  Transport window, 168
  trim tool editing before downbeat, 186–9
  Universe window, 164
  vocals editing, 183–5
  zero crossing edits, 185
  zooming and navigation, 164–5
  see also MIDI data editing

Emagic Logic Audio Platinum, 285–8
Emagic's virtual instruments:
  about Emagic's virtual instruments, 266
  Epic TDM, 270
  ESB TDM, 268–9
  HD Extension, 267
  Host TDM Enabler, 267–8
Emulator sampler, 7
Epic TDM, Emagic, 270
EQ plug-ins, 242
ESB TDM, Emagic, 268–9
Event List editing, 87
Exabyte as backup, 49–50
Expansion chassis, 28
Exporting:
  MIDI files, 330
  OMFI or AAF files from Pro Tools, 337–8

Fairlight, 7
File management:
  about file management, 320–1
  audio file size limits, 42
  Browser facilities, 320–1
  candidate files, 326
  Catalogs, 322–3
  Missing Files dialog, 324
  naming files, 138
  Project Browser, 323
  relinking missing files, 324–6
  Transfer files, 324
  unique file identifiers, 323–4
  Volume Browsers, 322
  Workspace Browser, 321–2
Final Cut Pro software, 36
Firewire hard drives, 45
Focusrite 24-channel controller, 16

GoLive software, 37
Grabber tool, 85–6, 93, 182–3, 200
Graphic editing, 85–6
Grid mode features, 174–5, 177
Groove Control with Stylus, 78–81
Groove Quantize command, 90–1
Groove Template Extraction, 180
Groups:
  Edit Groups List Key Focus, 216
  fader groups, 215–16
  Group List Key Focus feature, 216

Half-speed recording and playback, 125–6
Hard drives, 39–46, 348
  ATA/IDE drives, 39–40, 45
  and audio data storage requirements, 41
  audio file size limits, 42

Hard drives (*continued*)
  Average Seek Time, 39
  bandwidth considerations, 40–2
  compatibility guidelines, 39
  and data transfer rates, 41
  defragmenting, 39, 43
  failure problems, 43, 62–3
  Firewire, 45
  formatting, 42
  maintenance and repair, 43–4
  multichannel data transfer rates, 41–2
  Norton's Disk Doctor, 43
  permission settings, 63
  and Pro Tools requirements, 39
  SCSI drives, 39–40, 44
  zero free space problem, 62
  see also Backup systems
Hardware control/controllers:
  about hardware control, 307, 319
  history, 15–16
  see also Control|24; HUI (Human User
      Interface), Mackie; ProControl
Hardware faults/problems, 60
HD core systems/cards/interfaces:
  about HD core systems, 21
  96 I/O, 16, 24
  96i I/O, 24–5
  192 Digital I/O, 23
  192 I/O, 21–2
HD Extension, Emagic, 267
Headroom:
  clipping, 345
  and mixing, 213
HTDM (Host TDM) plug-ins, 232–3, 267–8
HUI (Human User Interface), Mackie:
  about HUI, 314–15, 319
  application, 315
  channel strips, 316–17
  Control Room section, 317
  features, 315
  history, 15–16
  monitoring facilities, 318
  need for, 30
  operation, 316
  rear panel connections, 318
  Switch Matrix section, 317
  talkback function, 318
  Transport controls, 317

Identify Beat, 144–7
IDVD software, 36
IEEE 1394 digital interface (Firewire), 45
Importing:
  existing audio files, 138–41
  from CD, 128–31
  Import Session Data, 331–2
  MIDI files, 330
  OMFI or AAF files, 338–40
  selecting session data, 334
  Time Code Mapping options, 333
  Track Offset options, 333
  track playlist options, 333–4
  tracks and track attributes, 330–3
Interfaces:
  about interfaces, 348
  legacy, 26–7
  third-party, 27
  see also HD core systems/cards/interfaces
Internet connection, with Mac OSX, 55
ITunes software, 36

Keyboards:
  commands and short cuts, 165–6
  custom, 28

LaCie 22-inch computer monitor, 35
Legacy interfaces, 26–7
Legacy Peripheral port, 27
Lexicon NuVerb card, 11
Linked sections, 177–8
Linn Drum, 7
Live, Ableton, 299–302
Logic Audio Platinum, Emagic, 285–8
Looping:
  about looping, 349
  Loop Playback, 219
  Loop Recording feature, 124–5

Mac, Pro Tools with, 17–18
Mac OSX installations, 51–60
  about Mac OSX, 51
  administrators, 54
  disk formatting/partitioning, 52–3
  Energy Saver Control Panel, 57
  folder arranging, 58
  Home folders, 54
  installing process, 53
  Internet connection, 55
  logging In, 56
  MIDI configuring, 66–8
  operating tips, 58–60
  OS9 to OSX upgrade, 51–2
  password forgotten?, 56–7
  Pro Tools issues, 57–8
  project file organization, 56
  screensaver problems, 58
  software updates, 56
  start-up problems, 61–2

user accounts, 53–4
see also MIDI data recording; Power Macs
MachineControl option, 28
Mackie 1604 mixer, 29
Mackie HUI *see* HUI (Human User Interface)
Markers, 126–7, 142–4
    Bar|Beat Markers, 147–51
    Markers window set-up, 160–1
MAS format plug-ins, MOTU, 284
Master Faders, 195, 213
Mastering:
    CDs, 209
    terminology, 207–8
Mbox, 16
Meter with Pro Tools, 111–12
Mezzo backup system, 46
Microsoft Office software, 37
MIDI:
    about MIDI, 349–50
    early sequencers, 7
MIDI + Audio Sequencers:
    Ableton Live, 299–302
    about MIDI + Audio Sequencers, 282, 302–3
    BIAS Deck, 297–9
    Cakewalk Sonar, 299
    Emagic Logic Audio Platinum, 285–8
    MOTU Digital Performer, 282–5
    Sony Acid Pro, 302–3
    Steinberg Cubase SX, 292–7
    Steinberg Nuendo, 288–92
MIDI data editing:
    about MIDI editing, 83–5
    Absolute/Bar|Beat referencing, 84–5
    drum patterns, 91–3
    Event List editing, 87
    Flatten Performance, 90
    Grabber tool, 85–6, 93
    graphic editing, 85–6
    Grid and Nudge values, 83, 98–102
    Groove Quantize command, 90–1
    MIDI Event List editor, 93, 95
    MIDI menu, 87–91
    note length trimming, 93–4
    patterns, working with, 96–102
    Pencil tool, 85–6
    Pro Tools resolution, 83
    Quantize command, 88–9
    Restore Performance, 90
    routing from ProTools to Reason, 277–8
    sample/tick-based operation, 84
    selection facilities, 96–9
    Shuffle mode, 102
    synthesizer tracking, 152–4
    synthesizing brass, 93–4

Trimmer tool, 86
MIDI data recording:
    about MIDI, 66
    Auxiliary Inputs, 75
    Device/Channel Selector, 70
    external synthesizers as sources, 77
    Groove Control with Stylus, 78–81
    Loop Record, 73–4
    Mac OSX configuring, 66–8
    MIDI Thru enabling, 71
    multiple destinations assigning, 76
    Patch Names Window, 69–71
    playing back, 74–6
    recording process, 72–3
    setting up, 68–71
    SysEx data into Pro Tools, 76–7
    track configuring, 68–9
    virtual instruments, 77–8, 81–2
MIDI I/O (peripheral), 26
MIDI project transfers, 329–30
    exporting MIDI files, 330
    importing MIDI files, 330
MIX Farm card, 14
Mixing:
    about mixing, 195
    analogue tape output, 208
    automation, 214
    Auxiliary Inputs, 195
    auxiliary routings, 222–6
    bounce to Disk, 207, 209–11
    DAT (Digital Audio Tape) output, 208
    digital media output, 208
    Dither plug-in, 211–12
    Dithered Mixer plug-ins, 214
    final mix, 229–31
    groups, 215–17
    headroom, 213
    Master Faders, 195
    and mastering, 207–8
    mix session setting up, 214–31
    Mix window, 31
    mixdown, 207
    mixing precision, 212–13
    mutes, writing, 220–1
    Output window for tracks and sends, 197–9
    panning the instruments, 228–9
    and recording to audio tracks, 211
    sends, 196–7, 224–6
    tracks, 195–6
    view choosing, 214
    vocals mixing, 226–8
    see also Automation
Monitoring with Pro Tools, 107–8
Monitors, computer, 35

MOTU Digital Performer, 282–5
Moving sessions between platforms, 340
MTC (MIDI Time Code), 350–1
Mute button, 200
'Mute Frees Assigned Voice', 123
Mutes, writing, 220–1

Native audio systems, 19–20
Native Instruments Studio Collection:
    B-4 plug-in, 264–5
    Battery percussion sampler, 266–7
    Pro-53 plug-in, 265–6
Neve Electronics, 6–7
Norton Utilities software, 37
    Norton's Disk Doctor, 43
Note timing correction, 180–3
Nuendo, Steinberg, 288–92
    Cubase SX comparison, 296–7
    Pro Tools comparison, 289
    use with Pro Tools hardware, 290–2

Offline media working, 32
OMF Interchange (OMFI):
    about OMFI, 335–6
    with Digital Performer file format, 285
    exporting from Pro Tools, 337–8
    importing into Pro Tools, 338–40
    preparation, 336
    transferring projects, 336–40
Opcode's Vision, 8
Orange Vocoder RTAS plug-in, 257–8
OSX *see* Mac OSX installations
Output windows, tracks and sends, 197–9
Overdubbing, 154–7

Page Scroll option, 162
Panning instruments, 228–9
Parameter RAM (PRAM), 64
Patterns, working with, 96–102
PCI expansion chassis, 13
PCs (personal computers):
    monitors, 35
    need for top models, 34–5
    Sony VAIO laptop range, 45
Pencil tool, 85–6, 168, 201
Performer Midi sequencing software, 8
Peripherals:
    MIDI I/O, 26
    PRE, 25–6
    SYNC I/O, 26
Pitch shifting, 151
Playback Cursor, 30–1
    locator, 163
Playlists, 116–17, 190–2, 203–4, 333–4

Plug-ins:
    Ableton Live Digidesign Edition, 254
    about plug-ins, 11–12, 232, 254
    Access Virus Indigo TDM Synthesizer, 255–6
    AmpliTube LE, 254
    AudioSuite, 232–3, 237–8
    automation facilities, 238
    Click, 248–9
    DeEsser, 249, 250
    delay problems, 236, 238
    HTDM (Host TDM), 232–3
    'inactive' facilities, 236–7
    inserting during playback, 252–3
    Multi-channel TDM, 245–8
    Multi-mono TDM and RTAS, 248–52
    optional, 253
    packs, 253
    Prosoniq Orange Vocoder RTAS plug-in, 257–8
    real-time, 234–6, 238
    RTAS (Real Time Audio Suite), 232–4, 238, 239–45
    SampleTank SE, 254
    Signal Generator, 249, 250
    T-RackS EQ, 254
    TDM, 232–4, 238, 245–8
    third-party, 253
    Trim, 249–52
    Waldorf Q TDM virtual instrument, 256–7
POW-r Dither plug-in, 241
Power Macs:
    battery-backed RAM, 64
    CDs unreadable, 64–5
    freezes, foibles and crashes, 60–1
    hardware faults/problems, 60
    Power Mac G5, 33–4
    PowerBook G4, 36
    see also Mac OSX installations
PRE (Digidesign peripheral), 25–6
Prism Sound ADA-8 interface, 27
Pro Tool MIX cards/systems, 14–15
    interface options, 15
Pro Tools:
    about Pro Tools, 1–5, 6–18, 19–20
    applications, 2–4
    beginnings, 6–10, 17–18
    with Macs, 17–18
    mixers, need for?, 29–30
    pre Pro Tools systems, 7–8
    Pro Tools III, 10–11
    Pro Tools TDM hardware, 20
    Pro Tools TDM software, 30–2
    Pro Tools|24, 14
    Pro Tools|HD systems *see* HD core systems/
        cards/interfaces

resolution, 83
Processing power, 2
ProControl:
    about ProControl, 307–8, 310
    DigiFaders, 308
    modules, 309
    rear panel, 309–10
    visual feedback, 309
Project Browser, 323
Project transfers *see* Exporting; Importing;
        Transferring projects
Propellerheads *see* Reason virtual instruments,
        Propellerheads
Prosoniq Orange Vocoder RTAS plug-in, 257–8
Punch-in, 108, 118
Punch-out, 118

Q-sheet software, 8–9
Quantization, 352
    Quantize command, 88–9
QuickPunch feature, 123–4
QuickTime movie playback, 11

RAM, Parameter RAM (PRAM), 64
Real-Time Pitch Processor TDM plug-ins, 245
Reason virtual instruments, Propellerheads,
        261–3
    DR.REX Loop layer, 261
    Matrix, 262
    ReBirth drum machine, 261
    RECycle software, 261
    with ReWire, 271–8
    routing MIDI from Pro Tools, 277–8
    setting up, 275–7
Recording:
    24-bit recording, 104
    about recording, 103–6
    anti-aliasing filters, 104, 105
    audio recording, 135–6, 211
    audiophile-quality, 104
    autosave, 134–5
    Bar|Beat Markers, 147–51
    'click' facilities, 113–15
    destructive record, 115
    digitizing audio, 103–4
    drum punch-ins, 109
    Edit window view options, 110–11
    file naming, 138
    half-speed recording and playback,   125–6
    Identify Beat, 144–7
    importing audio from CD, 128–31
    importing existing files, 138–41
    Loop Recording feature, 124–5
    Markers, 126–7, 142–4

meter, 111–12
monitoring, 107–8
new session opening, 131–4
overdubbing, 154–7
overloading problems, 106
playlists, 116–17
pre- and post-roll, 118
punch-in and -out, 118
quality, 105
QuickPunch feature, 123–4
real recording session, 131–57
Record Pause mode, 108
ruler tracks, 109
sampling rate, 105
shortcuts, 126
tempo, 111–13
tempo maps, 144–8
Time stamping, 127–8
topping and tailing, 136–8
Trim To Selection command, 137
User Time Stamp, 125
voices, channels and tracks, 118–23
    see also MIDI data recording
Region editing features, 173
    Region Separation, 180
Relative Grid Mode, 31
Resolution, Pro Tools, 83
Retrospect backup system, 46
Reverb One reverb plug-in, 16
ReWire:
    about ReWire, 271, 281
    auxiliary tracks caveat, 281
    looping playback, 281
    playing application sequences with Pro Tools,
        278–9
    with Reason, 271–8
    using/playing Ableton Live with Pro Tools,
        279–80
RTAS (Real Time Audio Suite) plug-ins:
    about RTAS plug-ins, 232–4, 238, 352
    D-Verb plug-in, 239
    delays:
        about delays, 242
        Long and Extra Long Mod Delay II, 243
        Low-Pass Filter (LPF) control, 242–3
        Medium Mod Delay II, 243
        Short Mod Delay II, 243
        Slap Delay II, 243
        tempo, meter, duration and groove, 243–4
    Dither plug-ins, 239–41
    dynamics processors, 242
    EQ plug-ins, 242
    POW-r Dither, 241
    real-time working, 238

Ruler tracks, 109

SampleTank SE plug-in, 254
Sampling, 353
    sampling rate, 105, 353
Screensaver problems, 58
Scrolling options, 162–3
Scrub tool, 167–8
SCSI, 353
    SCSI hard drives, 39–40, 44
Sections:
    linked sections, 177–8
    section arranging, 189–90
Selector tool, 167
Sends, 196–7
    output window for, 197–9
    Send window, 224–6
Session moving between platforms, 340
Short cuts, 126, 165–6, 196
Signal Generator plug-in, 249, 250
Slip mode button, 171
Smart tool, 167
SMPT (Society of Motion Picture and Television
    Engineers) synchrinizer/Time Code, 354
Snapshot automation, 206, 217–19
Software:
    Adobe Video Collection, 36–7
    Avid Xpress Pro, 36
    Final Cut Pro, 36
    GoLive, 37
    IDVD, 36
    iTunes, 36
    Microsoft Office, 37
    Norton Utilities, 37
    Tech Tool Pro, 37
    updates, 37–8
Sonar, Cakewalk, 299
Sonic Studio, 8
Sony Acid Pro, 302–3
Sony AIT as backup, 49
Sony VAIO laptop range, 35
Sound Designer II format, 31
Sound On Sound magazine, 56
Sound Tools, 9–10
Spectrasonics virtual instruments:
    about Spectrasonics virtual instruments,
        258
    Atmosphere, 259–60
    Stylus, 259
    Trilogy, 260–1
Spot mode, 192–4
Steinberg *see* Cubase SX, Steinberg; Nuendo,
    Steinberg
Studio Collection, Native Instruments, 264–6

Stylus virtual instruments with Groove Control,
    78–81, 259
Super DLT (Digital Linear Tape) as backup, 49
Surround mixing, 31
SYNC I/O (peripheral), 26
Synclavier, 7
Synthesizers:
    Access Virus Indigo TDM Synthesizer plug-in,
        255–6
    tracking, 152–4
SysEx data, recording into Pro Tools, 76–7

T-RackS EQ plug-in, 254
Tab to Transients button, 171–2
TDIF Format/Synch, 355
TDM hardware, 20
TDM plug-ins:
    about TDM plug-ins, 232–4, 238
    Real-Time Pitch Processor, 245
    TimeAdjuster, 246–8
TDM software, 30–2
Tech Tool Pro software, 37
Tempo with Pro Tools:
    Manual Tempo mode, 113
    and meter, 111–12
    setting default tempo, 112
    tempo maps, 144–8
Third-party interfaces:
    Apogee AD8000, 27
    PrismSound ADA-8, 27
Time Code Mapping options, 333
Time Compression/Expansion (TCE) plug-in,
    175–7
Time stamping, 127–8, 192–4
TimeAdjuster TDM plug-ins, 246–8
'Timeline Insertion Follows Playback', 162–3
Tools *see* Edit tools
Topping and tailing, 136–8
Tracks:
    Auxiliary input, 195–6
    importing, 330–3
    Master Faders, 195
    output window for, 197–9
    track counts, 20–1
    Track Offset options, 333
    track playlist options, 333–4
    see also Voices, channels and tracks
Transferring audio, 335–6
Transferring projects:
    about transferring projects, 327
    MIDI exports, 330
    MIDI imports, 330
    MIDI transfers, 329–30
    moving between PT LE and PT TDM, 340–2

moving sessions between platforms, 340
OMFI files, 336–40
preparation, 327–9
see also Exporting; Importing
Transients, tab to transients, 171–2
Transport window, 168
Trilogy virtual instruments, 260–1
Trim mode, and automation, 205–6
Trim plug-in, 249–52
Trim To Selection command, 137
Trim tool, 86, 137–8, 166–7
editing before downbeat, 186–9
Twenty-four-bit digital audio, 13–14

UMI sequencer, 7
Unity Session, Bitheadz, 263–4
Universe window, 164
User Time Stamp, 125

Virtual instruments, 255–70
Bitheadz Unity Session, 263–4
with MIDI, 77–8, 81–2
see also Digidesign virtual instruments;
Emagic's virtual instruments; Native
Instruments Studio Collection; Reason
virtual instruments, Propellerheads;
Spectrasonics virtual instruments

Viruses, 43
Vocals:
editing, 183–5
mixing, 226–8
Voicable tracks, 122–3
Voices, channels and tracks:
about channels and tracks, 118–19
'Mute Frees Assigned Voice', 123
Pro Tools|24 MIX systems, 119
Pro Tools|HD and HD Accel systems, 119–22
voicable tracks, 122–3
Volume Browsers, 322

Waldorf Q TDM virtual instrument plug-in, 256–7
Waves plug-ins, 12
Waves Renaissance compressor, 227
Workspace Browser, 321–2

Yamaha:
02R compact digital mixer, 12–13, 17
early digital mixers, 8

Zero crossing position/edits, 185–7
Zooming and navigation, 164–5
Zoom tool, 166

Also by Mike Collins

A Professional Guide to Audio Plug-ins and Virtual Instruments

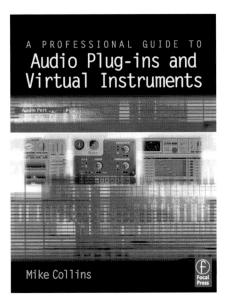

If you are about to purchase new software to enhance your existing system, wish to troubleshoot technical problems, need guidance on how to build up a collection of software to suit your needs, or simply want to know more about what's available, this is the book for you! Mike Collins has meticulously surveyed the major software available to provide a detailed overview of products and how they integrate into host platforms. Extensively colour-illustrated and packed with helpful tips and notes, this is a vital source of reference to keep by your side, whether you are a professional recording engineer or an amateur aspiring to professional results.

'Where keeping up with developments in software takes time, and time is money, this book is worth every penny. Keep it by your bed!'

Steve Parr, Vice-Chairman of the Music Producers Guild and co-owner of Hear No Evil Studios

'... really useful and unbiased guidance. Buy this book before you put your system together and it will help you to make truly informed decisions about what you need.'

Simon Boswell, Film Composer (*Shallow Grave*, *This Year's Love*, *A Midsummer Night's Dream*, *Warzone*)

Published: 2003
656 pages
530 colour illustrations
ISBN: 0240517067

For more details and to purchase a copy online, visit **www.focalpress.com**

Also by Mike Collins

Choosing and Using Audio and Music Software

A guide to the major software applications for Mac and PC.

This comprehensive reference features all the major audio software, including Pro Tools, Logic Audio, Cubase, Digital Performer, Sonar and Audition. If you need advice on which systems to purchase, which are most suitable for particular projects, and on moving between platforms mid-project, this book should be your one-stop reference. Mike Collins is a trainer and consultant who has been tackling these issues for years and his expert advice will save you time and money.

Each section covers a specific system, providing a handy overview of its key features and benefits, including help with setup. The many short tutorials provide both a source of comparison and means to get up to speed fast on any given software.

'Professional studio engineers are now expected to know all of the popular audio and MIDI software packages... This is just what the doctor ordered when you need to get up to speed quickly. The appendix on transferring projects between applications alone will make me buy this book!'

John Leckie, Producer (Radiohead, Simple Minds, Baaba Maal, Dr John)

'Three months ago I made some major changes in my home studio set-up, involving some serious shopping preceded by weeks of head-scratching and chin-stroking. If only this book had been available then, it would have saved me hundreds of phone calls and loads of angst and definitely resulted in a more informed choice!'

Chris 'Snake' Davis, Sax & Flute Player (Snake Davis Band, M-People)

Published: 2004
544 pages
252 colour illustrations
ISBN: 0240519213

For more details and to purchase a copy online, visit **www.focalpress.com**

 **Focal Press**   **www.focalpress.com**

Join Focal Press online
As a member you will enjoy the following benefits:

- browse our full list of books available
- view sample chapters
- order securely online

Focal eNews
Register for eNews, the regular email service from Focal Press, to receive:

- advance news of our latest publications
- exclusive articles written by our authors
- related event information
- free sample chapters
- information about special offers

Go to www.focalpress.com to register and the eNews bulletin will soon be arriving on your desktop!

If you require any further information about the eNews or www.focalpress.com please contact:

USA
Tricia Geswell
Email: t.geswell@elsevier.com
Tel: +1 781 313 4739

Europe and rest of world
Lucy Lomas-Walker
Email: l.lomas@elsevier.com
Tel: +44 (0) 1865 314438

Catalogue
For information on all Focal Press titles, our full catalogue is available online at www.focalpress.com, alternatively you can contact us for a free printed version:

USA
Email: c.degon@elsevier.com
Tel: +1 781 313 4721

Europe and rest of world
Email: j.blackford@elsevier.com
Tel: +44 (0) 1865 314220

Potential authors
If you have an idea for a book, please get in touch:

USA
editors@focalpress.com

Europe and rest of world
ge.kennedy@elsevier.com